Managing Enterprise Projects Using Project Online and Microsoft Project Server 2019

Dale Howard

José Marroig

Managing Enterprise Projects
Using Project Online and Microsoft Project Server 2019

Copyright © 2019 Projility, Inc.

Publisher: Projility, Inc.
Authors: Dale Howard and José Marroig
Technical Editor: Tom Henry

All rights reserved. You may not reproduce or transmit any part of this work in any form or by any means, electronic or mechanical, including photocopying, recording, or by any information storage or retrieval system, without the prior written consent of the copyright owner and the publisher.

We use trademarked names in this book in an editorial context and to the benefit of the trademark owner with no intention of infringement of the trademark.

Published and distributed by:

Projility, Inc.
2 Wisconsin Circle, Suite 700
Chevy Chase, MD 20815

(703) 448-6777

http://www.projility.com

We provide the information contained in this book on an "as is" basis, without warranty. Although we make every effort to ensure the accuracy of information provided herein, neither the authors nor the publisher shall have any liability to any person or entity with respect to any loss or damage caused or allegedly caused directly or indirectly by the information contained in this work.

Contents

About the Authors x
About the Technical Editor x
Introduction xi
About Projility xii
Downloading the Sample Files xiii
Installing the Project Template xiii

Module 01: Microsoft PPM Overview 1
    Project Online vs. Project Server 3
    Applying PPM Terminology to Project Online 3
    Understanding Project Online Terminology 4
        Enterprise Project 4
        Enterprise Resource 4
        Check In and Check Out 5
        Enterprise Resource Pool 5
        Enterprise Global 5
    Understanding the Project Communications Life Cycle 6

Module 02: Microsoft Project Overview 11
    Understanding the Ribbon and Quick Access Toolbar 13
    Using Views, Tables, Filters, and Groups 15
        Using Views 15
        Using Tables 19
        Using Filters 21
        Using Groups 26
    Creating a Project Web App Login Account 29
    Opening and Closing Enterprise Projects 34
    Understanding the Local Cache 39
        Adjusting Local Cache Settings 39
        Viewing Local Cache Contents 40
        Cleaning Up the Local Cache 42
        Resolving Local Cache Corruption 44
    Working with Offline Projects 45
        Offline Mode While Away from the Corporate Network 46
        Offline Mode When You Lose Connectivity 49

Module 03: Project Web App Overview 51
    Introducing Project Web App 53

Contents

Understanding the Default Project Web App Interface 53
    Navigating in Project Web App 54
Features and Functionality in Project Web App 56
    Using Project Web App Views 57
    Manipulating a Data Grid 61
    Printing the Data Grid 64
    Exporting the Data Grid to Excel 65

Module 04: Creating an Enterprise Project 67

Creating a New Enterprise Project 69
Creating a New Project Using an Enterprise Project Type 70
    Transferring a Proposed Project to a Project Manager 79
Creating a New Project Using Microsoft Project 83
Defining a New Enterprise Project 85
    Set the Project Start Date and Specify Enterprise Field Values 85
    Enter the Project Properties 87
    Display the Project Summary Task 88
    Set the Project Working Schedule 89
    Set Options Unique to the Project 93
    Save the Project 97
Creating Resource Engagements Using Generic Resources 99
    Setting Resource Utilization to Use Resource Engagements 100
    Building the Project Team with Generic Resources 100
    Creating the Resource Engagements 103
    Submitting the Resource Engagements 108

Module 05: Task Planning 111

Understanding the Task Planning Process 113
Understanding the Task Mode Setting 113
Using Basic Task Planning Skills 114
    Entering and Editing Tasks 114
    Moving Tasks 114
    Inserting Tasks 115
    Deleting Tasks 116
    Creating the Work Breakdown Structure (WBS) 118
    Creating Milestones 121
    Adding Task Notes 124
    Applying Cell Background Formatting 125
Using Task Dependencies 128
    Understanding Task Dependencies 128
    Setting and Removing Task Dependencies 130
    Using Lag Time with Dependencies 132

  Using Lead Time with Dependencies 133
 Setting Task Constraints and Deadline Dates 137
  Setting Constraints 137
  Understanding Flexible and Inflexible Constraints 139
  Understanding Planning Wizard Messages about Constraints 140
  Setting Deadline Dates 141
  Understanding Missed Constraints and Deadline Dates 143
 Applying Task Calendars 144
 Estimating Task Durations 147

Module 06: Resource Planning 149

 Understanding Enterprise Resources 151
 Building a Project Team 152
  Filtering Resources 154
  Grouping Resources 158
  Viewing Resource Information 158
  Viewing Resource Availability 159
  Adding Resources to Your Project Team 162
  Matching and Replacing Resources 162
  Using Proposed vs. Committed Booking 163
 Using Local Resources in a Project Team 165

Module 07: Assignment Planning 167

 Assigning Work Resources to Tasks 169
  Assigning Resources Using the Task Entry View 169
  Assigning Resources Using the Assign Resources Dialog 174
 Understanding Other Factors in Assignment Planning 177
  Understanding the Duration Equation 178
  Understanding Task Types 179
  Understanding Effort Driven Scheduling 183
 Assigning Cost Resources 189
  Adding Cost Resources to the Project Team 189
  Assigning a Budget Cost Resource 189
  Assigning an Expense Cost Resource 193
  Recommendations for Using Cost Resources 195
 Assigning Material Resources to Tasks 197
 Understanding Resource Overallocation 199
  Locating and Analyzing Resource Overallocations 200
 Leveling Overallocated Resources 204
  Using a Leveling Methodology 204
  Setting Leveling Options 205
  Leveling an Overallocated Resource 208

  Viewing Leveling Results *208*
  Clearing Leveling Results *210*
  Setting Task Priority Numbers *210*
 Creating Resource Engagements for Human Resources **213**

Module 08: Project Execution **219**

 Analyzing the Critical Path **221**
  Viewing the "Nearly Critical" Path *223*
 Working with Project Baselines **224**
  Saving a Project Baseline *225*
  Backing Up a Baseline *226*
  Clearing the Project Baseline *227*
 Understanding Publishing **228**
  Publishing an Enterprise Project *229*
  Setting Custom Permissions for a Project *230*
 Working with Project Deliverables **235**
  Adding a New Deliverable in Microsoft Project *235*
  Editing and Deleting Deliverables in Microsoft Project *239*
  Viewing Deliverables in Project Web App *242*
  Adding a New Deliverable Dependency *244*
  Updating the Baseline and Republishing the Project *247*

Module 09: Tracking Time and Task Progress **249**

 Tracking Time and Task Progress in Project Web App **251**
 Accessing the Timesheet Page **251**
  Understanding the Timesheet Page *253*
  Best Practices for Setting Up the Timesheet Page *255*
 Using the Timesheet Page **259**
  Entering Time in the Timesheet *260*
  Adding a New Row to a Timesheet *261*
  Reassigning a Task to a Fellow Team Member *268*
  Removing a Timesheet Line *269*
  Submitting a Timesheet for Approval *270*
  Recalling a Submitted Timesheet *273*
  Deleting a Timesheet *274*
  Responding to a Rejected Timesheet *275*
  Submitting Future Planned PTO *275*
 Using the Tasks Page **278**
  Understanding the Tasks Page *278*
  Understanding the Assignment Details Page *280*
  Reporting Progress from the Tasks Page *283*
  Adding Tasks to the Tasks Page *289*

Removing a Task 292
Reassigning a Task to a Fellow Team Member 293
Manually Entering Task Progress in the Microsoft Project Schedule 294
Entering Progress at the Task Level 295
Entering Progress at the Resource Assignment Level 296

Module 10: Approving Time and Task Progress 299

Understanding the Project Update Cycle 301
Viewing Pending Approvals on the Home Page 301
Accessing the Approval Center Page 302
Understanding the Approval Center Page 302
Best Practices for Setting Up the Approval Center Page 304
Processing Task Updates 306
Reviewing Pending Task Updates 309
Rejecting Task Updates 311
Approving Task Updates 313
Creating Rules for Auto Approving Task Updates 314
Rescheduling Incomplete Work from the Past 320
Updating Expense Cost Resource Information 326
Publishing the Latest Schedule Changes 328
Axioms for Success with Tracking Progress 328

Module 11: Variance Analysis, Plan Revision, and Reporting 331

Understanding Variance 333
Understanding Variance Types 333
Calculating Variance 333
Understanding Actual vs. Estimated Variance 334
Analyzing Project Variance 335
Analyzing Date Variance 335
Analyzing Work Variance 337
Analyzing Cost Variance 338
Revising a Project Plan 341
Potential Problems with Revising a Plan 341
Using a Change Control Process 343
Inserting New Tasks in a Project 343
Updating the Baseline in Your Project 345
Baselining Only Selected Tasks 345
Backing up the Current Baseline Data 347
Project Reporting Overview 349
Project Reporting Using the Timeline View 349
Adding Tasks to the Timeline View 350
Arranging Tasks in the Timeline View 350

Contents

  Formatting the Timeline View *351*
 Project Reporting Using the Dashboard Reports **353**
  Customizing a Chart *354*
  Customizing a Table *356*
 Project Reporting Using Power BI Reports **359**
  Using Ad Hoc Filtering *368*
  Using Natural Language Queries *370*

Module 12: Working with the Project Site 373

 Understanding the Project Site **375**
  Who Has Access to the Project Site? *375*
 Navigating to the Project Site **375**
  Navigating from the Microsoft Project Schedule *375*
  Navigating from Project Web App *376*
 Managing Project Risks **379**
  Creating a New Risk *379*
  Working with Existing Risks *383*
 Managing Project Issues **389**
  Creating a New Issue *390*
  Viewing and Editing Existing Issues *392*
 Viewing Your Assigned Issues and Risks **392**
 Managing Project Documents **393**
  Accessing the Documents Library *393*
  Creating a New Document in a Document Library *394*
  Creating a New Folder in a Documents Library *395*
  Uploading Templates and Documents to a Document Library *396*
  Renaming a Document *399*
  Sharing a Document *401*
  Checking Out a Document for Editing *402*
  Editing a Document *403*
  Working with the Version History of a Document *407*
  Subscribing to E-Mail Alerts about a Document *411*
 Attaching Risks, Issues, and Documents to Tasks **412**
 Using Other Features in the Project Site **419**
  Using OneNote Notebooks *419*
  Using the Tasks Page *420*
  Using the Newsfeed Web Part *422*

Module 13: Working in the Project Center 425

 Using the Project Center **427**
  Using the Projects Ribbon in the Project Center *428*
  Using Project Center Views *428*

Understanding Show/Hide Options *430*
Understanding Indicators *431*
Using the Open Menu Button *432*
Working With Projects in the Project Center **434**
Opening Projects in Microsoft Project from the Project Center *434*
Creating a Master Project from the Project Center *435*
Checking In a Project from the Project Center *436*
Setting Custom Permissions for a Project *438*
Working with Detailed Project Views *441*
Editing Projects in Project Web App **446**
Creating a New Project in Project Web App *446*
Understanding the PWA Task Planning Process *449*
Understanding the PWA Resource Planning Process *451*
Understanding the PWA Assignment Planning Process *452*
Finalizing Your PWA Project *453*
Closing and Reopening a PWA Project for Editing *454*

Module 14: Working in the Resource Center 457

Using the Resource Center 459
Applying Resource Center Views *460*
Selecting and Deselecting Resources *461*
Viewing Resource Availability 463
Viewing Resource Assignments 471
Working with Resource Engagements 473
Viewing Pending Resource Engagements *473*
Viewing Resource Availability *476*
Responding to Pending Resource Engagements *476*

Module 15: Managing Personal Settings 481

Personal Settings Overview 483
Managing Alerts and Reminders for Yourself 484
Managing Alerts and Reminders for Your Resources 487
Managing My Queued Jobs 490
Managing Delegates 491
Acting as Delegate 495

Module 16: Index 497

About the Authors

Dale Howard is the Director of Education for Projility. He has used Microsoft Project since version 4.0 for Windows 95 and every version of the Microsoft PPM tool since its first release as Project Central in 2000. He is the co-author of 21 books on Microsoft Project, Project Server, and Project Online. He is currently one of only 26 Microsoft Project MVPs in the entire world and one of only 4 in the United States. Dale has been married to the former Mickey Cobb, whose name is mentioned as a resource in a sample project several times in this book

José Marroig is the founder and chief executive officer of Projility. Under José's leadership, Projility has become one of a handful of leaders in the project portfolio management (PPM) technology products and services market. As a successful operations and strategy executive, with more than a decade of experience in the Project Management & Technology arena, he has managed every aspect in both the sales and system development life cycle. He is responsible for the strategic development, growth, and management of the company's business lines and consulting practices. Jose's career has included successful stops at Deloitte, Xpedior, and Arthur Anderson. He received a Bachelor of Science (BS) in Information Technology & Business Administration from Marquette University and a Master's in Information Systems from The George Washington University.

About the Technical Editor

Tom Henry has been involved with Microsoft Project and Portfolio Management (PPM) since graduating from Bournemouth University in the UK. Working first in London and then moving to the US, Tom has seen how project managers interact with Microsoft Project and Project Online from using different methodologies such as Prince 2 in the UK to PMI in the USA. Tom has also worked with Agile teams looking to utilize Microsoft Project to manage projects using Scrum, Kanban, and other iterative methods. Tom believes that after seeing so many different ways to manage a project, always remember the golden rule: "A project schedule is a model of a project, therefore, it should be significantly less complex than the project itself so that you can use it to manage your project schedule more efficiently" Tom prides himself on being able to understand his clients' needs and to translate those needs into simplified PPM solutions.

Introduction

Having both worked for many years with Microsoft's project management tools, this book is truly a "labor or love" for us. For Dale, this is his 22nd book on Microsoft Project and the Microsoft PPM solution, while for José, this book is his first.

Our book focuses on the use of Microsoft Project 2019, Project Online, and Project Server 2019. In this book, we use the term "Project Server 2019" sparingly to avoid repetition. Therefore, every time you see the words "Project Online" in this book, we refer to both Project Online and Microsoft Project Server 2019, since the functionality is nearly identical between these two tools.

We organize the content in this book in a logical manner, beginning with teaching you the basics about Microsoft Project and Project Online, and then teaching you how to use these tools effectively to manage your projects. Modules 01-03 provide you with an overview of the Microsoft PPM tools, Microsoft Project, and Project Web App. Modules 04-07 teach you how to create a new project, then to complete task planning, resource planning, assignment planning, and leveling of overallocated resource.

Modules 08-11 teach you the steps you need to follow to take the project into the Execution stage, how to use the *Timesheet* and *Tasks* pages for tracking task progress, how to approve or reject task updates, along with the steps required for completing the project updating life cycle each reporting period. Modules 12-15 teach you everything else you need to know about Project Online, including how to use the Project Site for team collaboration, how to use the *Project Center* and *Resource Center* pages, and how to specify your personal settings for Project Web App.

Throughout this book, we provide a generous number of Notes, Warnings, and Best Practices. Notes call your attention to important additional information about a subject. Warnings help you avoid the most common problems experienced by others. Best Practices provide recommendations for best uses of the tool based on our years of field experience.

With this book, we believe that you can become more effective at using Microsoft Project and the Microsoft PPM tools. If you have questions about the book or are interested in our professional services, please contact us at **info@projility.com**.

Dale Howard and José Marroig

About Projility

Projility is a Microsoft Project Portfolio Management (PPM), Project Online, Power BI and Dynamics PSA services leader and product innovator. Our consulting services and Hammerhead solutions have helped organizations around the world achieve the following:

- Improved resource utilization
- Increased visibility, insight and control of project, programs, and portfolios
- Reduced cost of deployment and management of PPM systems
- Increased on-time and on-budget projects
- Delivery of true portfolio optimization and strategic execution management capability

The Projility combination of process expertise and implementation capability is unique in the market, providing good practices that have resulted in a 98% on-time, on-budget track record of success for hundreds of Fortune 1000 and Public Sector clients since the firm's inception in 2005.

Projility's mantra is "Expert solutions for effective change." Organizations around the world look to Projility for rapid, comprehensive, and high value Microsoft-based solution implementations, and the unique ability to drive operational change while reducing impact to end users, based upon our implementation methodology and project-based approach.

Our clients come to us for a number of reasons ranging from digital transformations, organization process improvements, end user training, or operational growth. We continue to improve on our services and strive to ensure that our clients are able to achieve their target business outcomes and objectives.

Downloading the Sample Files

The Hands On Exercises in this book require you to use multiple sample files that accompany this book. Before you attempt to work the first exercise, please download and unzip the contents of the **Managing 2019 Student Files.zip** file to a folder of your own choice. You can find the ZIP file at the following URL:

https://www.projility.com/managing2019

Installing the Project Template

After unzipping the student sample files to your desired location, you need to ask your Project Online or Project Server 2019 application administrator to install one of the sample projects as an enterprise project template. Please ask your app admin to complete the following steps:

1. Launch Microsoft Project 2019 and connect to Project Web App.
2. Open the **PPM Training Template.mpp** sample file.
3. Click the **File** tab and then click the **Save As** tab in the *Backstage*.
4. In the *Project Web App* section of the *Save As* page, confirm that your Project Online or Project Server 2019 instance is selected (if not, select it).
5. At the top of the *Save As* page, click the **Save** button.
6. In the *Save to Project Web App* dialog, enter the name **PPM Training** in the **Name** field.
7. In the dialog, click the **Type** pick list and select the **Template** item.
8. In the dialog, *do not select* any values in the *Custom fields* section, and then click the **Save** button.
9. In the *Save As Template* dialog, select *every checkbox* and then click the **Save** button.
10. In the confirmation dialog about enterprise calendars, click the **Yes** button.
11. When the save job is 100% complete, close and check in the *PPM Training* enterprise template.

After saving the PPM Training enterprise template in your Project Online or Project Server 2019 system, your application administrator should also create a new Enterprise Project Type (EPT) in Project Web App for use with this book. When creating this new EPT on the *Add Enterprise Project Type* page of PWA, your app admin should enter the name **PPM Training for PMs** in the **Name** field, and should select the **PPM Training** template in the **Project Plan Template** pick list. Your app admin should set all other values for the new EPT in accordance with your organization's use of Project ID numbers, SharePoint workflows, Project Detail Pages (PDPs), Departments, Project Site creation, synchronization setting with the Project Site, and the template for the Project Site. After completing these steps, you should be able to work every Hands On Exercise in the book.

Module 01

Microsoft PPM Overview

Learning Objectives

After completing this module, you will be able to:

- Understand the difference between Project Online and Project Server
- Apply Project Portfolio Management (PPM) terminology to Project Online
- Understand terminology exclusive to Project Online
- Understand the project communications life cycle in Project Online

Inside Module 01

Project Online vs. Project Server 3
Applying PPM Terminology to Project Online 3
Understanding Project Online Terminology 4
    Enterprise Project 4
    Enterprise Resource 4
    Check In and Check Out 5
    Enterprise Resource Pool 5
    Enterprise Global 5
Understanding the Project Communications Life Cycle 6

Project Online vs. Project Server

Microsoft currently offers two versions of its powerful Project Portfolio Management (PPM) tool: Project Online and Project Server 2019. Project Server 2019 is Microsoft's eighth generation version of the PPM tool. Previous versions included Project Central, released in 2000, as well as Project Server 2002, 2003, 2007, 2010, 2013, and 2016. Project Online was introduced in 2013 and incorporated into Microsoft Office 365.

To a project manager, the differences between Project Online and Project Server 2019 are difficult to discern, as the user interfaces are nearly identical. However, there are several major differences between these two PPM tools:

- Project Online is a **cloud-based product**, with all of its software installed and hosted within Microsoft's data centers. Microsoft staff maintain the software, and regularly apply software patches and updates. Project Server 2019 is an **on-premise product** that must be installed on the customer's own servers, or in cloud based infrastructure owned by the customer. The customer's own IT staff must maintain the software and must manually apply software patches and updates.

- Project Online is a **subscription-based product**. Organizations pay a monthly fee per user, based on the user's role in the organization. Project Server 2019 is a **license-based product**. Organizations must pay for a Project Server 2019 license for each running instance of the software, along with a Client Access License (CAL) for each user accessing the system.

- Project Online natively provides data access for reporting using only **OData**. Project Server 2019 provides multiple methods for data access for reporting, including **SSRS, OData, and OLAP cubes**.

> **Information**: As you read this book, keep in mind that I am using Project Online, but every feature I discuss about Project Online is also available in Project Server 2019.

Applying PPM Terminology to Project Online

In the world of enterprise project management, you hear terms like project, program, and portfolio. How do these terms apply to your organization's project management environment? According to the Project Management Institute (PMI), a **project** is "a temporary endeavor undertaken to create a unique product, service, or result." A project is *temporary*, meaning that it has a beginning and an end. A project is *unique*, meaning that it is something that your organization has not done before, or never done quite this way before.

For the purposes of this book, a **program** is "a collection of related projects" and a **portfolio** is "a collection of programs and/or projects within a business unit or across an entire enterprise." The concept of a portfolio is flexible, depending on the size of the company. A smaller organization may have a single portfolio of projects, whereas a larger business may have numerous departmental or line-of-business portfolios, each containing its own set of programs and projects. Regardless of the way a business conceives these terms, you can model them in Project Online using the portfolio hierarchy diagram shown in Figure 1 - 1.

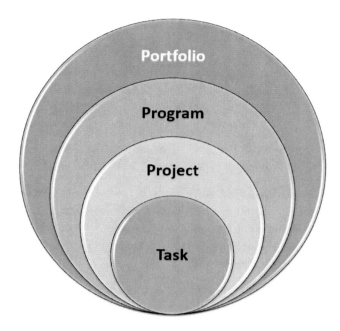

Figure 1 - 1: Portfolio hierarchy diagram

Understanding Project Online Terminology

Two terms that you must understand in the context of Project Online are **enterprise project** and **enterprise resource**. Very specific criteria determine whether a project is an enterprise or non-enterprise project, and whether a resource is an enterprise or local resource. In addition to these two important terms, there are additional Project Online terms of which you should be aware. I discuss each of these terms individually.

Enterprise Project

A project is an **enterprise project** when one of the following two conditions is true:

- You create the project using Microsoft Project while connected to Project Online and save the project in the Project Online database.

- You create and save the project using Project Web App, the web-based user interface for Project Online.

Any project not stored in the Project Online database, such as a project saved as a .MPP file on your hard drive or a network share, is termed a **non-enterprise project** or a **local project**.

 Information: Projility highly recommends using enterprise projects wherever possible to provide visibility for your organization's executives and business leaders about the status of every project in your portfolio and the utilization of enterprise resources across all projects.

Enterprise Resource

An **enterprise resource** is any resource stored in the centralized Enterprise Resource Pool in Project Online. Although your application administrator creates enterprise resources in the system, you can use them as team members in your own enterprise projects. If a resource exists in an enterprise project but does not exist in the Enterprise Resource Pool, then this resource is termed a **local resource**, meaning that it is local to that particular project only.

 Information: Projility highly recommends utilizing enterprise resources wherever possible so that your organization's executives, portfolio managers, and resource managers can analyze capacity and availability information for all resources across all enterprise projects.

Check In and Check Out

For a project manager, the terms **check in** and **check out** apply to enterprise projects. When you open an enterprise project for editing, Project Online checks out the project. While your project is checked out, no one else can open and edit the project. When you close a project, should always check in the project. Once the project is checked in, others can edit the project if they have the proper permissions in the system.

Enterprise Resource Pool

The **Enterprise Resource Pool** is the collection of all resources that may be required for staffing enterprise projects. This resource pool may contain Work resources, Cost resources, and Material resources. Work resources are typically people but may also include equipment. You can use Cost resources in your own project to specify a budget for the project and to track non-labor related costs. You can use Material resources to track the utilization of consumable items.

The resources in the Enterprise Resource Pool usually include resource attributes that you can use for filtering and grouping, and which your organization can use for resource reporting and resource management. Resource attributes might include the resource's role, location, skills, and department, for example.

Your organization's application administrator manages the Enterprise Resource Pool, although your organization's resource managers may participate in the resource management process. You, as a project manager, can use any of these resources in your own projects based on your resource staffing needs.

Enterprise Global

Every time you launch Microsoft Project and connect to Project Online, the system opens a copy of two global files in the background: the **Global.mpt** file and the **Enterprise Global** file. Each Global file contains a library of project objects such as views, tables, filters, and groups.

The Global.mpt file is stored on your hard drive and contains a standard set of objects included by default with the Microsoft Project desktop client. The *Gantt Chart* view and the *Entry* table are two of these default objects. The Enterprise Global file is stored in the Project Online database and functions as your organization's "library" of custom enterprise objects. Your organization's application administrator manages the Enterprise Global file and is responsible for creating custom enterprise objects such as enterprise views, tables, filters, and groups.

When you launch Microsoft Project and connect to Project Online, the system opens the Global.mpt file from your hard drive, opens the Enterprise Global file from the server, and then combines the two into a single "cached" global file for the current session. This gives you access to all of the default objects in the Global.mpt file and all of the custom enterprise objects in the Enterprise Global file.

Understanding the Project Communications Life Cycle

The core functionality in Project Online provides a cyclical assignment and update process between project managers and team members. This cycle is the heart of the system's work and resource management system. Work assignments flow from the Microsoft Project schedule to the resources performing the work, and resources report progress data back to the project schedule. This project communication cycle flows through the following steps:

1. The project manager saves the project schedule to Project Online, as shown in Figure 1 - 2. At this point, executives *cannot see* the project information in the *Project Center* page of Project Web App (PWA) or in PWA reports, and project team members *cannot see* their task assignments in the *Tasks* or *Timesheet* pages of PWA.

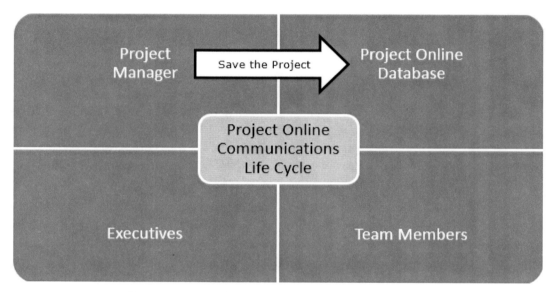

Figure 1 - 2: Save the project schedule to Project Online

2. The project manager publishes the project schedule to Project Online, as shown in Figure 1 - 3. At this point, team members can now see their task assignments in the *Tasks* and *Timesheet* pages in Project Web App. If your organization's application administrator enabled the *Notifications* feature in Project Online, team members also receive an e-mail message notifying them about their tasks in the newly published project. In addition, on the first publish of the project, Project Online also creates the Project Site for the project (not shown). The Project Site is where the project manager, team members, and executives collaborate together to manage project-related items such as risks, issues, and documents.

 Information: When a project manager publishes a project schedule to Project Online, this action makes the project data visible to executives in multiple locations in Project Web App. They can see the project data in the *Project Center* page and in detailed *Project* views. They can see resource data about the project's team members in the *Resource Assignments* and *Capacity Planning* pages. They can also see project and resource data in custom reports created by the organization's report authors.

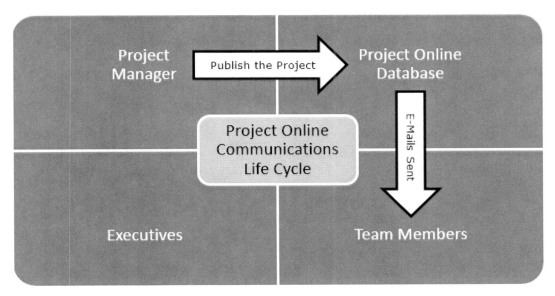

Figure 1 - 3: Publish the project – Team Members can see their task assignments

3. On the last day of each reporting period, team members submit their time and task progress for the tasks on which they worked during the current reporting period, as shown in Figure 1 - 4. If the application administrator enabled the *Notifications* feature in Project Online, the project manager receives an e-mail message notifying him/her about the pending task updates for the project. Once submitted, the task updates are visible to the project manager in the *Approval Center* page of PWA. At this point, Project Online does not apply the task updates to the project schedule until the project manager formally approves the updates.

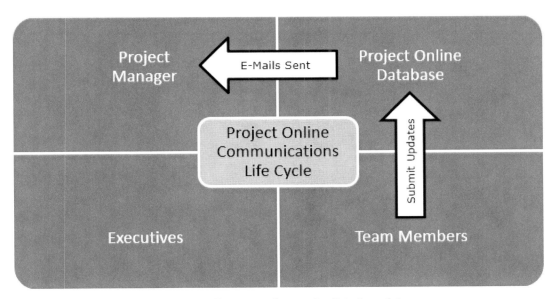

Figure 1 - 4: Team members submit task updates

4. The project manager reviews each set of task updates from project team members, as shown in Figure 1 - 5. The project manager can individually approve or reject each task update in the *Approval Center* page in PWA. Once approved, Project Online applies the task updates to the project schedule. At this point, the project manager can open the project schedule in Microsoft Project, analyze variance, revise the project schedule as needed, and then publish the project schedule.

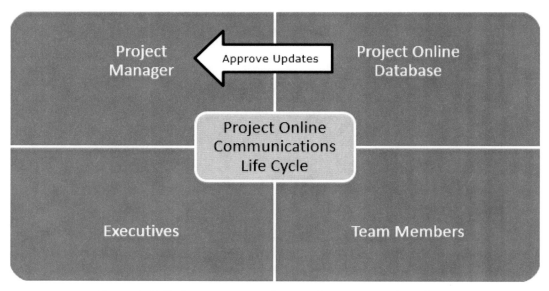

Figure 1 - 5: Project manager reviews and approves task updates

5. After accepting or rejecting each task update, the project manager saves and publishes the latest schedule changes from Microsoft Project to Project Online, as shown in Figure 1 - 6. Although not indicated in the figure, this action makes the latest schedule changes visible to executives in the *Project Center* page in PWA and in PWA reports, and to team members in the *Timesheet* and *Tasks* pages of PWA.

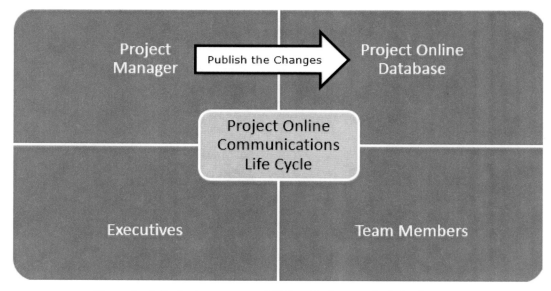

Figure 1 - 6: Project manager publishes the latest schedule changes

6. At any time, executives can view data about all published projects in the organization's portfolio, as shown in Figure 1 - 7. Project Web App provides numerous locations for viewing this information, including the *Project Center* page, the *Resource Center* page, and in custom reports created by your application administrator or reporting specialist.

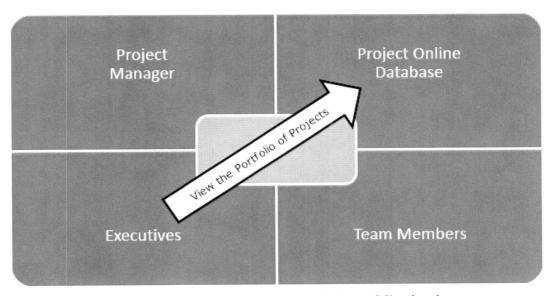

Figure 1 - 7: Executives view the organization's portfolio of projects

Module 02

Microsoft Project Overview

Learning Objectives

After completing this module, you will be able to:

- Customize the Ribbon and Quick Access Toolbar using a customization file
- Apply the appropriate views, tables, filters, and groups to display the data you want to see
- Create your Project Web App login dialog
- Open, check out, close, and check in enterprise projects
- Understand and work with the local Cache
- Work with an enterprise project in Offline mode

Inside Module 02

Understanding the Ribbon and Quick Access Toolbar	13
Using Views, Tables, Filters, and Groups	15
Using Views	*15*
Using Tables	*19*
Using Filters	*21*
Using Groups	*26*
Creating a Project Web App Login Account	29
Opening and Closing Enterprise Projects	34
Understanding the Local Cache	39
Adjusting Local Cache Settings	*39*
Viewing Local Cache Contents	*40*
Cleaning Up the Local Cache	*42*
Resolving Local Cache Corruption	*44*
Working with Offline Projects	45
Offline Mode While Away from the Corporate Network	*46*
Offline Mode When You Lose Connectivity	*49*

Understanding the Ribbon and Quick Access Toolbar

At the top of the application window, Microsoft Project displays the **Ribbon** and the **Quick Access Toolbar**. The software displays in the Ribbon across the top of the entire application window, while it displays the Quick Access Toolbar in the far upper-left corner of the application window. Use the Ribbon to access the most commonly used commands and buttons. Use the Quick Access Toolbar to access the commands you use most frequently.

By default, the software includes six ribbons, including the *Task, Resource, Report, Project, View, Help,* and *Format* ribbons. To display any ribbon, click the relevant ribbon tab. For example, click the **Resource** tab to display the *Resource* ribbon shown in Figure 2 - 1.

Figure 2 - 1: Resource ribbon

The *Task* ribbon contains the commands you need for working with tasks, while the *Resource* ribbon contains the commands you need for working with resources. The *Report* ribbon offers you three sets of commands for working with reporting on an individual project. The *Project* ribbon contains the commands you need for working with high-level project information. The *View* ribbon contains commands that allow you to manipulate your project by applying items such as views, tables, filters, or groups. The *Help* ribbon offers what the name implies: help with using the software. The *Format* ribbon is a context sensitive ribbon that contains commands you can use to format the view currently applied.

By default, the Quick Access Toolbar contains only three buttons: *Save, Undo,* and *Redo*. To add your favorite buttons to this toolbar, click the **Customize Quick Access Toolbar** pick list button at the right end of the toolbar, and select your favorite command from the pick list.

Microsoft Project allows you to customize both the Ribbon and the Quick Access Toolbar. Although the process for manually customizing these two objects is beyond the scope of this book, the fastest and easiest way to customize them is to import a customization file that contains a set of customizations for the Ribbon and/or the Quick Access Toolbar. The following Hands On Exercise teaches you how to import a customization file.

 Warning: If you accidentally or intentionally double-click any ribbon tab, Microsoft Project automatically **collapses** the Ribbon. Although this action gives you more vertical screen space in the application window, it also forces you to click a ribbon tab each time you want to click a command on that ribbon. To expand the Ribbon so that you can see all of the commands, double-click on any ribbon tab. Alternatively, you can also right-click on any ribbon tab and **deselect** the **Collapse the Ribbon** item on the shortcut menu.

Module 02

Hands On Exercise

Exercise 2 - 1

Customize your Ribbon and Quick Access Toolbar by importing a customization file.

Warning: In order to work the Hands On Exercises in this book, please follow the directions in the *Downloading the Sample Files* section on page xiii in the introduction to the book. **You must download and unzip the sample files before proceeding with this exercise.**

1. Launch Microsoft Project.
2. Click the **File** tab and then click the **Options** tab in the *Backstage*.
3. In the **Project Options** dialog, click the **Customize Ribbon** tab.
4. In the lower-right corner of the *Project Options* dialog, click the **Import/Export** pick list button and select the **Import customization file** item on the menu.
5. In the *File Open* dialog, navigate to the folder where you saved the sample files that accompany this book.
6. In the dialog, select the **Projility PPM Ribbon and QAT Customizations.exportedUI** file, and then click the **Open** button.
7. In the confirmation dialog, click the **Yes** button.
8. Click the **OK** button to close the *Project Options* dialog.
9. Right-click on the **Task** ribbon tab and select the **Show Quick Access Toolbar Below the Ribbon** item on the shortcut menu.

Information: To reset the customizations made to the ribbon and Quick Access Toolbar, click the **File** tab and then click the **Options** tab in the *Backstage*. In the *Project Options* dialog, click the **Customize Ribbon** tab. In the lower-right corner of the *Project Options* dialog, click the **Reset** pick list button and select the **Reset all customizations** item on the menu. Click the **Yes** button in the confirmation dialog. Microsoft Project resets both the ribbon and the Quick Access Toolbar to their default configurations.

Your copy of Microsoft Project now includes a new *Definition* ribbon, a new *Microsoft PPM* ribbon, and the customized Quick Access Toolbar displayed below the Ribbon. Use commands on the *Definition* ribbon to properly define each new project you create. Use the commands in the *Microsoft PPM* ribbon to work with Project Online or Project Server. The customized Quick Access Toolbar contains a collection of commonly used buttons, such as the *Zoom In* and *Zoom Out* buttons, for example. Figure 2 - 2 shows the new *Microsoft PPM* ribbon and the customized Quick Access Toolbar.

Microsoft Project Overview

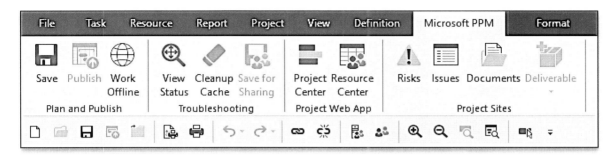

Figure 2 - 2: Custom Ribbon and Quick Access Toolbar

Using Views, Tables, Filters, and Groups

Figure 2 - 3 displays the complete Microsoft Project Data Model as it affects views, tables, filters, and groups. Notice that the software recognizes three separate and distinct types of data: **Task** data, **Resource** data, and **Assignment** data. *Task* and *Resource* data include their own unique sets of views, tables, filters, and groups. *Assignment* data is actually a joined set of data about a task and the resources assigned to it. Because *Assignment* data is so unique, Microsoft Project includes two special views for displaying it: the *Task Usage* view (which displays task data and resource assignments) and the *Resource Usage* view (which displays resource data and task assignments). When you apply the *Task Usage* view, you can then apply task tables, filters, and groups with the view. When you apply the *Resource Usage* view, you can then apply resource tables, filters, and groups with the view.

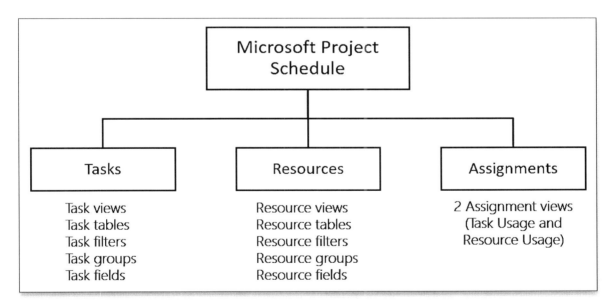

Figure 2 - 3: Microsoft Project Data Model

Using Views

Microsoft Project includes a number of built-in task views and resource views. Software users often define the term "view" as "different ways of looking at my project data." Although this definition is correct, Microsoft Project formally defines a view as:

View = Table + Filter + Group + Screen

In the Microsoft Project definition of a view, the **Table** displays the columns you see on the left side of the view, the **Filter** defines which rows you see, the **Group** organizes the rows into groups with common attributes or characteristics, and the **Screen** determines what appears on the right side of the view. Specifically, the *Screen* determines whether you see some type of a Gantt chart screen, a timephased grid screen, or no screen at all on the right side of the view.

To apply a view, use any of the following methods:

- Click the **Task** tab to display the *Task* ribbon. In the *View* section of the *Task* ribbon, click the **Gantt Chart** pick list button and select any view on the list. You can also select the **More Views** item at the bottom of the list and then apply a view in the *More Views* dialog. Figure 2 - 4 shows the views available on the *Gantt Chart* pick list.

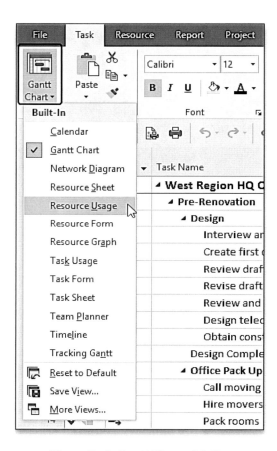

Figure 2 - 4: Gantt Chart pick list
on the Task ribbon

- Click the **Resource** tab to display the *Resource* ribbon. In the *View* section of the *Resource* ribbon, click the **Team Planner** pick list button and select any view on the list. You can also select the **More Views** item at the bottom of the list and then apply a view in the *More Views* dialog.

Microsoft Project Overview

- Click the **View** tab to display the *View* ribbon. Click any pick list button in the *Task Views* or *Resource Views* sections of the *View* ribbon and select a view on the list. On any *view* pick list, you can also select the **More Views** item at the bottom of the list and then apply a view in the *More Views* dialog.

When you select the **More Views** item at the bottom of any view pick list, Microsoft Project displays the *More Views* dialog shown in Figure 2 - 5. This dialog displays every default and custom view. Select any view in the dialog and then click the **Apply** button.

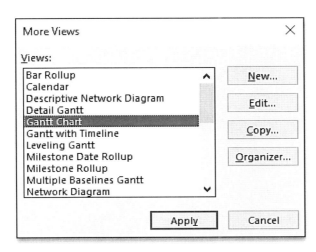

Figure 2 - 5: More Views dialog

In Microsoft Project, a **Single view** is any view that fills the entire screen. You commonly use Single views such as the *Gantt Chart* and *Resource Sheet* views.

In Microsoft Project, a **Combination view** (also known as a **Split view**) is any view that contains two views, displayed in a split-screen format with each view tiled in its own pane. The most commonly used Combination views are the *Task Entry* view and the new *Gantt with Timeline* view. To apply a Combination view, click the **View** tab to display the *View* ribbon. In the *Split View* section of the *View* ribbon, select the **Timeline** checkbox to apply the *Gantt with Timeline* view, or select the **Details** checkbox to apply the *Task Entry* view, such as shown in Figure 2 - 6. Deselect the selected checkbox to return to a Single view.

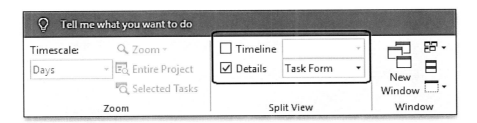

Figure 2 - 6: Split view section of the View ribbon

Module 02

Information: The fastest way to apply the *Gantt with Timeline* view is to right-click anywhere in the chart portion of the *Gantt Chart* view and select the **Show Timeline** item on the shortcut menu. The fastest way to apply the *Task Entry* view is to right-click anywhere in the chart portion of the *Gantt Chart* view and select the **Show Split** item on the shortcut menu. To return to the default *Gantt Chart* view, right-click again in the chart portion of the *Gantt Chart* view and **deselect** either the **Show Timeline** or **Show Split** item on the shortcut menu.

Hands On Exercise

Exercise 2 - 2

Explore commonly used views in Microsoft Project.

1. Open the **Project Navigation 2019.mpp** sample file from the folder where you saved the sample files that accompany this book.

2. Click the **Task** tab to display the *Task* ribbon.

3. In the *View* section of the *Task* ribbon, use the **Gantt Chart** pick list button to apply and study the following views:

    - **Tracking Gantt**
    - **Task Usage**

4. Click the **Resource** tab to display the *Resource* ribbon.

5. In the *View* section of the *Resource* ribbon, use the **Team Planner** pick list button to apply and study the following views:

    - **Resource Sheet**
    - **Resource Usage**

6. Click the **View** tab to display the *View* ribbon.

7. In the *Task Views* section of the *View* ribbon, click the **Gantt Chart** pick list button and apply the **Gantt Chart** view.

Information: When you click the *Gantt Chart* pick list button, Microsoft Project displays a pick list of the most commonly used Gantt-based views. If you click the *Gantt Chart* button instead of the pick list button, the software applies the most recently used Gantt-based view, which in this exercise would be the *Tracking Gantt* view and not the *Gantt Chart* view.

8. In the *Split View* section of the *View* ribbon, select the checkboxes to individually apply the following Combination views:

18

- **Gantt with Timeline**
- **Task Entry**

9. In the *Split View* section of the *View* ribbon, ***deselect*** the selected checkbox to return to the default **Gantt Chart** view.

Using Tables

Microsoft Project includes a number of built-in task tables and resource tables. By definition, a table is a "collection of columns" and the name of each table often describes the type of columns in the table. For example, the task *Cost* table contains columns showing the cost data associated with each task in the project. Because the software displays tables as a component of views, you can only use task tables with task views and can only use resource tables with resource views. To apply any table in the current view, use either of the following methods:

- Click the **View** tab to display the *View* ribbon. In the *Data* section of the *View* ribbon, click the **Tables** pick list button and select any table on the list. You can also select the **More Tables** item at the bottom of the Tables pick list. Figure 2 - 7 shows the list of available tables on the *Tables* pick list in the *Data* section of the *View* ribbon.

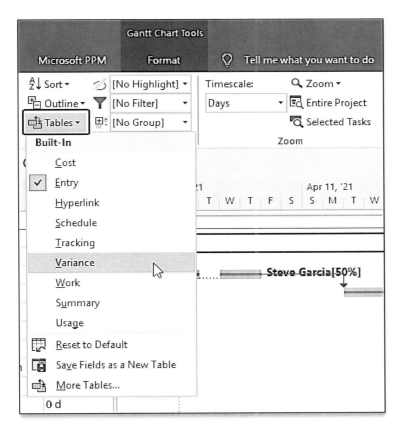

Figure 2 - 7: Tables pick list on the View ribbon

- Right-click the **Select All** button and then select any table on the shortcut menu, such as shown in Figure 2 - 8. You can also select the **More Tables** item at the bottom of the shortcut menu.

Figure 2 - 8: Right-click on the Select All button

When you select the **More Tables** item at the bottom of the **Tables** pick list or shortcut menu, Microsoft Project displays the *More Tables* dialog shown in Figure 2 - 9. This dialog displays every default and custom table for the type of view applied currently. This means if you have a task view applied, the *More Tables* dialog displays the complete list of task tables. If you have a resource view applied, the *More Tables* dialog shows the complete list of resource tables. Select any table in the dialog and then click the **Apply** button.

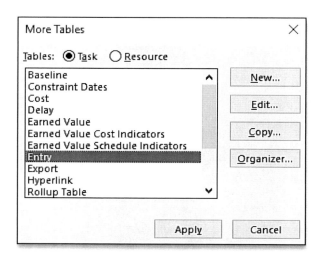

Figure 2 - 9: More Tables dialog

Hands On Exercise

Exercise 2 - 3

Explore commonly used task tables in Microsoft Project.

1. Apply the **Gantt Chart** view, if necessary.
2. Click the **View** tab to display the *View* ribbon, if necessary.
3. In the *Data* section of the *View* ribbon, click the **Tables** pick list and select the **Cost** table.
4. Drag the split bar to the right edge of the **Remaining** column.

Examine all of the columns included in the *Cost* table. Notice that every column in the *Cost* table displays task cost information.

5. In the *Data* section of the *View* ribbon, click the **Tables** pick list and select the **Work** table.

Examine all of the columns included in the *Work* table. Notice that every column in the *Work* table displays task work information.

6. Right-click the **Select All** button and select the **Schedule** table on the shortcut menu.

Examine all of the columns included in the *Schedule* table. Notice that the last column in the *Schedule* table is *Total Slack*. Microsoft Project calculates this column and uses the *Total Slack* information to determine whether each task is a Critical task. Critical tasks have a *Total Slack* value of *0 days*, while non-Critical tasks have a *Total Slack* value greater than *0 days*.

7. Right-click the **Select All** button again and select the **Entry** table on the shortcut menu.
8. Drag the split bar to the right edge of the **Duration** column.

Using Filters

Microsoft Project includes a large number of built-in task filters and resource filters. Filters allow you to display or highlight specific data rows according to your filter criteria. Because the software applies filters within views, you can only use task filters with task views and resource filters with resource views. The software offers you three ways to apply filters:

- Apply a filter as a **standard filter**.
- Apply a filter as a **highlight filter**.
- Use **AutoFilter**.

Using Standard Filters

When you apply a filter as a **standard filter**, Microsoft Project displays only the data rows that meet your filter criteria. To apply a standard filter, click the **View** tab to display the *View* ribbon. In the *Data* section of the *View* ribbon, click the **Filter** pick list and select a filter, such as shown in Figure 2 - 10. You can also select **the More Filters** item near the bottom of the list.

Notice that the *Filter* pick list allows you to choose from a list of most commonly used filters, plus you can choose additional filtering options. These additional options include *Clear Filter* to remove filtering criteria, *New Filter* to create a new filter from scratch, *More Filters* to display the *More Filters* dialog, and *Display AutoFilter* to turn on the AutoFilter feature in the data grid. By the way, Microsoft Project enables the *AutoFilter* feature by default in all task and resource tables.

Any filter name that ends with an ellipsis character (...) is an interactive filter. This means that when you apply the filter, Microsoft Project prompts you in one or more dialogs for your filter criteria, and then it applies the filter using the criteria you supply. Interactive filters on the *Filter* pick list include the *Date Range*, *Task Range*, and *Using Resource* filters.

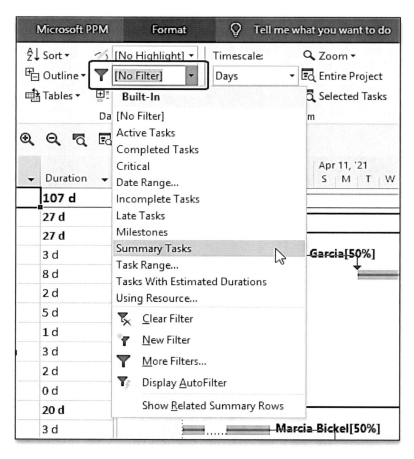

Figure 2 - 10: Apply a Standard Filter

If you click the **More Filters** item near the bottom of the **Filter** pick list, the software displays the *More Filters* dialog shown in Figure 2 - 11. This dialog displays all available default and custom filters. After selecting any filter in the dialog, click the **Apply** button to apply the filter as a standard filter. You can also click the **Highlight** button to apply the filter as a highlight filter.

Microsoft Project Overview

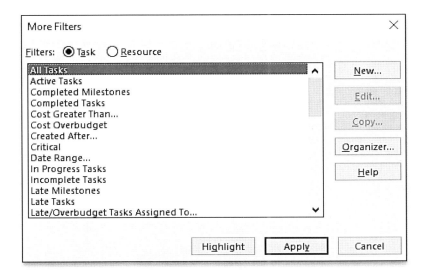

Figure 2 - 11: More Filters dialog

To remove a standard filter and reapply the default [No Filter] filter, click the **Filter** pick list again and choose either the **[No Filter]** item or the **Clear Filter** item on the pick list.

 Information: Press the **F3** function key on your computer keyboard to quickly apply the **[No Filter]** filter in any task or resource view.

Using Highlight Filters

When you apply any filter as a **highlight filter**, Microsoft Project displays *all rows* in the project, but *highlights* only the rows that meet the filter criteria using yellow cell background formatting. To apply any filter as a highlight filter, click the **View** tab. In the *Data* section of the *View* ribbon, click the **Highlight** pick list, and then select a highlight filter.

 Information: The list of filters is identical on the *Highlight* pick list and the *Filter* pick list.

The *Highlight* pick list allows you to choose from a list of the most commonly used filters, and provides other filtering options such as *Clear Highlight* to remove the highlight filter, *New Highlight Filter* to create a new highlight filter, and *More Highlight Filters* to display the *More Filters* dialog. If you click the **More Highlight Filters** item near the bottom of the **Highlight** pick list, Microsoft Project displays the *More Filters* dialog shown previously in Figure 2 - 11. Remember that you can apply any filter as a highlight filter in the *More Filters* dialog by selecting the filter and then clicking the **Highlight** button.

To remove a highlight filter and reapply the default *[No Filter]* filter, click the **Highlight** pick list again and choose either the **[No Highlight]** item or the **Clear Highlight** item on the pick list.

23

Information: Press the **F3** function key on your computer keyboard to quickly apply the **[No Highlight]** filter in any task or resource view.

Using AutoFilter

AutoFilter is your third filtering option. Interestingly enough, Microsoft Project actually offers you *two ways* to AutoFilter your project. To use either AutoFilter method, click the **AutoFilter** pick list arrow button in the column header of any column on which you want to filter, and the software displays the *AutoFilter* menu. The AutoFilter menu contains filtering options that are appropriate for the type of data displayed in the column.

Notice the *AutoFilter* menu for the *Duration* column shown in Figure 2 - 12. To apply an *AutoFilter* using the first method, *deselect* the **(Select All)** checkbox, select the checkboxes for one or more items on the menu, and then click the **OK** button. The software displays all rows that meet your AutoFilter criteria.

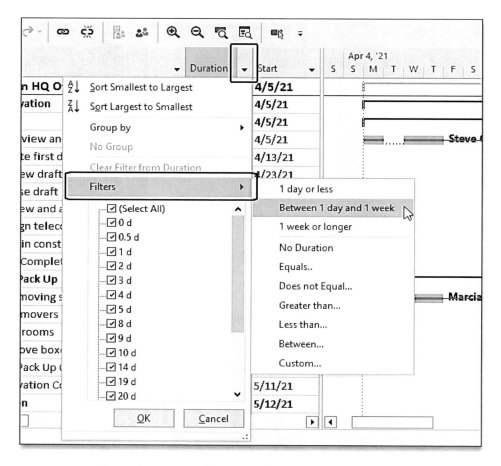

Figure 2 - 12: AutoFilter menu for the Duration column

 Warning: When you apply an AutoFilter by selecting more than one checkbox on the *AutoFilter* menu, Microsoft Project applies your filter criteria using a Boolean "Or" filter. This means that if you select the names of two resources in the AutoFilter menu for the Resource Names column, the software displays the tasks assigned to *either* of the selected resources.

Microsoft Project Overview

To apply an *AutoFilter* using the second method, click the **Filters** item on the **AutoFilter** menu and select one of the built-in filters on the flyout menu. Notice in the *AutoFilter* menu for the *Duration* column shown previously in Figure 2 - 12 that the *Filters* flyout menu contains three default filters, including the *1 day or less*, the *Between 1 day and 1 week*, and the *1 week or longer* items. Select one of the default filters on the flyout menu and the software displays all rows that meet your *AutoFilter* criteria.

When you apply an *AutoFilter* to a column, Microsoft Project displays a "funnel" indictor instead of the pick list button in that column header. For example, notice the "funnel" indicator in the *Duration* column shown in Figure 2 - 13 to indicate that I applied an *AutoFilter* to the *Duration* column.

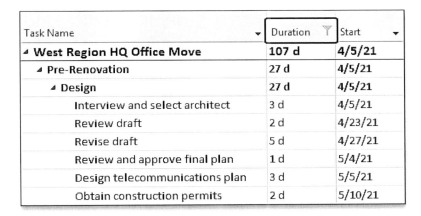

Figure 2 - 13: AutoFilter applied to the Duration column

To remove the current *AutoFilter* applied, click the funnel button in the column header and then select the **Clear Filter from** _____ item on the menu. Alternately, you can select the (**Select All**) checkbox on the **AutoFilter** menu and then click the **OK** button.

Information: Press the **F3** function key on your computer keyboard to remove the current *AutoFilter* and reapply the **[No Filter]** filter in any task or resource view.

Hands On Exercise

Exercise 2 - 4

Explore standard filters, highlight filters, and AutoFilter in Microsoft Project.

1. Apply the **Gantt Chart** view and **Entry** table, if necessary.
2. Click the **View** tab to display the *View* ribbon, if necessary.

25

3. In the *Data* section of the *View* ribbon, click the **Filter** pick list and select the **Using Resource...**filter.

4. In the *Using Resource* dialog, click the **Show tasks using** pick list and select **Russ Powell**, and then click the **OK** button.

Notice that Microsoft Project displays all tasks assigned to Russ Powell in different sections of the project, as indicated by their respective summary tasks.

5. Press the **F3** function key on your computer keyboard to clear the standard filter.

6. In the *Data* section of the *View* ribbon, click the **Highlight** pick list and select the **Using Resource...**filter again.

7. In the *Using Resource* dialog, click the **Show tasks using** pick list and select **Russ Powell**, and then click the **OK** button.

Scroll down through the project and notice that Microsoft Project displays all tasks in the project, but highlights the tasks assigned to Russ Powell using the yellow cell background color.

8. Press the **F3** function key on your computer keyboard to clear the highlight filter.

9. Drag the split bar to the right edge of the **Resource Names** column.

10. Click the **AutoFilter** pick list button in the **Resource Names** column header.

11. In the *AutoFilter* menu, *deselect* the **(Select All)** checkbox, select the checkboxes for **Marcia Bickel** and **Russ Powell**, and then click the **OK** button.

12. Drag the split bar to the right edge of the **Duration** column.

Notice that Microsoft Project displays the tasks assigned to either Marcia Bickel or to Russ Powell.

13. Press the **F3** function key on your computer keyboard to clear the *AutoFilter*.

Using Groups

Microsoft Project offers a small selection of built-in task groups and resource groups. You use groups to organize tasks or resources into groups with similar attributes or characteristics. Because the software applies groups within views, you must use task groups with task views and resource groups with resource views.

To apply a group to your project data, click the **View** tab to display the *View* ribbon. In the *Data* section of the *View* ribbon, click the **Group By** pick list and select a group, as shown in Figure 2 - 14. You can also select **the More Groups** item near the bottom of the list.

Notice that the *Group By* pick list allows you to choose from a list of most commonly used groups, plus you can choose additional grouping options. These additional options include *Clear Group* to remove the applied grouping, *New Group By* to create a new group from scratch, *More Groups* to display the *More Groups* dialog, and *Maintain Hierarchy in Current Group* to display the Work Breakdown Structure (WBS) of summary tasks for the tasks in each group.

Microsoft Project Overview

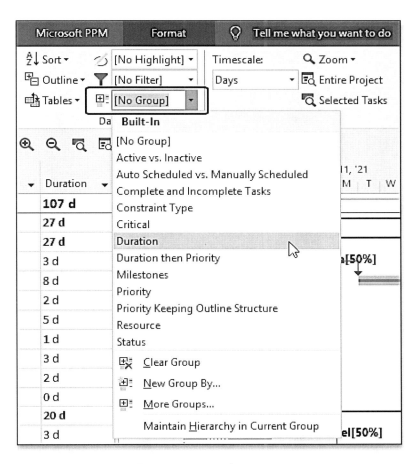

Figure 2 - 14: Apply a group

When you apply a group, Microsoft Project organizes the tasks or resources into groups with similar attributes or characteristics. For example, Figure 2 - 15 shows *Gantt Chart* view with the *Resource* group applied. Notice that the software organized the tasks into groups according to the resources assigned to each task.

If you select the **More Groups** item on the **Group By** pick list, the software displays the *More Groups* dialog shown in Figure 2 - 16. This dialog displays all available default and custom groups. In the *More Groups* dialog, select any default or custom group and then click the **Apply** button.

27

Module 02

	ⓘ	Task Mode ▾	Task Name		Duration ▾
			▷ Resource Names: No Value		87d
			▲ Resource Names: A1 Movers		89d
14	✓ 🗎	➡	Pack rooms		4 d
15	✓	➡	Remove boxes to storage		3 d
67		➡	Furniture delivery and setup		1 d
71	🗎	➡	Retrieve boxes from storage		1 d
72		➡	Unpack		4 d
			▲ Resource Names: Amy McKay		1d
36		➡	Install appliances		1 d
			▲ Resource Names: Amy McKay,Gene Cain		5d
26		➡	Frame new walls		5 d
			▲ Resource Names: Amy McKay,Gene Cain,Terry Madison		20d
21	✓	➡	Strip walls		3 d
24	✓ 🗎	➡	Remove asbestos in ceiling		5 d
27		➡	Put up dry wall		2 d
			▲ Resource Names: Bob Siclari		7d
48		➡	Install pipes		5 d
49		➡	Install sink and faucets		1 d
50		➡	Connect appliances		1 d

Figure 2 - 15: Task list with Resource group applied

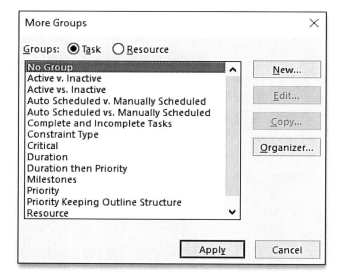

Figure 2 - 16: More Groups dialog

To remove a group and reapply the default *[No Group]* group, click the **Group By** pick list and select either the **[No Group]** item or the **Clear Group** item on the pick list.

28

Microsoft Project Overview

 Information: Press the **Shift + F3** key combination on your computer keyboard to remove the current group and reapply the **[No Group]** group

 Hands On Exercise

Exercise 2 - 5

Explore groups in Microsoft Project.

1. Apply the **Gantt Chart** view and **Entry** table, if necessary.
2. Click the **View** tab to display the *View* ribbon, if necessary.
3. In the *Data* section of the *View* ribbon, click the **Group By** pick list and select the **Duration** group.

Scroll through the project and notice that the *Duration* group organizes tasks into groups according their *Duration* value. To most project managers, this grouping is not useful.

4. Press the **Shift + F3** key combination on your computer keyboard to remove the current group.
5. In the *Resource Views* section of the *View* ribbon, click the **Resource Usage** pick list button and select the **Resource Usage** view.
6. Click the **Group By** pick list button and select the **Assignments Keeping Outline Structure** group.

As you scroll through the list of resources, notice that this group displays the Work Breakdown Structure (WBS) for every task assignment in the *Resource Usage* view. To most project managers, this grouping is particularly useful!

7. Close but do not save the **Project Navigation 2019.mpp** sample file.

Creating a Project Web App Login Account

Before you can use Microsoft Project with Project Online or Project Server, you must create a login account that connects Microsoft Project to Project Web App. To create a login account in Microsoft Project, complete the following steps:

1. Launch Microsoft Project and create a new blank project from the *Start* page, if necessary.
2. Click the **File** tab and then click the **Info** tab in the *Backstage*. The software displays the *Info* page, such as shown in Figure 2 - 17.

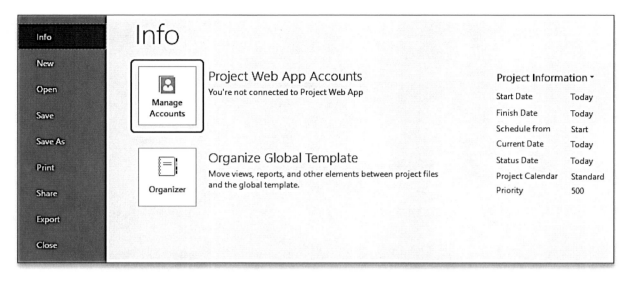

Figure 2 - 17: Click the Manage Accounts button

3. In the *Project Web App Accounts* section of the *Info* page, click the **Manage Accounts** button. The software displays the *Project Web App Accounts* dialog shown in Figure 2 - 18.

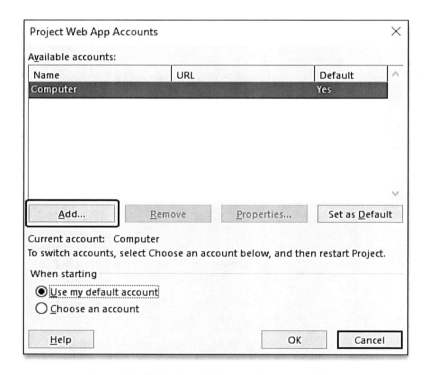

Figure 2 - 18: Project Web App Accounts dialog

4. In the *Project Web App Accounts* dialog, click the **Add** button. The software displays the *Account Properties* dialog shown in Figure 2 - 19.

Microsoft Project Overview

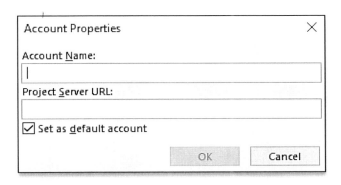

Figure 2 - 19: Account Properties dialog - blank

5. In the *Account Properties* dialog, enter a friendly name such as **Project Online** in the **Account Name** field.

6. In the *Account Properties* dialog, enter the URL of your organization's Project Web App in the **Project Server URL** field.

 Information: If you do not know the URL of your organization's Project Web App, you should contact your application administrator to obtain this important information.

7. In the *Account Properties* dialog, leave the **Set as default account** checkbox selected. Figure 2 - 20 shows the completed *Account Properties* dialog.

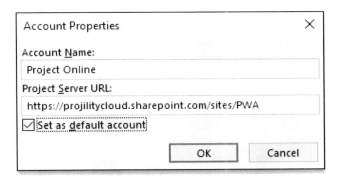

Figure 2 - 20: Account Properties dialog - completed

8. In the *Account Properties* dialog, click the **OK** button. The software displays your new login account in the *Available accounts* section of the *Project Web App Accounts* dialog, such as shown in Figure 2 - 21.

9. In the lower left corner of the *Project Web App Accounts* dialog, select the **Choose an account** option, and then click the **OK** button.

31

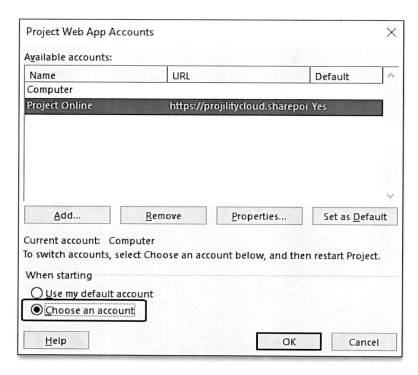

Figure 2 - 21: Project Web App Accounts dialog
displays new login account

10. Exit Microsoft Project and then launch the software again. The system displays the *Login* dialog with your default account selected in the *Profile* field.

11. In the *Login* dialog, leave the default **Profile** value selected and then ***deselect*** the **Load Summary Resource Assignments** checkbox, such as in Figure 2 - 22.

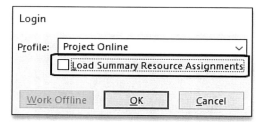

Figure 2 - 22: Login dialog

 Information: The *Load Summary Resource Assignments* option determines how Microsoft Project displays resource assignment information when you open an enterprise project and apply the *Resource Usage* view. If you select this option, the *Resource Usage* view displays resource assignments for the current project plus a collapsed summary line named *Other projects and commitments*. When you expand this summary line, you see a list of every other enterprise project in which the resource is assigned to tasks. If you deselect this option, the *Resource Usage* view displays resource assignments only for the current project. Because of how this option impacts the process of leveling overallocated resources, Projility strongly recommends that you ***deselect*** the *Load Summary Resource Assignments* option.

Microsoft Project Overview

12. In the *Login* dialog, click the **OK** button.

When Microsoft Project successfully connects to Project Web App, it displays a "globe" indicator at the left end of the *Status* bar at the bottom of the application window. If you float your mouse pointer over this "globe" indicator, the software displays a *Connected to _____* tooltip with the name of the login account you previously created, such as shown in Figure 2 - 23.

Figure 2 - 23 Successful connection to Project Web App

Hands On Exercise

Exercise 2 - 6

Create your Project Web App login account in Microsoft Project, if you have not done so already.

1. Obtain the URL of your organization's Project Web App from your application administrator, if necessary.
2. Launch Microsoft Project and create a new blank project from the *Start* page, if necessary.
3. Click the **File** tab and then click the **Info** tab in the *Backstage*.
4. In the *Project Web App Accounts* section of the *Info* page, click the **Manage Accounts** button.
5. In the *Project Web App Accounts* dialog, click the **Add** button.
6. In the *Account Properties* dialog, do the following:
    - Enter a friendly name in the **Account Name** field.
    - In the **Project Server URL** field, enter the URL of your organization's Project Web App.
    - Leave the **Set as default account** checkbox selected.
    - Click the **OK** button.
7. In the *Project Web App Accounts* dialog, select the **Choose an account** option, and then click the **OK** button.
8. Exit Microsoft Project and then launch the software again.
9. In the *Login* dialog, leave the default **Profile** value selected and then *deselect* the **Load Summary Resource Assignments** checkbox.
10. In the *Login* dialog, click the **OK** button.

Opening and Closing Enterprise Projects

To open an enterprise project using Microsoft Project, click the **File** tab and then click the **Open** tab in the *Backstage*. The software displays the *Open* page with the *Recent* option selected, such as shown in Figure 2 - 24. The *Projects* list on the right side of the page displays the list of enterprise projects with which you worked previously, listed in chronological order beginning with the most recently-opened project. You can see in Figure 2 - 24 that I previously opened five enterprise projects, the most recent of which is the *Chi Rho Software Marketing Campaign* project.

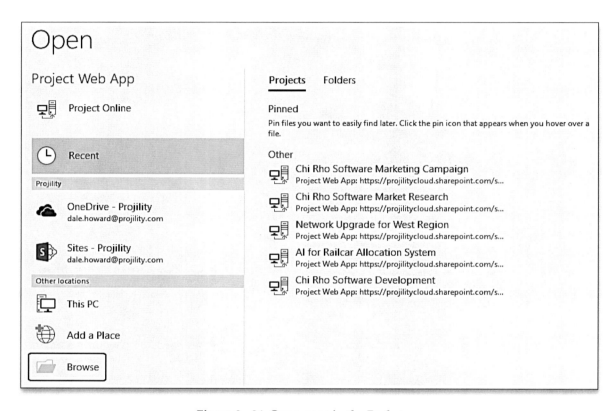

Figure 2 - 24: Open page in the Backstage

To open a recently used enterprise project, click the name of the project in the *Projects* list. To open a project not listed in the *Projects* section of the page, click the **Browse** button at the bottom of the page. The software displays the *Open* dialog. Notice in the *Open* dialog shown in Figure 2 - 25 lists the same five previously-opened enterprise projects displayed in the *Projects* section of the *Open* page shown previously in Figure 2 - 24.

 Information: The fastest and most direct way to display the *Open* dialog is to click the **Open** button on your Quick Access Toolbar, allowing you to bypass the *Open* page in the *Backstage* entirely.

Microsoft Project Overview

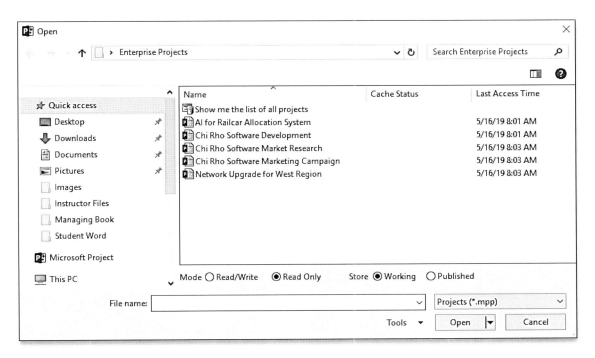

Figure 2 - 25: Open dialog displays recent projects only

To open a recent project, click the name of the project in the *Open* dialog and then click the **Open** button. To open a project not listed in the *Open* dialog, double-click the **Show me the list of all projects** link at the top of the project list. Microsoft Project updates the *Open* dialog with the list of all projects you have permission to access, such as shown in Figure 2 - 26.

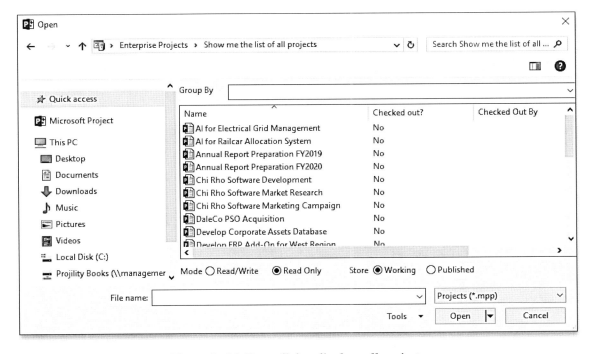

Figure 2 - 26: Open dialog displays all projects

35

Information: The system controls your project permissions based on the security Groups to which your application administrator has added you in Project Web App. The default permissions for members of the *Project Managers* group allow you to open and edit only the projects that belong to you. Depending on how your application administrator has customized security permissions, you may also be able to open and edit projects that belong to other project managers.

When viewing the list of all projects you have permission to access, the *Open* dialog includes a *Group By* pick list above the list of projects, such as shown in Figure 2 - 27. The *Group By* pick list allows you to apply grouping to the projects shown in the dialog using any default or custom enterprise *Project* field that contains a lookup table. If you want to group the projects shown in the dialog, click the **Group By** pick list and select a custom enterprise field from the list.

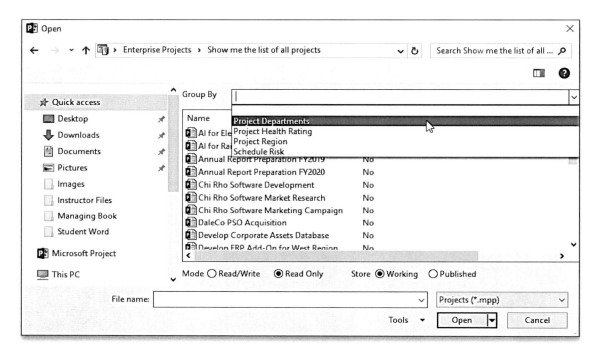

Figure 2 - 27: Open dialog – Group By pick list

Figure 2 - 28 shows the *Open* dialog with the list of projects grouped by the *Project Departments* field. Grouping on this field allows a project manager to group the projects by their respective department. In the *Open* dialog shown in Figure 2 - 28, you can see two projects that belong to the *Finance* department and multiple projects that belong to the *HR* department.

Microsoft Project Overview

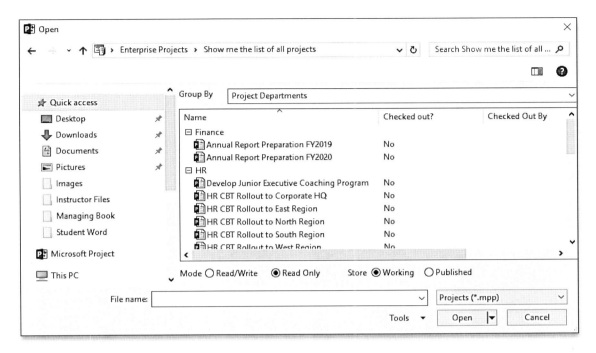

Figure 2 - 28: Open dialog with projects grouped by the Department field

To open a project from the list of all projects, click the name of the project in the *Open* dialog and then click the **Open** button. Regardless of which method you use to open an enterprise project, Microsoft Project always opens the project in *Read-Only* mode by default. The software indicates this by displaying a READ-ONLY gold banner across the top of the project, such as for the project shown in Figure 2 - 29. To check out the project so that you can work with it in *Read/Write* mode, click the **Check Out** button. When you check out an enterprise project, the system does not allow a fellow project manager or application administrator to edit and save your project, but they can open the project in *Read-Only* mode to view it.

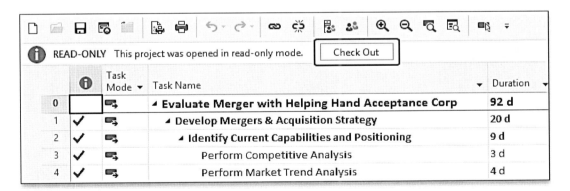

Figure 2 - 29: Check out an enterprise project

When you finish working with an enterprise project, you need to close and check in the project. Click the **File** tab and then click the **Close** tab in the *Backstage* or click the **Close** button in your Quick Access Toolbar. Microsoft Project displays the confirmation dialog shown in Figure 2 - 30. Click the **Yes** button in the confirmation dialog to initiate the check-in process.

Figure 2 - 30: Check in a project

After clicking the *Yes* button, it is a good idea to wait 5-10 seconds to allow Microsoft Project and Project Web App to finish the check-in process. You can actually see the state of the check-in process by displaying the *Open* dialog immediately after clicking the *Yes* button. Notice the *Checkin Pending* message in the *Open* dialog shown in Figure 2 - 31, indicating that the check-in process is underway for this project. Click the **Cancel** button to close the *Open* dialog. After you finish checking in the enterprise project, you can safely exit Microsoft Project.

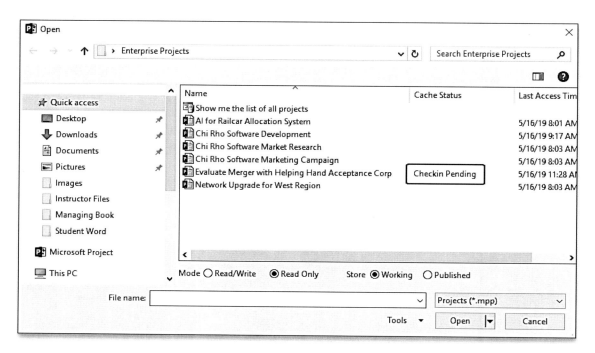

Figure 2 - 31: Checkin Pending message in the Open dialog

 Warning: ***Do not*** close an enterprise project by clicking the **Close** button (large **X** button) in the upper right corner of the Microsoft Project application window as this can cause your enterprise project to get "stuck" in a checked-out state. When this happens, the system will only allow you to open the project Read-only, which prevents you from saving and publishing the project.

Understanding the Local Cache

When you save a new enterprise project to Project Web App, Microsoft Project saves a complete copy of your project file in a special location on your hard drive known as the **Cache**, and then it "spools" all the project data to the Project Web App database. Each time you edit your project and save the changes, the software again saves a complete copy of your project in the Cache, but "spools" only the changes to the database.

When you open an enterprise project, Microsoft Project opens the copy from your Cache and then synchronizes this local copy with the enterprise project data stored in the database. Because the Cache works in the background, it makes the process of opening and saving enterprise projects much faster, even over a Wide Area Network (WAN) connection.

When you save an enterprise project in Microsoft Project, you can see the status of the Cache "spooling" operation near the right end of the *Status* bar at the bottom of the application window. When the system completes the "spooling" operation, it displays a *Save completed successfully* message in the *Status* bar.

Adjusting Local Cache Settings

To view and adjust the default settings for the Cache, click the **File** tab and then click the **Options** tab in the *Backstage*. In the *Project Options* dialog, click the **Save** tab. Microsoft Project displays the default settings in the *Cache* section of the *Save* page in the *Project Options* dialog, as shown in Figure 2 - 32.

The options in the *Cache* section of the *Project Options* dialog allow you to change two settings for the Cache: the size limit and its location. To adjust the size limit of the Cache, enter a new value in megabytes in the **Cache size limit (MB)** field. To change the location of the Cache, click the **Browse** button to the right of the *Cache location* field and select a new file location. Click the **OK** button when finished.

Information: By default, Microsoft Project stores the Cache data in the following folder on your laptop or PC:

C:\Users\YourUserID\AppData\Roaming\Microsoft\MS Project\16\Cache\

Warning: If your job requires you to travel, and you need to take enterprise projects with you while traveling, *do not* specify a Cache location that is outside your laptop's hard drive, such as on a network share. Microsoft Project needs to use the Cache to work with Offline projects, which allows you to check out and modify enterprise projects while you are away from your corporate network

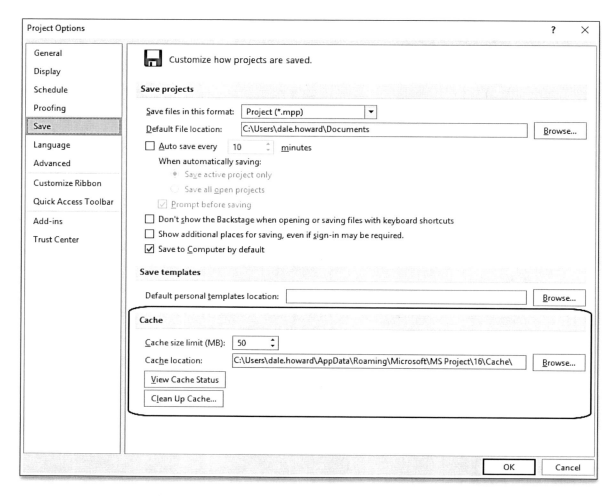

Figure 2 - 32: Cache settings in the Project Options dialog

Viewing Local Cache Contents

To view the contents of the Cache, click the **File** tab and then the **Options** tab in the *Backstage*. In the *Project Options* dialog, click the **Save** tab. In the *Cache* section of the *Save* page in the *Project Options* dialog shown previously in Figure 2 - 32, click the **View Cache Status** button. Alternately, you can also click the **View Cache Status** button in the *Troubleshooting* section of the custom *Microsoft PPM* ribbon. Using either method, Microsoft Project displays the *Active Cache Status* dialog shown in Figure 2 - 33.

The *Active Cache Status* dialog contains two pages: the *Status* and *Errors* pages. The *Status* page shows the status of recent Cache activities, such as saving or publishing a project. Notice in Figure 2 - 33 that I successfully published one enterprise project and successfully checked in three other projects.

Click the **Errors** tab to display the *Errors* page of the dialog. This page allows you to see Cache errors that occurred during a save or publish operation. The system displays the *Errors* page of the dialog shown in Figure 2 - 34. Notice that there are no errors shown on the *Errors* page of the *Active Cache Status* dialog. By the way, saving and publishing errors in the Cache are very rare with either Project Online or Project Server.

Microsoft Project Overview

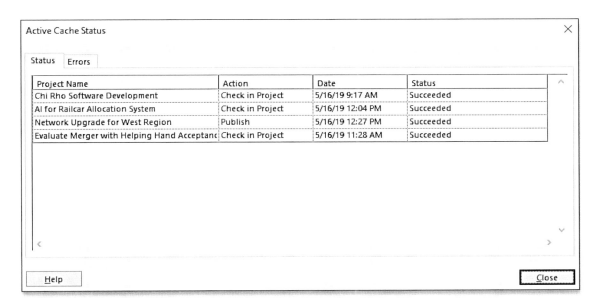

Figure 2 - 33: Active Cache Status dialog – Status page

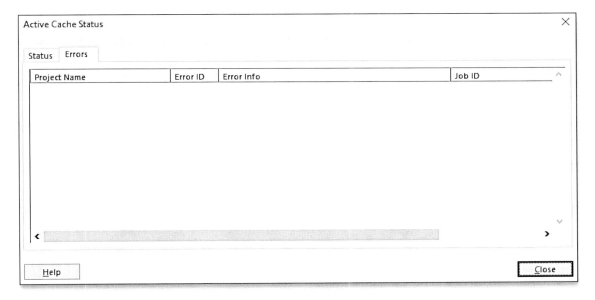

Figure 2 - 34: Active Cache Status dialog – Errors page

 Information: If you see errors of any kind on the *Errors* page of the *Active Cache Status* dialog, contact your application administrator immediately for help.

When finished, click the **Close** button to close the *Active Cache Status* dialog. Click the **OK** or **Cancel** button to close the *Project Options* dialog as well.

41

Cleaning Up the Local Cache

At some point, you may want to clean up the Cache to remove projects with which you no longer work, such as completed or cancelled projects, or to remove a project that is "stuck" in a checked-out state. To remove projects from the Cache, click the **File** tab and then the **Options** tab in the *Backstage*. In the *Project Options* dialog, click the **Save** tab. In the *Cache* section of the *Save* page in the *Project Options* dialog shown previously in Figure 2 - 32, click the **Cleanup Cache** button. Alternately, you can also click the **Cleanup Cache** button in the *Troubleshooting* section of the custom *Microsoft PPM* ribbon. Using either method, Microsoft Project displays the *Clean Up Cache* dialog shown in Figure 2 - 35.

The *Clean Up Cache* dialog displays statistics about the Cache in the *Cache Details* section at the top of the dialog. When you click the **Project Filter** pick list in the middle of the dialog, you can select one of two filters:

- **Projects not checked out to you** – Use this filter to clean up completed or cancelled projects.
- **Projects checked out to you** – Use this filter to clean up projects "stuck" in a checked-out state.

By default, Microsoft Project initially applies the *Projects not checked out to you* filter each time you display the *Clean Up Cache* dialog. Notice in Figure 2 - 35 that the system lists five projects not checked out to me currently.

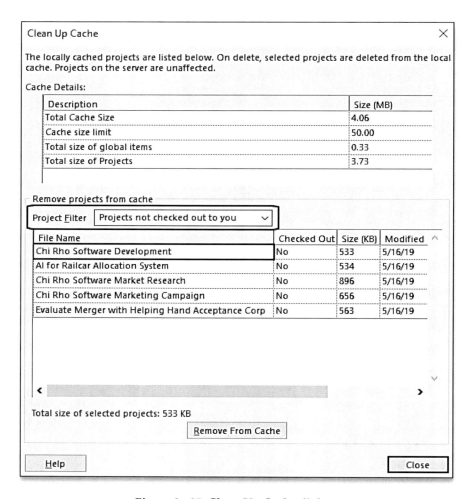

Figure 2 - 35: Clean Up Cache dialog

Microsoft Project Overview

To remove a completed or cancelled project from the Cache, select the name of the project in the *Clean Up Cache* dialog, and then click the **Remove from Cache** button. Microsoft Project removes the project from the Cache without displaying a confirmation dialog.

To remove a project from the Cache that is "stuck" in a checked-out state, click the **Project Filter** pick list and select the **Projects checked out to you** filter. Figure 2 - 36 shows the *Clean Up Cache* dialog with the *Projects checked out to you* filter applied. Notice that the *Clean Up Cache* dialog shows one enterprise project "stuck" in a checked-out state, which is the *Network Upgrade for West Region* project.

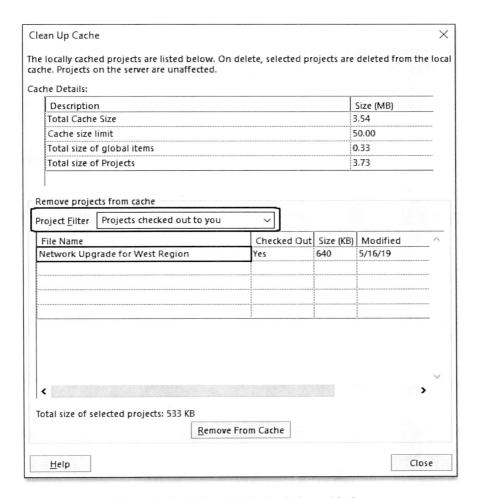

**Figure 2 - 36: Clean Up Cache dialog with the
Projects checked out to you filter applied**

To remove a project from the Cache that is "stuck" in a checked-out state, select the project and then click the **Remove From Cache** button. Microsoft Project displays the warning dialog shown in Figure 2 - 37.

43

Module 02

Figure 2 - 37: Remove from Cache confirmation dialog

If you are absolutely certain the project is "stuck" in a checked-out state, click the **Yes** button in the confirmation dialog to remove the project from the Cache. Click the **Close** button to close the *Clean Up Cache* dialog when you finish cleaning up enterprise projects in the Cache. Click the **OK** or **Cancel** button to close the *Project Options* dialog as well.

Information: Deleting a project from the Cache does not delete the project from the Project Web App database. Instead, this operation removes only the Cached version of the project from your hard drive. To open a project deleted from the Cache, you must double-click the **Show me the list of all projects** item in the *Open* dialog and open the project from the list of all projects you have permission to access.

Warning: If your organization's application administrator deletes a project from the Project Web App database, the system ***does not*** delete the project from your Cache. This means that the project continues to appear in the *Open* dialog, but you are not able to open it. You should remove the project from your Cache to avoid confusion and frustration.

Resolving Local Cache Corruption

Even though Microsoft has made improvement to Project Online and Project Server, it is still possible for the Cache to become corrupted. If you see unusual behavior in your enterprise projects, and suspect that a corrupted Cache is the cause, you can safely delete the Cache files by completing the following steps:

1. Exit Microsoft Project completely.

2. Click the **Start** button, type **Run**, and then select the **Run** application in the *Apps* section of the *Start* menu.

3. Enter the following command in the **Open** field of the *Run* dialog, as shown in Figure 2 - 38:

%appdata%\Microsoft\MS Project\16\Cache

44

Microsoft Project Overview

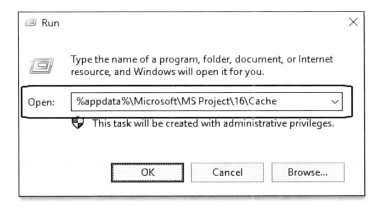

Figure 2 - 38: Run dialog

4. Click the **OK** button in the *Run* dialog.

The Windows system launches the File Explorer application and navigates to the Cache folder on your hard drive. You may see only one folder in the Cache folder, or you may see multiple folders, such as shown in Figure 2 - 39.

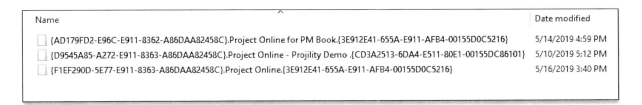

Figure 2 - 39: Multiple folders in the Cache folder

5. Select all of the folders in the *Cache* folder and then press the **Delete** key on your computer keyboard to delete all of them.

The next time you launch Microsoft Project and connect to Project Web App, the system rebuilds the Cache, giving you a fresh and uncorrupted Cache. To open any enterprise project initially after deleting the Cache, you must double-click the **Show me the list of all projects** item in the *Open* dialog and open the project from the list of all projects you have permission to access.

Working with Offline Projects

One of the features available in Microsoft Project with Project Server or Project Online is working with enterprise projects in **Offline mode** when you are not connected to the Internet through your company network. You may need to work with Microsoft Project in Offline mode in situations such as the following:

- You need to travel away from your corporate network, such as on a business trip or out in the field, and you need to work with an enterprise project while you are away.

- You temporarily lose network connectivity while you have an enterprise project open, and you are not able to save your project changes to Project Web App.

45

Module 02

Offline Mode While Away from the Corporate Network

If you are a traveling project manager and need to take an enterprise project with you on the road, complete the following steps to work with the project in Offline mode:

1. Before you leave your office, and while you are still connected to your corporate network, launch Microsoft Project, and connect to Project Web App.

2. Open and check out the enterprise project with which you want to work while you are away from your corporate network.

3. Save the enterprise project one final time, if necessary.

4. Click the **File** tab and then click the **Close** tab, or you can click the **Close** button in your Quick Access Toolbar. Microsoft Project displays a confirmation dialog such as the one show in Figure 2 - 40.

Figure 2 - 40: Check In confirmation dialog

5. In the confirmation dialog, click the **No** button to close the project and to intentionally leave it in a checked-out state.

6. Exit Microsoft Project.

Completing the first set of steps loads the latest version of the enterprise project into your Cache but leaves the project in a checked-out state so that no one can modify it while you are working with it in Offline mode. While you are away from your corporate network, such as while you are flying on an airliner, you can work with the project in Offline mode by completing the following steps:

1. Launch Microsoft Project.

2. In the *Login* dialog shown in Figure 2 - 41, leave your Project Web App login account selected in the **Profile** pick list. *Do not* select the *Computer* account in the *Profile* pick list!

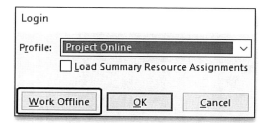

Figure 2 - 41: Login dialog

3. In the *Login* dialog, click the **Work Offline** button.

46

Microsoft Project Overview

Microsoft Project launches in Offline mode without connecting to Project Web App. If you float your mouse pointer over the "globe" indicator at the left end of the *Status* bar, the software displays a *Working Offline* tooltip, such as shown in Figure 2 - 42.

Figure 2 - 42: Working in Offline mode

4. Click the **File** tab, click the **Open** tab in the *Backstage*, and then click the **Browse** button. You can also click the **Open** button in the Quick Access Toolbar.

Microsoft Project displays the *Open* dialog with a list of projects in your Cache, as shown in Figure 2 - 43. Notice that the *Cache Status* column displays a *Checked Out* message for the *Network Upgrade for West Region* project, indicating that I can work with this project in Offline mode. Notice also that the dialog does not display any message in the *Cache Status* column for the other four projects, indicating that these projects are in a checked-in state.

 Warning: If the *Cache Status* column **does not** display a *Checked Out* message for a project in your Cache, this means that you cannot work with it in Offline mode. If you attempt to open the project, Microsoft Project will open it in Read-Only mode, which means that you cannot to save your changes to the Cache. Remember that you can only work with projects in Offline mode if they are in a checked-out state.

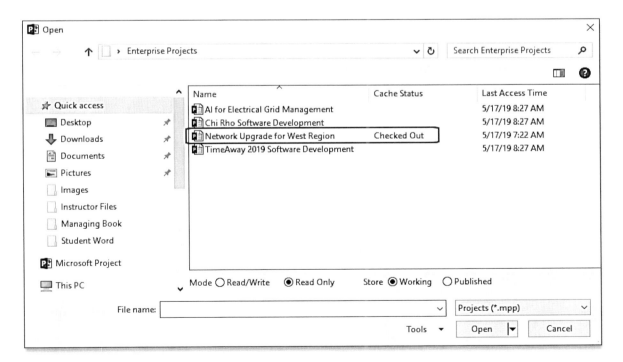

Figure 2 - 43: Open dialog shows projects in the Cache

47

Module 02

5. In the *Open* dialog, select any project that displays a *Checked Out* message in the *Cache Status* column and then click the **Open** button.

 Information: While working with a project in Offline mode, you can perform many activities that you can do while working in Online mode, such as adding and editing tasks, setting task dependencies, etc. However, you cannot perform any activity that requires an active connection to Project Web App, such as using the *Build Team* dialog to add enterprise resources to your project team. If a feature is not available for use in Offline mode, Microsoft Project disables ("grays out") that feature.

6. Edit your project, as necessary, and save the changes to the Cache. Notice the message near the right end of the *Status* bar shown in Figure 2 - 44. This message indicates that Microsoft Project is in Offline mode and cannot synchronize the project data to Project Web App.

Figure 2 - 44: Cannot synchronize data to server message

7. When you finish editing your project, click the **File** tab and then click the **Close** tab in the *Backstage* to close the project. You can also click the **Close** button in the Quick Access Toolbar.

8. In the *Close* dialog shown in Figure 2 - 45, leave the **Save** and **Keep it checked out** options selected, and then click the **OK** button.

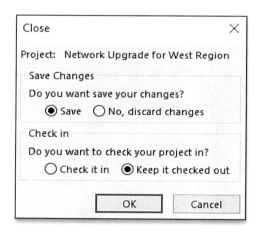

Figure 2 - 45: Close dialog

9. Exit Microsoft Project.

When you return to your corporate network, complete the following steps to synchronize the changes you made to your enterprise project while working in Offline mode:

1. Launch Microsoft Project

2. In the *Login* dialog, leave your Project Web App account selected in the **Profile** pick list, and then click the **OK** button to connect to Project Web App.

3. Click the **File** tab, click the **Open** tab in the *Backstage*, and then click the **Browse** button. You can also click the **Open** button in the Quick Access Toolbar.

4. In the *Open* dialog, select the project with which you worked in Offline mode, and then click the **Open** button.

5. Click the **File** tab and then click the **Save** tab in the *Backstage*, or click the **Save** button in the Quick Access Toolbar to synchronize your project changes to Project Web App.

Microsoft Project confirms the success of the project synchronization process by displaying a confirmation message near the right end of the *Status* bar, such as shown in Figure 2 - 45.

Figure 2 - 45: Project data synchronized

Offline Mode When You Lose Connectivity

If you lose network or Internet connectivity while you have an enterprise project open in Read/Write mode, Microsoft Project cannot save your project changes to Project Web App. So that you do not lose your project changes, you can work with the software in Offline mode.

 Information: While writing this module, I actually needed to work with an enterprise project in Offline mode. Due to construction work in my neighborhood, I lost network connectivity several times one afternoon. Each time this occurred, I quickly changed to Offline mode and did not lose any changes to my enterprise project.

To work with Microsoft Project in Offline mode when you lose network connectivity, complete the following steps:

1. Click the "globe" indicator at the left end of the *Status* bar and select the **Work Offline** item on the menu as shown in Figure 2 - 46.

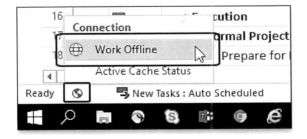

Figure 2 - 46: Work in Offline mode

2. Click the **File** tab and then click the **Save** tab in the *Backstage* or click the **Save** button in the Quick Access Toolbar.

Module 02

Microsoft Project indicates that it cannot synchronize the project changes to Project Web App by displaying the message in the *Status* bar shown previously in Figure 2 - 44.

3. When network connectivity returns, click the "globe" indicator at the left end of the *Status* bar and select the **Connect to Server** item on the menu as shown in Figure 2 - 47

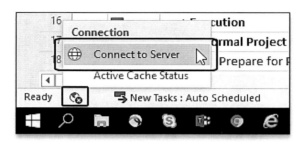

Figure 2 - 47: Connect to Server

4. Click the **File** tab and then click the **Save** tab in the *Backstage* or click the **Save** button in the Quick Access Toolbar.

Microsoft Project confirms the success of saving the project to Project Web App by displaying a *Save completed successfully* message at the right end of the *Status* bar.

Module 03

Project Web App Overview

Learning Objectives

After completing this module, you will be able to:

- Understand the Project Web App interface
- Navigate in Project Web App
- Use the features and functionality of Project Web App

Inside Module 03

Introducing Project Web App	**53**
Understanding the Default Project Web App Interface	**53**
Navigating in Project Web App	*54*
Features and Functionality in Project Web App	**56**
Using Project Web App Views	*57*
Manipulating a Data Grid	*61*
Printing the Data Grid	*64*
Exporting the Data Grid to Excel	*65*

Introducing Project Web App

Project Web App, also known as PWA, is the web-based user interface to Project Online and Project Server. PWA is a central point to access and manage nearly all project and resource data. While PWA offers many features, the core functionality is as follows:

- **Project Information:** Use the *Project Center* page to view both high-level and detailed project information.
- **Resource Information:** Use the *Resource Center* to manage your organization's project resources, and to navigate to other PWA pages to view resource assignments and resource availability.
- **Time and Task Progress:** PWA offers the *Timesheet* page for organizations that want to track all types of time spent on project work, non-project work, and nonworking time. For organizations that need a less challenging approach, the *Tasks* page offers two simpler methods of tracking only task progress.
- **Collaboration:** Once it is published, every project has its own SharePoint Project Site for managing project-related items such as risks, issues, and documents. PWA automatically manages the links between each project and its associated Project Site.
- **Portfolio Management and Project Governance:** PWA provides advanced capabilities to allow for the selection and oversight management of all projects in the system. This system includes business driver management, cost constraint analysis, resource constraint analysis, and governance workflow.
- **Application Administration:** PWA provides a rich set of administration tools necessary to manage a complex project and portfolio management information system.

To use Project Online effectively requires your organization to use a combination of Microsoft Project, Project Web App (PWA), and the SharePoint Project Site. The following table shows the recommended toolset for each role in your organization:

Project Managers	Team Members	Executives	Portfolio Managers	Resource Managers
Microsoft Project Project Web App Project Site	Project Web App Project Site	Project Web App Project Site	Project Web App	Project Web App

Keep in mind that your organization's Project Online application administrator needs to use both Microsoft Project and Project Web App to set up and manage the system effectively.

Understanding the Default Project Web App Interface

To navigate to Project Web App (PWA), launch your preferred web browser and enter the URL of your organization's Project Web App. Project Online displays the *Home* page of PWA. Each organization customizes the *Home* page according to the PPM needs of the organization. The customizations done to the *Home* page shown in Figure 3 - 1 are very common, but may differ from what you see on your own organization's *Home* page of PWA.

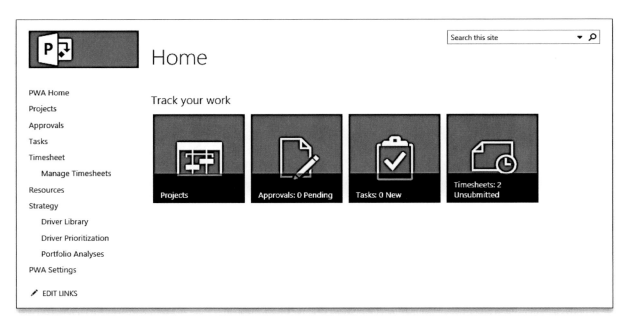

Figure 3 - 1: Home page of Project Web App

Navigating in Project Web App

The *Home* page of PWA consists of two parts: the *Quick Launch* menu on the left and the *Track your work* carousel at the top. The *Quick Launch* menu consists of a series of links you can use to navigate to different locations in PWA. The links you see in the *Quick Launch* menu depend on your security permissions in PWA and also depend upon how your application administrator configured the menu. Following is the list of all possible default links you may see in the *Quick Launch* menu in the *Home* page of your organization's Project Web App. Keep in mind that your application administrator may have added additional non-default links not mentioned below:

- Click the **Projects** link to navigate to the *Project Center* page where you can view your project portfolio.

- Click the **Approvals** link to navigate to the *Approval Center* page where you can approve pending task updates from your project team members.

- Click the **Tasks** link to navigate to the *Tasks* page and view all of the tasks assigned to you across all projects.

- Click the **Timesheet** link to navigate to the *Timesheet* page and view your timesheet for the current reporting period.

- Click the **Issues and Risks** link to navigate to the Issues and Risks page where you can view issues and risks assigned to you in all of your projects.

- Click the **Resources** link to navigate to the *Resource Center* page where you can view and work with resource information.

- Click the **Status Reports** link to navigate to the *Status Reports* page where you can create or respond to a *Status Report* request.

- The *Strategy* section contains three links that are ***only visible*** to portfolio managers and application administrators by default. Click the **Driver Library** link to navigate to the *Driver Library* page and view a list of your organization's business drivers for projects. Click the **Driver Prioritization** link to navigate to the *Driver Prioritization* page and see your organization's priorities for business drivers. Click the **Portfolio**

Analyses link to navigate to the *Portfolio Analyses* page where you can create Portfolio Analyses to manage your organization's project intake process.

- Click the **Reports** link to navigate to the *PWA Reports* page where you can see custom Excel reports for leaders in your organization.

- Click the **Server Settings** link to navigate to the **PWA Settings** for page.

The *Quick Launch* menu shown previously in Figure 3 - 1 contains two custom links our application administrator created for the writing of this book: the *PWA Home* link and the *Manage Timesheets* link. I can click the **PWA Home** link to return to the *Home* page of PWA. I can the **Manage Timesheets** link to display the *Manage Timesheets* page where I can create, edit, and delete my timesheets as needed. The *PWA Settings* link is the default link *Server Settings* link, but with a custom name. I can click the **PWA Settings** link to navigate to the *PWA Settings* page.

The *Track your work* carousel consists of a series of "live tiles" that you can click to navigate to different locations in Project Web App. Again, please remember that the tiles you see in your organization's *Track your work* carousel depend on how your application administrator configured it. Following is the list of all possible default tiles you may see in the *Track your work* carousel on the *Home* page of your organization's PA:

- Click the **Projects** tile to navigate to the *Project Center* page where you can view your project portfolio.

- The *Approvals* tile displays your total number of pending approvals. Click the **Approvals** tile to navigate to the *Approval Center* page where you can approve pending task updates from your project team members.

- The *Tasks* tile displays the total number of new tasks assigned to you. Click the **Tasks** tile to navigate to the *Tasks* page and view all of the tasks assigned to you across all projects.

- The *Timesheets* tile displays the total number of your unsubmitted timesheets. Click the **Timesheets** tile to navigate to the *Manage Timesheets* page where you can create, edit, and delete timesheets.

- The *Team Timesheets* tile displays the total number of unsubmitted timesheets from you and the resources who report to you. Click the **Team Timesheets** tile to navigate to the *Timesheet Approval History* page to view the history of timesheet submittal for your resources.

- The *Issues* tile displays the number of active issues assigned to you in your projects. Click the **Issues** tile to navigate to the *Issues and Risks* page where you can view all active risks and issues assigned to you.

- The *Risks* tile displays the number of active risks assigned to you in your projects. Click the **Risks** tile to navigate to the *Issues and Risks* page where you can view all active risks and issues assigned to you.

- Click the **Status Reports** tile to navigate to the *Status Reports* page where you can create or respond to a *Status Report* request.

- Click the **Reports** tile to access the *PWA Reports* page where you can see custom Excel reports for leaders in your organization.

Features and Functionality in Project Web App

Every Project Web App page that contains a data grid also includes a ribbon with one or more ribbon tabs at the top of the page. By default, the ribbon is collapsed on most PWA pages, such as the *Project Center* and *Resource Center* pages, for example. To expand the ribbon on any page, use one of the following actions:

- Click anywhere in the data grid to automatically expand the ribbon.
- Click the ribbon tab to manually expand the ribbon.

Information: In Project Web App, there is no way to "pin" a ribbon tab to keep the ribbon expanded. This differs with applications in the Microsoft Office in which you can "pin" a ribbon to keep it expanded or "unpin" the ribbon to collapse it.

Figure 3 - 2 shows the *Project Center* page. Notice that I expanded the *Projects* ribbon at the top of the page.

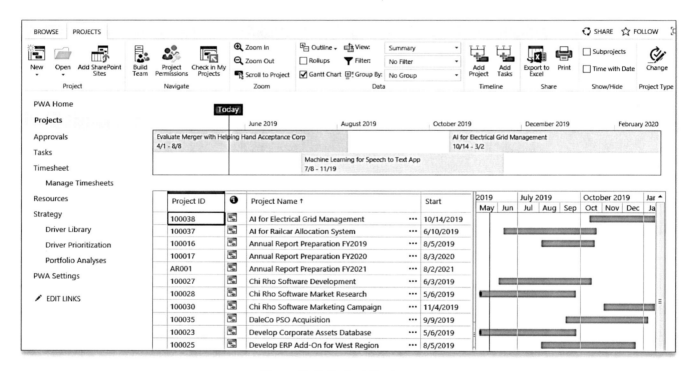

Figure 3 - 2: Project Center page

Some pages contain a ribbon with more than one ribbon tab at the top of the page. For example, Figure 3 - 3 shows the *Schedule* page that Project Web App displays when I clicked the name of a project in the *Project Center* data grid. Notice that the *Schedule* page contains a ribbon with four ribbon tabs: the *Project*, *Page*, *Task*, and *Options* tabs. You can use the commands on these four ribbons to manipulate how PWA displays the project data, and even to open and check out the project for editing using the web-based user interface. For example, you can click the **Options** tab and then select the **Project Summary Task** checkbox to display the Project Summary Task (aka Row 0 or Task 0) in the project schedule shown on this page. On the *Task* ribbon, you can click the **Edit** button to check out the project for editing, use the **Save** button to save your changes, use the **Publish** button to publish your changes, and use the **Close** button to close and check in the project at the end of the editing session.

Project Web App Overview

 Warning: Because of the many limitations imposed by the web-based user interface for editing projects, Projility recommends that project managers avoid editing projects in Project Web App. Instead, always use Microsoft Project to plan, edit, and manage your projects.

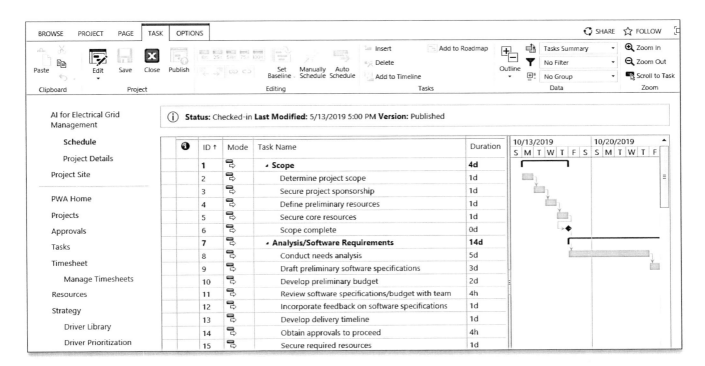

Figure 3 - 3: Schedule page for a selected project

Using Project Web App Views

Figure 3 - 4 shows the *Resource Center* page with the *Resources* ribbon expanded. Notice that the *Data* section of the *Resources* ribbon contains a *View* pick list, a *Filter* pick list, and a *Group By* pick list. You can use any of these three pick lists to manipulate how Project Web App displays the data in the data grid.

The default view displayed on the *Resource Center* page is the *All Resources* view. Notice that this view displays every resource and applies grouping on the *Type* column to organize the resources into *Cost*, *Material*, and *Work* groups of resources.

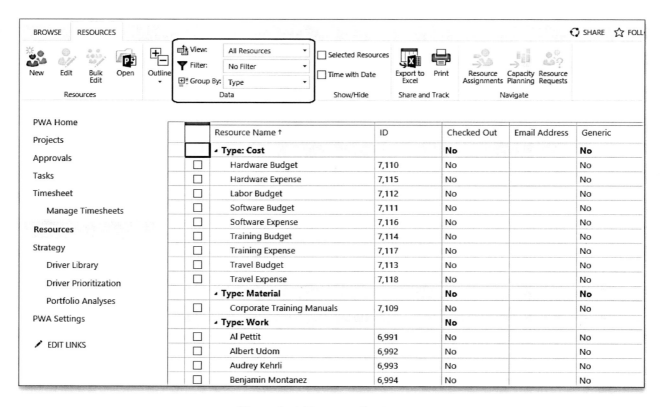

Figure 3 - 4: Resource Center page

To apply a different view, click the **View** pick list and select one of the views in the list, such as shown the *Work Resources* view shown in Figure 3 - 5. As the name of the view implies, the *Work Resources* view displays only *Work* resources.

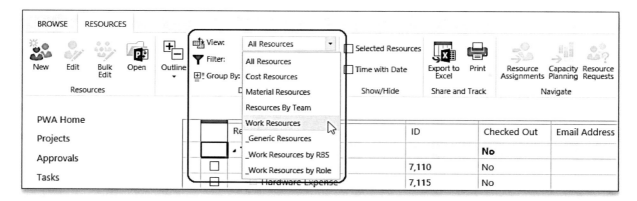

Figure 3 - 5: Click the View pick list

 Information: In Figure 3 - 5 shown previously, notice the three views whose names are prefixed with the underscore character, as these views are custom views. The *_Generic Resources* view displays only generic Work resources. The *_Work Resources by RBS* view displays only Work resources, grouped by their RBS values. The *_Work Resources by Role* view displays only Work resources, grouped by their Role values. Our application administrator created these custom views based on the resource management needs of our organization.

Project Web App Overview

To apply a filter to the data displayed in any view, click the **Filter** pick list, and select the **Custom Filter** item, as shown in Figure 3 - 6.

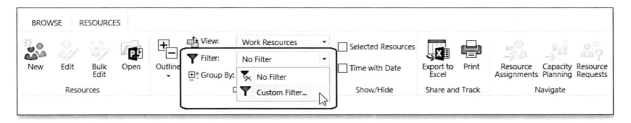

Figure 3 - 6: Click the Filter pick list

Project Web App displays the *Custom Filter* dialog shown in Figure 3 - 7.

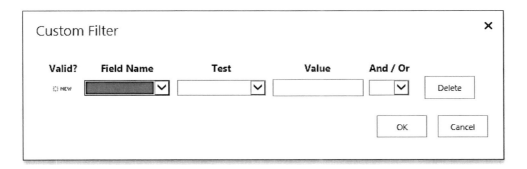

Figure 3 - 7: Custom Filter dialog

To create a custom filter, click the **Field Name** pick list and select one of the fields included in the current view. Click the **Test** pick list and select one of the filtering tests. Manually type a value in the **Value** field. Optionally select a value in the **And / Or** pick list. When you enter valid filtering criteria, the *Custom Filter* dialog displays a green checkmark indicator in the *Valid* field, such as shown in Figure 3 - 8. Click the **OK** button to apply the filter.

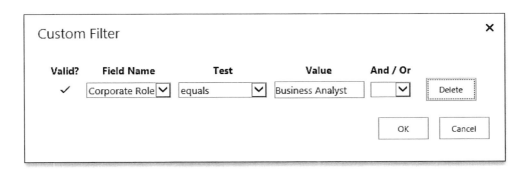

Figure 3 - 8: Valid filtering criteria in the Custom Filter dialog

When you apply a custom filter in any PWA view, the *Filter* field in the ribbon displays a *Custom Filter...* value. The system continues to apply the custom filter each time you return to the PWA page in which you applied the filter. To remove a custom filter, click the **Filter** pick list and select the **No Filter** item.

59

Warning: Project Web App allows you to create and apply custom filters, but it **does not** allow you to save your custom filters. Once you have removed a custom filter from a view, you must manually recreate it to reapply the filter. If you find yourself constantly creating a particular filter, you could request that your application administrator adjust the configuration of the view to include your custom filter.

To apply grouping using a single field to a view in any Project Web App page, click the **Group By pick** list and select one of the fields included in the current view, such as the **Resource Departments** field shown in Figure 3 - 9. The system organizes the data into groups using the selected field.

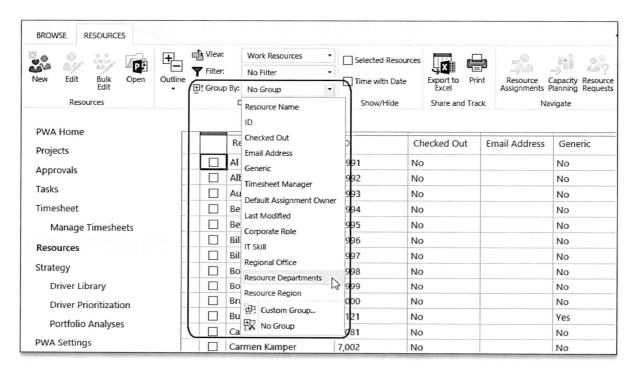

Figure 3 - 9: Click the Group By pick list

To apply custom grouping using multiple fields, click the **Group By** pick list and select the **Custom Group** item. The system displays the *Group Fields* dialog shown in Figure 3 - 10.

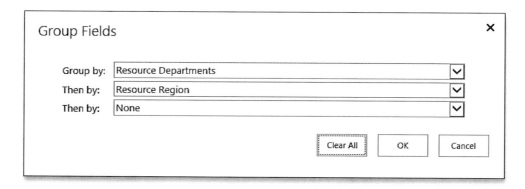

Figure 3 - 10: Group Fields dialog

In the *Group Fields* dialog, click the **Group by** pick list and select one of the fields included in the current view. Click the first **Then by** pick list and select a field, then optionally click the second **Then by** pick list and select another field. Figure 3 - 11 shows the *Group fields* dialog with grouping applied to the *Resource Departments* field and then to the *Resource Region* field. Click the **OK** button when finished to apply the custom grouping.

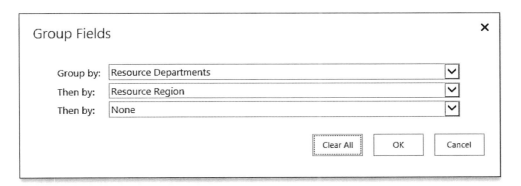

Figure 3 - 11: Group Fields dialog with custom grouping

 Information: When attempting to apply filtering or grouping in a PWA view, if you do not see the field on which you want to filter or group, ask your organization's Project Online application administrator to add the necessary field to the view. Remember that you can only filter or group on fields included in the view.

When you apply grouping on a single field in any Project Web App view, the *Group By* field displays the name of the field you selected. When you apply custom grouping using multiple fields, the *Group By* field displays a *Custom Group...* value. Regardless of which type of grouping you applied, the system continues to apply your grouping each time you return to the PWA page in which you applied the grouping. To remove the grouping applied to a view on any PWA page, click the **Group By** pick list and select the **No Group** value.

 Warning: Project Web App allows you to create and apply a custom group using multiple fields, but it **does not** allow you to save your custom group. Once you have removed the custom group from a view, you must manually recreate it to reapply the custom group. If you find yourself constantly creating a particular group, you could request that your application administrator adjust the configuration of the view to include your custom group.

Manipulating a Data Grid

As previously mentioned, a number of Project Web App pages contain a data grid that displays task, resource, or assignment data. Some data grids, such as the *Project Center* page, include a vertical split bar separating the grid into two sections; other pages contain a single grid only, such as the *Resource Center* page. For example, notice that the *Project Center* page shown previously in Figure 3 - 2 consists of two sections: the project list on the left side of the split bar and the *Gantt Chart* pane on the right side. To work with the data in a data grid most effectively, it is important to know how to take the following actions:

- **Move the Split Bar** – You can move the split bar in a data grid by floating your mouse pointer anywhere over the split bar itself. When the mouse pointer changes from a single arrow to a double-headed arrow, click and hold the mouse button to "grab" the split bar, and then drag it to the new position on the screen. Figure 3 - 12 shows the mouse pointer floating over the split bar, ready for me to move the split bar.

Figure 3 - 12: Move the split bar

- **Change Column Widths** – To change the width of any column in the data grid, float your mouse pointer anywhere on the right edge of the column header. The mouse pointer changes from a single arrow to a double-headed arrow. Click and hold to "grab" the right edge of the column, and drag the edge of the column to the proper width. Notice in Figure 3 - 13 that I am changing the width of the *Project Name* column.

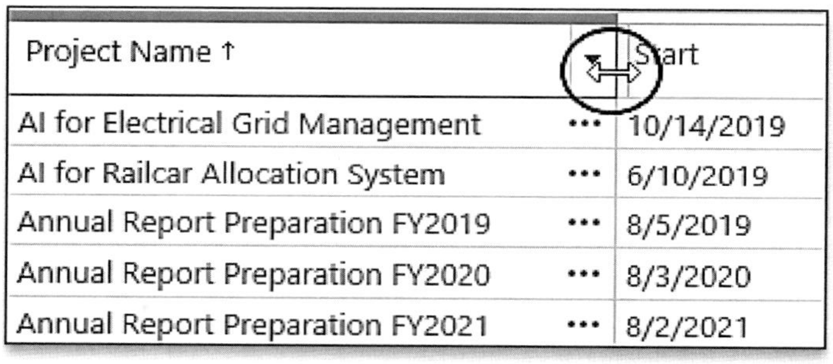

Figure 3 - 13: Change the column width

- **Move a Column** – To move any column in the grid, click and hold the column header of the column you want to move to "grab" the column, then "drag and drop" the column to its new position in the grid. Notice in Figure 3 - 14 that I am ready to drag the *% Complete* column to a new position.

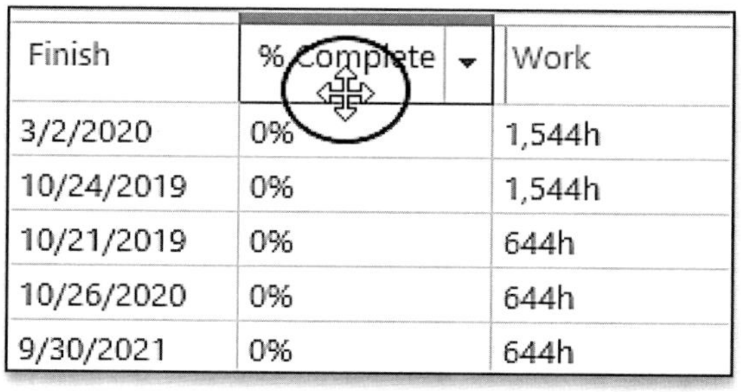

Figure 3 - 14: Move a column

- **Hide a Column** – To hide any column, float your mouse pointer over the column header of the column you want to hide, and then click the pick list button that appears in the column header. On the pick list, select the **Hide Column** item. Notice in Figure 3 - 15 that I am preparing to hide the *Project ID* column.

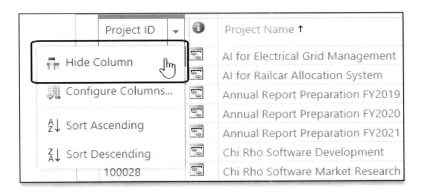

Figure 3 - 15: Hide a column

- **Unhide a Column** – To unhide a column, float your mouse pointer over any column header, click the pick list button, and select the **Configure Columns** item on the pick list shown previously in Figure 3 - 15. PWA displays the *Configure Columns* dialog shown in Figure 3 - 16. In the *Configure Columns* dialog, select the checkboxes for the columns you want to unhide and then click the **OK** button.

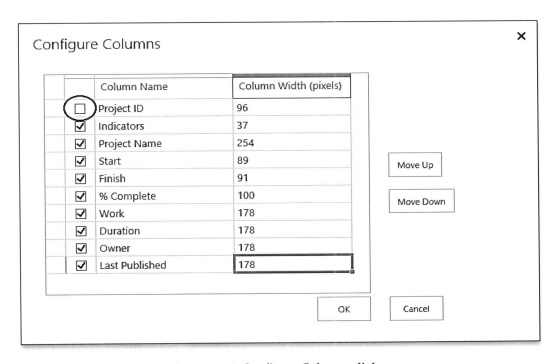

Figure 3 - 16: Configure Columns dialog

63

Module 03

Information: Notice that *Configure Columns* dialog shown in previously in Figure 3 - 16 also allows you to set column widths and to change the column display order. To change the width of any column, enter the width in pixels in the **Column Width** field for that column. To change the display order of columns, select any column and use the **Move Up** and **Move Down** buttons. Click the **OK** button when finished.

- **Sort Columns** – To sort the data in the grid, float your mouse pointer over the column header of the column that you want to sort, click the pick list button, and then select either the **Sort Ascending** or **Sort Descending** item, such as shown in Figure 3 - 17.

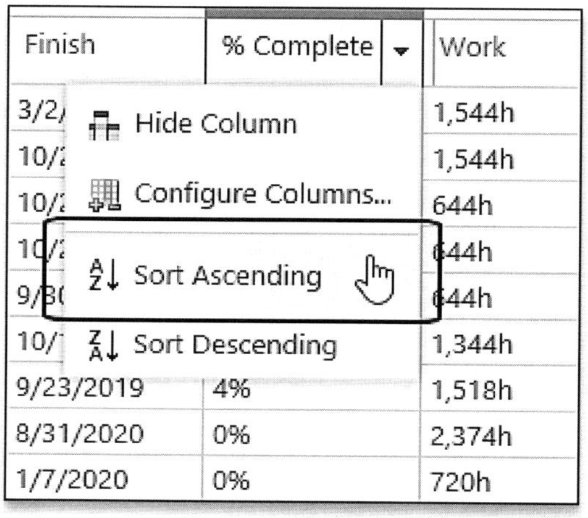

Figure 3 - 17: Sort the data in the grid

Information: Project Web App automatically saves any changes you make to the layout of a data grid (such as column order, column width, etc.) so that the layout of the grid reappears the next time you return to the page. These changes affect the current user only and do not affect the grid layout of any other users.

Printing the Data Grid

Project Web App allows you to print a data grid or to export the data grid information to Microsoft Excel. To print a data grid, click the **Print** button in the ribbon at the top of the PWA page. The system displays a duplicate data grid on a new web browser tab and then displays the *Print* dialog, as shown for the *Resource Center* page in Figure 3 - 18. To print the data grid currently displayed, click the **Print** button in the *Print* dialog.

Information: If you want to modify the layout of the data grid before printing, press the **Cancel** button in the *Print* dialog, then modify the data grid layout as needed. For example, if you attempt to print a *Project Center* view, the system displays ***every column*** included in the view in the duplicate data grid, along with the *Gantt Chart* pane on the right side of the view. In the secondary data grid, change column widths as needed and then position the split bar for printing. Press the **Control + P** keyboard shortcut to display the *Print* dialog, then click the **Print** button to print the modified data grid.

Project Web App Overview

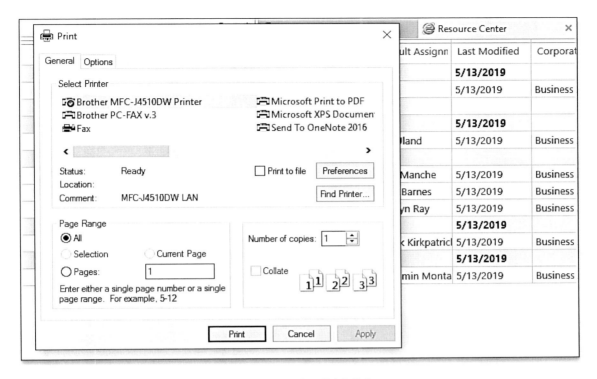

Figure 3 - 18: Print Grid dialog

Exporting the Data Grid to Excel

You can export a data grid to a Microsoft Excel workbook by clicking the **Export to Excel** button in the ribbon at the top of the PWA page. The system displays the *File Download* dialog at the bottom of the page, such as shown in Figure 3 - 19.

Figure 3 - 19: File Download dialog

In the *File Download* dialog, click the **Open** button to open the file in Microsoft Excel. Figure 3 - 20 shows a portion of the workbook exported from the *Project Center* data grid to Excel. Once the grid data is exported to an Excel workbook, you can print the workbook or use any of Excel's data analysis features.

65

Module 03

	A	B	C	D	E	F
1	Project ID	Indicators	Project Name	Start	Finish	% Complete
2	100038	This is an	AI for Electrical Grid Management	10/14/2019	3/2/2020	0%
3	100037	This is an	AI for Railcar Allocation System	6/10/2019	10/24/2019	0%
4	100016	This is an	Annual Report Preparation FY2019	8/5/2019	10/21/2019	0%
5	100017	This is an	Annual Report Preparation FY2020	8/3/2020	10/26/2020	0%
6	AR001	This is an	Annual Report Preparation FY2021	8/2/2021	9/30/2021	0%
7	100027	This is an	Chi Rho Software Development	6/3/2019	10/17/2019	0%
8	100028	This is an	Chi Rho Software Market Research	5/6/2019	9/23/2019	4%
9	100030	This is an	Chi Rho Software Marketing Campaign	11/4/2019	8/31/2020	0%
10	100035	This is an	DaleCo PSO Acquisition	9/9/2019	1/7/2020	0%
11	100023	This is an	Develop Corporate Assets Database	5/6/2019	9/24/2019	3%
12	100025	This is an	Develop ERP Add-On for West Region	8/5/2019	12/20/2019	0%
13	100036	This is an	Develop Junior Executive Coaching Program	8/5/2019	1/3/2020	0%
14	100034	This is an	Evaluate Merger with Helping Hand Acceptance Corp	4/1/2019	8/8/2019	40%
15	100015	This is an	HR CBT Rollout to Corporate HQ	6/1/2020	10/22/2020	0%
16	100008	This is an	HR CBT Rollout to East Region	9/9/2019	1/28/2020	0%
17	100007	This is an	HR CBT Rollout to North Region	6/3/2019	10/24/2019	0%
18	100013	This is an	HR CBT Rollout to South Region	12/2/2019	4/23/2020	0%
19	100014	This is an	HR CBT Rollout to West Region	3/2/2020	7/23/2020	0%

Figure 3 - 20: Project Center data grid exported to Excel

Module 04

Creating an Enterprise Project

Learning Objectives

After completing this module, you will be able to:

- Create a new enterprise project in Project Web App
- Transfer an enterprise project to a different project manager
- Create a new enterprise project in Microsoft Project
- Properly define a new enterprise project using a six-step method
- Create Resource Engagements using Generic resources

Inside Module 04

Creating a New Enterprise Project	**69**
Creating a New Project Using an Enterprise Project Type	**70**
Transferring a Proposed Project to a Project Manager	*79*
Creating a New Project Using Microsoft Project	**83**
Defining a New Enterprise Project	**85**
Set the Project Start Date and Specify Enterprise Field Values	*85*
Enter the Project Properties	*87*
Display the Project Summary Task	*88*
Set the Project Working Schedule	*89*
Set Options Unique to the Project	*93*
Save the Project	*97*
Creating Resource Engagements Using Generic Resources	**99**
Setting Resource Utilization to Use Resource Engagements	*100*
Building the Project Team with Generic Resources	*100*
Creating the Resource Engagements	*103*
Submitting the Resource Engagements	*108*

Creating a New Enterprise Project

As the name implies, **Enterprise Project Types (EPTs)** determine the types of enterprise projects that that users can create in your organization. For example, in the Project Online instance used for the writing of this book, our application administrator created the following EPTs:

- *Annual Report Preparation* for projects created by the *Finance* department
- *Database Upgrade or Migration* for projects created by the *IT* department
- *Market Research* for projects created by the *Marketing and Sales* department
- *Marketing Campaign* for projects created by the *Marketing and Sales* department
- *Merger or Acquisition Evaluation* for projects created by the *Legal* department
- *Network Infrastructure* for projects created by the *IT* department
- *Software Development* for projects created by the *IT* department
- *Training Rollout* for projects created by the *HR* department

Each Enterprise Project Type consists of the following:

- An enterprise project template used to create the project in Microsoft Project
- An optional SharePoint workflow to guide the approval process for new proposed projects
- Project Detail Pages (PDPs) displayed for the EPT
- Department associated with the EPT
- *Side Creation* option that determines when the system creates the Project Site for the associated project
- *Synchronization* options that determine whether the system synchronizes user permissions with the associated Project Site when the project manager publishes the project
- An optional Project Site template used to create the Project Site for the project in SharePoint

As a project manager, you do not have permission in Project Online to create Enterprise Project Types, as this responsibility belongs solely to your application administrator. For you as a project manager, it is enough to know whether Enterprise Project Types exist in your Project Online instance, and if so, how to use them to create enterprise projects in your organization.

Depending on the PPM intake process in your organization, you may need to create new enterprise projects using two different methods. These methods include:

- **Create the project from an Enterprise Project Type in Project Web App** – Use this method to create a new proposed project that is subject to a SharePoint workflow approval process. You may also use this process if your organization mandates that all projects must be created using an Enterprise Project Type (EPT) from the *Project Center* page in PWA.
- **Create the project using Microsoft Project** – Use this method if your project is already approved or does not need approval, or if your organization does not use EPTs to create new projects.

I discuss the first method in this main topical section and discuss the second method in the next topical section.

Creating a New Project Using an Enterprise Project Type

When you create a new enterprise project using an Enterprise Project Type from the *Project Center* page in Project Web App, you can submit the project as a proposed project and submit it to a SharePoint workflow approval process. To create a proposed project and submit it to a SharePoint workflow for approval, complete the following steps:

1. Launch your preferred web browser and navigate to the *Home* page of Project Web App.

2. Click the **Projects** link in the *Quick Launch* menu or click the **Projects** tile in the *Track your work* carousel. PWA displays the *Project Center* page.

3. Click the **Projects** tab at the top of the *Project Center* page to expand the *Projects* ribbon.

4. In the *Project* section of the *Projects* ribbon, click the **New** pick list button and select one of the Enterprise Project Types (EPTs) listed in the menu, such as shown in Figure 4 - 1.

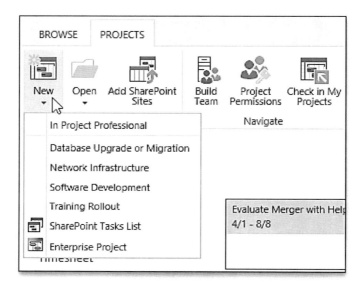

Figure 4 - 1: New pick list menu

 Information: Our application administrator created the Enterprise Project Types (EPTs) that appear in the *New* pick list shown in Figure 4 - 1. If your application administrator created EPTs for your organization, you will see them when you click the *New* pick list button on the *Project Center* page of your organization's Project Web App. The only default EPTs in Project Online, by the way, are the *Enterprise Project* and *SharePoint Tasks List* items.

When you select one of the EPTs on the *New* pick list, Project Web App displays the first Project Detail Page (PDP) of the intake process, such as the *IT Intake* page shown in Figure 4 - 2. Depending on how your application administrator configured your organization's intake process, you may be required to fill in one or more PDPs as part of the process.

Creating an Enterprise Project

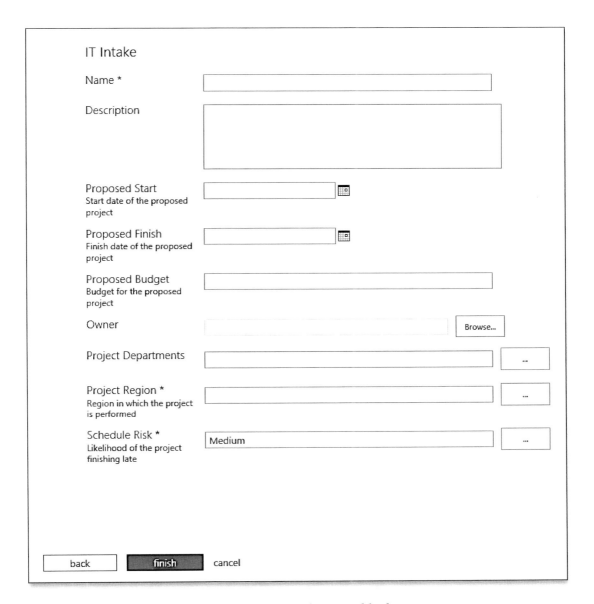

Figure 4 - 2: IT Intake page - blank

The *IT Intake* page, shown previously in Figure 4 - 2, includes three required fields indicated by the asterisk character (*) after the field name: the *Name, Project Region,* and *Schedule Risk* fields. The system requires you to enter or select a value in all required fields before you can save the information in the form in PWA.

5. Enter or select information in each of the required and optional fields, such as in the completed *IT Intake* page shown in Figure 4 - 3.

Module 04

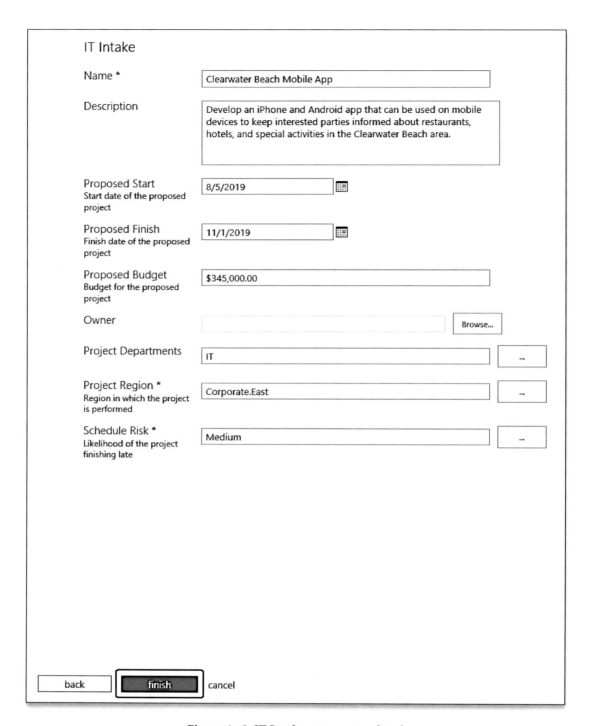

Figure 4 - 3: IT Intake page – completed

 Information: ***Do not*** enter a name in the *Owner* field. Even though this field is currently blank, Project Web App will automatically specify you as the *Owner* of the proposed project when you save the it in the system.

6. Click the **Finish** button at the bottom of the page to save the project information to Project Web App.

When you click the *Finish* button, Project Web App saves the project as an enterprise project, publishes the project for the first time, creates a Project Site in SharePoint for the project, and then displays the *Workflow Stage Status* page, such as the one shown in Figure 4 - 4.

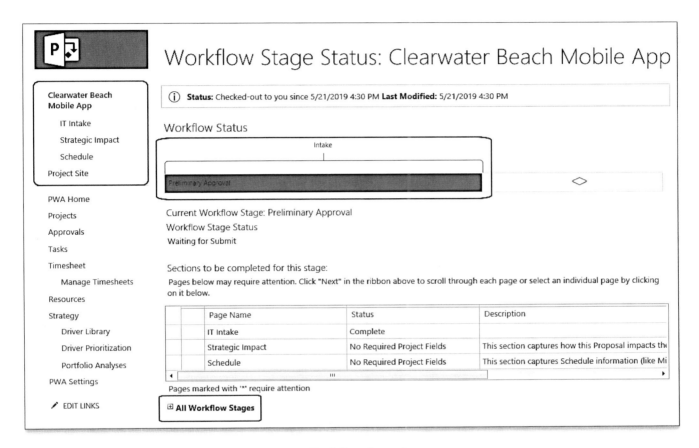

Figure 4 - 4: Workflow Stage Status page

In the *Workflow Stage Status* page, such shown previously in Figure 4 - 4, notice the following:

- The *Quick Launch* menu contains a special "drill down" section at the top of the menu which includes links to additional Project Detail Pages associated with the proposed project. Depending on how your application administrator configured your Project Web App, you may see a variation of the links shown in the "drill down" section of your organization's *Quick Launch* menu.

- The *Workflow Status* visualization at the top of the page reveals that the proposed project is in the *Intake* phase and the *Preliminary Approval* stage.

- There is an *All Workflow Stages* section at bottom of the page. This section is always collapsed by default when you display the *Workflow Stage Status* page.

Module 04

At the bottom of the page, click the **All Workflow Stages** section name to expand the section. This section contains the information that its name implies: it displays every *Stage* in the SharePoint workflow associated with the project, such as shown in Figure 4 - 17. Notice that the SharePoint workflow associated with my proposed project contains three *Stages* during the *Intake* phase: the *Preliminary Approval, Resource Engagements,* and *Final Approval* stages. Notice in the figure that the workflow is currently in the *Preliminary Approval* stage and that its *State* is *In Progress (Waiting for Input)*. This information indicates that I need to complete another Project Detail Page before I can submit the proposed project to the SharePoint workflow approval process. Depending on how your application administrator configured your Project Web App, you may see a different set of *Phases* and *Stages* in your organization's workflows.

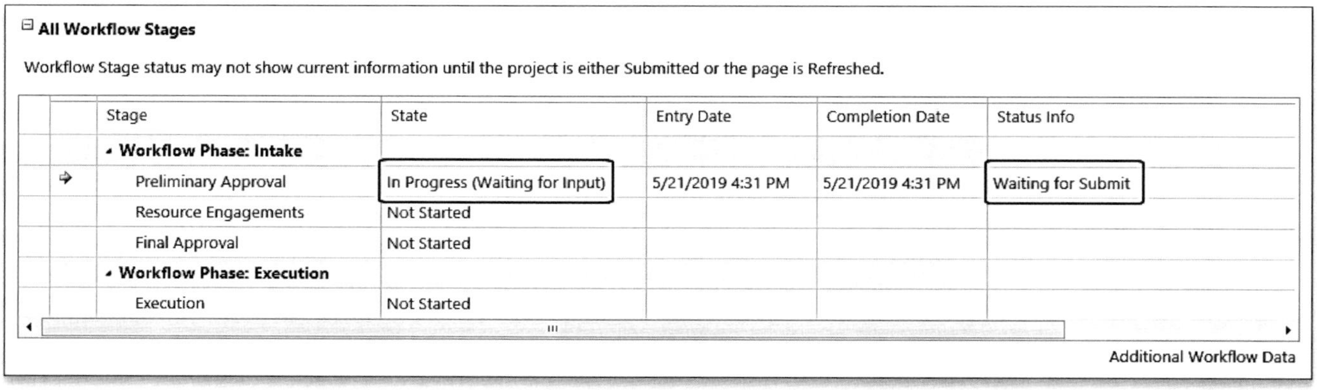

Stage	State	Entry Date	Completion Date	Status Info
Workflow Phase: Intake				
Preliminary Approval	In Progress (Waiting for Input)	5/21/2019 4:31 PM	5/21/2019 4:31 PM	Waiting for Submit
Resource Engagements	Not Started			
Final Approval	Not Started			
Workflow Phase: Execution				
Execution	Not Started			

Figure 4 - 5: All Workflow Stages section expanded

7. In the "drill down" section of the *Quick Launch* menu, click the **Strategic Impact** link.

Project Web App displays the *Strategic Impact* page, such as the one shown in Figure 4 - 6. This page contains our organization's business drivers, as developed by our executives and business leaders. You use this page to "rate" your proposed project again your organization's business drivers. Your executives can use the "rating" of each proposed project during the project intake process to determine which proposed projects to approve, and which ones to reject or delay.

8. Select a value other than *No Rating* for each of the business drivers shown on the *Strategic Impact* page.

Creating an Enterprise Project

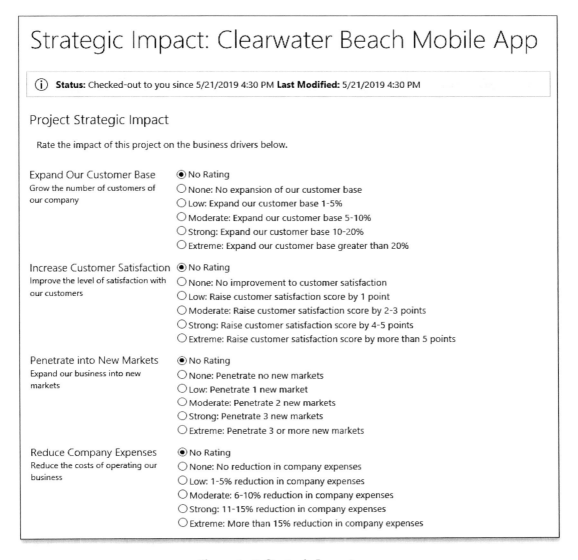

Figure 4 - 6: Strategic Impact page

9. In the *Project* section of the *Project* ribbon shown in Figure 4 - 7, click the **Save** button to save the changes to the *Strategic Impact* page.

Figure 4 - 7: Project ribbon

10. In the *Workflow* section of the *Project* ribbon shown previously in Figure 4 - 7, click the **Submit** button to submit the proposed project to the SharePoint workflow approval process.

11. In the confirmation dialog shown in Figure 4 - 8, click the **OK** button. Project Web App saves any unsaved changes to the proposed project, then submits it to the workflow approval process.

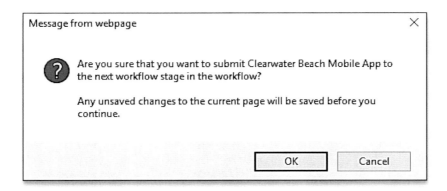

Figure 4 - 8: Confirmation dialog

When you click the *OK* button, Project Web App sends an e-mail message to all members of the workflow approvers group. Members of this group are usually executives or portfolio managers, and they can individually select whether to accept or reject the proposed project. If the proposed project is accepted, Project Web App updates the *Workflow Stage Status* page to show the current *Stage* of the workflow, such as shown in Figure 4 - 9 in the *All Workflow Stages* section of the page.

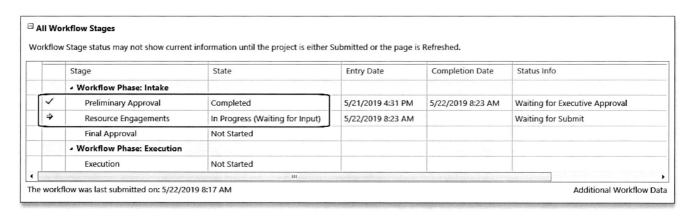

Figure 4 - 9: Current Stage updated in the All Workflow Stages section

Notice in Figure 4 - 9 shown previously that the *State* of the *Preliminary Approval* stage is *Completed*, and the *State* of the *Resource Engagements* stage is *In Progress (Waiting for Input)*. This means that the next step for a project manager is to create Resource Engagements in the Microsoft Project schedule to forecast the resource needs for this project. I discuss how to create these Resource Engagements later in this module.

12. When you finish submitting the proposed project to the SharePoint workflow, click the **Close** button in the *Project* section of the *Project* ribbon as shown in Figure 4 - 10.

Creating an Enterprise Project

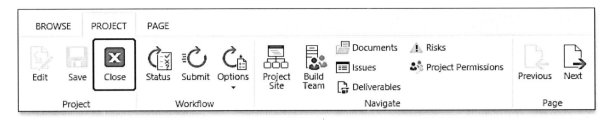

Figure 4 - 10: Close the proposed project

13. In the *Close* dialog shown in Figure 4 - 11, leave the **Check it in** option selected, and then click the **OK** button.

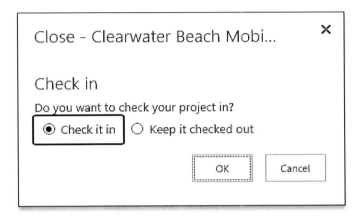

Figure 4 - 11: Confirmation dialog

Project Web App checks in the proposed project and then returns you to the *Project Center* page.

 Warning: If you close your web browser without closing and checking in the proposed project, the project will be left in a checked out state. This means you **will not** be able to open the project for editing in Microsoft Project. If you closed your web browser prematurely and forgot to check in the project, reopen your web browser and navigate to the *Project Center* page in PWA. In the *Project Center* data grid, click the name of your proposed project to return to the *Workflow Stage Status* page of the project. Click the **Project** tab to expand the *Project* ribbon, then click the **Close** button to close and check in the project properly.

77

Module 04

 Hands On Exercise

Exercise 4 - 1

> **Important Note:** You should work this exercise if your organization requires you to create new enterprise projects from the *Project Center* page in Project Web App and to optionally to submit each new project to a SharePoint workflow process for approval.

Create a new project using the PPM Training EPT and submit it to a SharePoint workflow for approval.

1. Launch your preferred web browser and navigate to the *Home* page of your organization's Project Web App.
2. Click the **Projects** link in the *Quick Launch* menu to navigate to the *Project Center* page.
3. At the top of the *Project Center* page, click the **Projects** tab to expand the *Projects* ribbon.
4. In the *Project* section of the *Projects* ribbon, click the **New** pick list button and select the Enterprise Project Type named **PPM Training for PMs**.
5. In the **Name** field of the first Project Detail Page (PDP), enter the name **Training YourFirstName YourLastName Develop ERP Class**, such as in the following example:

    Training Mickey Cobb Develop ERP Class

6. Fill in a brief description of the project in the **Description** field.
7. In the **Proposed Start** field, enter the date of the *first Monday of next month*. For example, if the current month is May, enter the date of the first Monday in June in the *Proposed Start* field.
8. In the **Proposed Finish** field, enter the date of the *last Friday of the month three months from now*. For example, if the current month is May, enter the date of the last Friday of August in the *Proposed Finish* field.
9. *Do not* enter a value in the **Owner** field.
10. Enter or select a value in every other field on the first PDP and then click the **Finish** button at the bottom of the page.
11. When PWA displays the *Workflow Stage Status* page for the new proposed project, click the **Strategic Impact** link in the "drill down" menu at the top of the *Quick Launch* menu.

> **Note:** If your "drill down" menu does not include a *Strategic Impact* link, click the links for every other PDP required by your organization, and enter the information on each of these PDPs in place of step #12 below, and then continue with step #13.

12. Select a value other than *No Rating* for each of the business drivers shown on the *Strategic Impact* page.

13. In the *Project* section of the *Project* ribbon, click the **Save** button.

14. In the *Workflow* section of the *Project* ribbon, click the **Submit** button.

15. Click the **OK** button in the confirmation dialog to submit the new project to a SharePoint workflow approval process.

16. Scroll down to the bottom of the *Workflow Stage Status* page, expand the **All Workflow Stages** section, and then examine the information shown in this section of the page.

17. Click the **Project** tab to expand the *Project* ribbon.

18. In the *Project* section of the *Project* ribbon, click the **Close** button.

19. In the *Close* dialog, leave the **Check it in** option selected, and then click the **OK** button to close and check in the project.

Transferring a Proposed Project to a Project Manager

In many organizations, it is not the project manager who creates new proposed projects from the *Project Center* page in Project Web App. Instead, people such as portfolio managers or business leaders create the new proposed projects. When executives give initial approval for a new proposed project, a project manager should be named for the project, and the project manager is responsible for creating Resource Engagements in the project. The process of "handing off" the new proposed project from its original creator to a project manager is a two-step process:

1. The creator of the proposed project must open the project in Project Web App and designate the project manager as the new *Owner* of the project. Doing so gives the new Owner full permissions to open, edit, save, and publish the proposed project.

2. The designated project manager must open the project in Microsoft Project and specify him/herself as the *Status Manager* for each task in the project schedule. By the way, the *Status Manager* is the person who must approve pending task updates submitted by assigned team members in the project.

Setting the New Project Owner

To designate a project manager as the new *Owner* of the proposed project, the original project creator (or a portfolio manager or an application administrator) ***must complete*** the following steps:

1. Launch your preferred web browser and navigate to the *Home* page of Project Web App.

2. Click the **Projects** link in the *Quick Launch* menu to navigate to the *Project Center* page.

3. In the *Project Center* data grid, click the name of the project to be transferred to the new project manager.

4. In the "drill down" section at the top of the *Quick Launch* menu, click the link for the Project Detail Page (PDP) that is the first page of your organization's intake process for new projects. In my case, I need to click the *IT Intake* link to display the *IT Intake* PDP, as shown in Figure 4 - 12.

 Information: If your organization uses the default Project Detail Pages in Project Web App for capturing information about new projects, click the **Project Details** link in the "drill down" section of the *Quick Launch* menu.

Module 04

Figure 4 - 12: Click the
IT Intake link

5. Click the **Project** tab to expand the *Project* ribbon.

6. In the *Project* section of the *Project* ribbon, click the **Edit** button to check out the project for editing.

7. Click the **Browse** button to the right of the **Owner** field, such as shown in Figure 4 - 13.

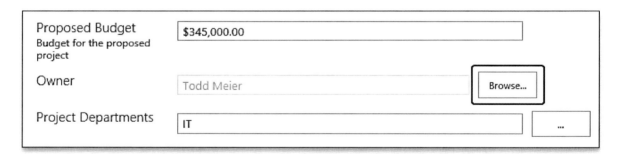

Figure 4 - 13: Click the Browse button for the Owner field

8. In the *Pick Resource* dialog, select the name of the project manager and then click the **OK** button, such as shown in Figure 4 - 14.

 Information: The *Pick Resource* dialog displays the names of every user who has permission to manage enterprise projects in Project Web App. This includes users who are members of the *Project Managers*, *Portfolio Managers*, and *Administrators* security groups in PWA.

Creating an Enterprise Project

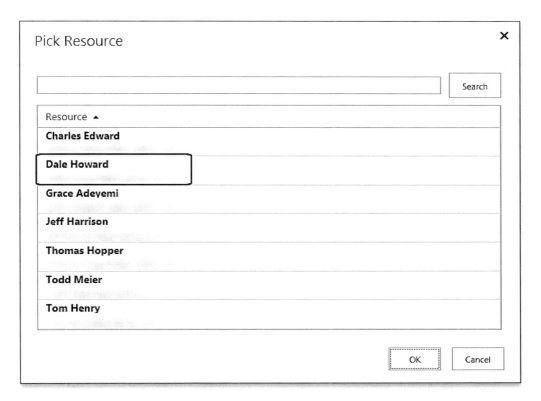

Figure 4 - 14: Pick Resource dialog

9. In the *Project* section of the *Project* ribbon, click the **Save** button to save and publish the latest changes to the project.

10. In the *Project* section of the *Project* ribbon, click the **Close** button.

11. In the *Close* dialog, leave the **Check it in** option selected, and then click the **OK** button.

Setting the Status Manager for Every Task

At this point, the designated project manager has permission to open and edit the proposed project. The project manager must now designate him/herself as the *Status Manager* for every task in the project schedule by completing the following steps:

1. Launch Microsoft Project and connect to Project Web App in the *Login* dialog.

2. Click the **File** tab, click the **Open** tab in the *Backstage*, and then click the **Browse** button. Alternately, you can click the **Open** button on the Quick Access Toolbar.

3. In the *Open* dialog, double-click the **Show me the list of all projects** link at the top of the project list.

4. Select the name of the new project and then click the **Open** button.

5. In the *READ-ONLY* gold banner at the top of the project, click the **Check Out** button.

6. Right-click on the **Duration** column header and select the **Insert Column** item on the shortcut menu, as shown in Figure 4 - 15.

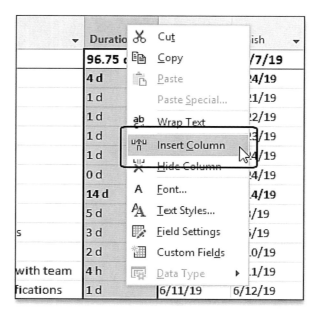

Figure 4 - 15: Insert a new column

7. In the list of available task columns, select the **Status Manager** column.

The *Status Manager* field designates the name of the person who approves task updates submitted by team members in the project. The *Status Manager* field currently displays the name of the original project creator for every detailed task in the project schedule, but needs to display the name of the new project manager instead.

8. Select the **Status Manager** cell for the first detailed task (not a summary task), click the pick list button in the cell, and select the name of the new project manager, such as shown in Figure 4 - 16.

		Task Mode	Task Name	Status Manager	Duration
	0	⏵	▲ Clearwater Beach Mobile App	Dale Howard	96.75 d
	1	⏵	▲ Scope	Dale Howard	4 d
	2	⏵	Determine project scope	Todd Meier	1 d
	3	⏵	Secure project sponsorship	Dale Howard / Todd Meier	1 d
	4	⏵	Define preliminary resources	Todd Meier	1 d
	5	⏵	Secure core resources	Todd Meier	1 d
	6	⏵	Scope complete	Todd Meier	0 d
	7	⏵	▲ Analysis/Software Requirements	Dale Howard	14 d
	8	⏵	Conduct needs analysis	Todd Meier	5 d
	9	⏵	Draft preliminary software specifications	Todd Meier	3 d
	10	⏵	Develop preliminary budget	Todd Meier	2 d
	11	⏵	Review software specifications/budget with team	Todd Meier	4 h
	12	⏵	Incorporate feedback on software specifications	Todd Meier	1 d
	13	⏵	Develop delivery timeline	Todd Meier	1 d
	14	⏵	Obtain approvals to proceed	Todd Meier	4 h
	15	⏵	Secure required resources	Todd Meier	1 d
	16	⏵	Analysis complete	Todd Meier	0 d

Figure 4 - 16: Select the new Status Manager for a task

Creating an Enterprise Project

9. Click and hold the **Fill Handle** in the lower right corner of the selected *Status Manager* cell and then drag down to the last task in the project schedule to fill the name of the new project manager in the *Status Manager* cell for every task.

> **Information**: You can use a couple of keyboard shortcuts to speed up the "fill" process in a large project. Select the *Status Manager* cell containing the name of the new project manager, then press **Control + Shift + Down-Arrow** to select all the *Status Manager* cells down to the last task in the project schedule. Press **Control + D** to "fill" the name of the new project manager into the *Status Manager* cell for every selected task.

10. Right-click on the **Status Manager** column and select the **Hide Column** item on the shortcut menu.
11. Click the **File** tab and then click the **Save** tab in the *Backstage*, or click the **Save** button on your Quick Access Toolbar.
12. Click the **File** tab and then click the **Publish** button on the *Info* page in the *Backstage*, or click the **Publish** button on your Quick Access Toolbar.
13. Click the **File** tab then click the **Close** tab in the *Backstage*, or click the **Close** button on your Quick Access Toolbar.
14. In the confirmation dialog, click the **Yes** button to check in the enterprise project.

> **Warning**: The set of steps documented in the *Setting the Status Manager for Every Task* topical section ***must*** be performed by the new project manager for the project. These steps ***cannot*** be performed on behalf of the new project manager by the original project creator or even by an application administrator.

Creating a New Project Using Microsoft Project

If your project is already approved or does not need approval, or if your organization does not create new projects from the *Project Center* page in PWA, you can create new projects in Microsoft Project by completing the following steps:

1. Launch Microsoft Project and connect to Project Web App using the *Login* dialog.
2. Click the **File** tab and then click the **New** tab in the Backstage.
3. At the top of the *New* page, click the **Enterprise** link. Microsoft Project displays the list of your organization's enterprise project templates, such as those shown in Figure 4 - 17.

83

Module 04

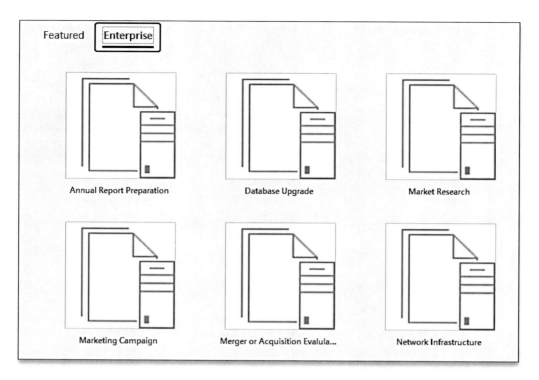

Figure 4 - 17: Enterprise project templates

4. Click the icon for the template from which you want to create a new enterprise project.

Microsoft Project creates a new unsaved project from the enterprise project template and sets the current date as the *Start* date of the project.

 Information: If your organization has not yet created any enterprise templates, you can start from a blank project template by clicking the **Blank Project** icon on the *New* page in the *Backstage*, or by clicking the **New** button in your Quick Access Toolbar.

Hands On Exercise

Exercise 4 - 2

> **Important Note:** You should **only** work this exercise if your organization requires you to create new enterprise projects using Microsoft Project. If you previously completed Exercise 4 - 1, skip this exercise entirely.

Create a new project in Microsoft Project using an enterprise template.

1. Launch Microsoft Project and connect to Project Web App in the Login dialog.
2. Click the **File** tab and then click the **New** tab in the Backstage.
3. At the top of the *New* page, click the **Enterprise** link.
4. Click the **PPM Training** template to create a new enterprise project from this template.

Defining a New Enterprise Project

After creating a new project from Project Web App or in Microsoft Project, the project manager should properly define the project by completing a six-step process recommended by Projility. These six steps are:

1. Set the project *Start* date and specify enterprise field values.
2. Enter the project properties.
3. Display the Project Summary Task (aka Row 0 or Task 0).
4. Set the working schedule for the project using a calendar.
5. Set project options unique to this project.
6. Save the project using your organization's naming convention.

I discuss each of these steps individually, showing how to perform the steps both using and not using the custom *Definition* ribbon.

Set the Project Start Date and Specify Enterprise Field Values

When you define a new project using Microsoft Project, you must set the *Start* date of the project. When you set the *Start* date, you allow the software to calculate an estimated *Finish* date based on the information you enter during the task, resource, and assignment planning processes. To enter the *Start* date for your new enterprise project, complete the following steps:

1. Click the **Project** tab to display the *Project* ribbon.

2. In the *Properties* section of the *Project* ribbon, click the **Project Information** button. The software displays the *Project Information* dialog shown in Figure 4 - 18.

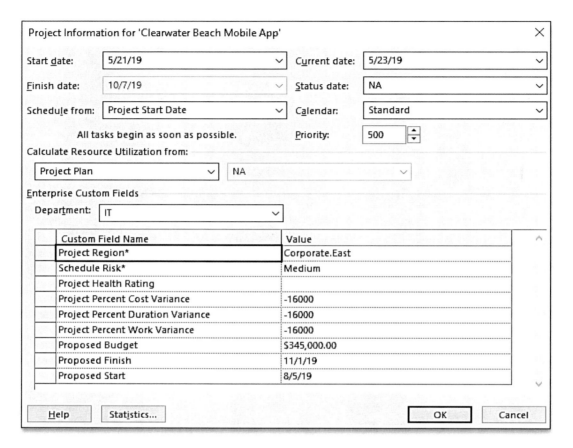

Figure 4 - 18: Project Information dialog

 Information: You can also set the *Start* date of the project by clicking the **Set Project Start Date** button in your custom *Definition* ribbon.

3. In the *Project Information* dialog, click the **Start date** pick list and select the starting date of the project in the calendar date picker.

The bottom half of the dialog contains the *Enterprise Custom Fields* section. After setting the *Start* date of your project, you should also enter or select values in all custom fields included in this section of the dialog. For example, notice in Figure 4 - 18 shown previously that I need to supply a value in the *Project Health Rating* field, as the value for this field is currently blank.

4. Select a value in the **Department** pick list.

 Information: When you select a value in the *Department* pick list, Microsoft Project may refresh the dialog and display additional fields that are specific to the department you selected.

Creating an Enterprise Project

5. Select or enter values for each of the required and optional enterprise custom fields displayed in the data grid. The system indicates required fields with an asterisk character (*) to the right of the field name. If you do not supply a value in every required field, Microsoft Project does not allow you to save your project to Project Web App.

 Information: In the *Enterprise Custom Fields* section of the dialog, notice that you cannot enter or select a value in any field that contains a formula, such as the *Project Percent Cost Variance* field, for example.

6. Click the **OK** button when completed.

Enter the Project Properties

Although you may be in the habit of skipping the *Properties* settings when creating a new document in Microsoft Word or a new workbook in Microsoft Excel, you should set the properties for each of your new enterprise projects. Once set, the *Properties* information in an enterprise project displays automatically in multiple locations throughout the project, such as in the headers and footers of views and reports you print. To set the properties information for your new enterprise project, complete the following steps:

1. Click the **File** tab and then click the **Info** tab in the *Backstage*.

2. In the upper right corner of the *Info* page, click the **Project Information** pick list and select the **Advanced Properties** item as shown in Figure 4 - 19.

 Information: You can also set the properties for the new project by clicking the **Set Project Properties** button in the custom *Definition* ribbon.

Figure 4 - 19: Click the Advanced Properties item

87

The system displays the *Properties* dialog for the new project, as shown in Figure 4 - 20.

Figure 4 - 20: Properties dialog

3. Click the **Summary** tab, if not already displayed, and then enter values in the following five fields used by Microsoft Project and Project Web App:

   - Title
   - Author
   - Manager
   - Company
   - Comments

4. Click the **OK** button when finished.

Display the Project Summary Task

The Project Summary Task, also known as Row 0 or Task 0, is the highest-level summary task in your project. The Project Summary Task summarizes or "rolls up" all task values in the entire project. For example, the value in the *Duration* column for the Project Summary Task represents the duration of the entire project, while the values in the *Work* and the *Cost* columns represent the total work and total cost for the entire project.

Creating an Enterprise Project

By default, Microsoft Project *does not* automatically display the Project Summary Task in any new blank project, so you must display it manually. To display the Project Summary Task in your new enterprise project, complete the following steps:

1. Apply any task view, such as the *Gantt Chart* view.
2. Click the **Format** tab to display the **Format** ribbon.
3. In the *Show/Hide* section of the **Format** ribbon, select the **Project Summary Task** checkbox.

Information: You can also display the Project Summary Task for the new project by clicking the **Display Project Summary Task** button in the custom *Definition* ribbon.

Microsoft displays the Project Summary Task (Row 0 or Task 0) as the first task in your new enterprise project, as shown in Figure 4 - 21.

	Task Mode	Task Name	Duration
0		▲ **Clearwater Beach Mobile App**	96.75 d
1		▲ Scope	4 d
2		Determine project scope	1 d
3		Secure project sponsorship	1 d
4		Define preliminary resources	1 d
5		Secure core resources	1 d
6		Scope complete	0 d

Figure 4 - 21: Project Summary Task

4. Widen the *Task Name* column, if necessary, to "best fit" the task name of the Project Summary Task.
5. If you widen the *Task Name* column, drag the split bar to the right side of the *Duration* column, as needed.

When you display the Project Summary Task, Microsoft Project uses the text you entered in the *Title* field of the *Properties* dialog as the task name of the Project Summary Task. The system also uses the text you entered in the *Comments* field of the *Properties* dialog as the body of the note for the Project Summary Task.

Set the Project Working Schedule

Setting the project working schedule is an ***optional step***, and is only necessary if your project schedule ***does not*** follow the working schedule shown on the enterprise *Standard* calendar. For example, suppose that you have project in which most of the work will occur in Brazil, and the project schedule will need to follow a Brazilian working schedule. In a situation such as this, you would need to use an alternate calendar with Brazilian national holidays specified on the calendar.

To determine which alternate enterprise calendars are available in your Project Online or Project Server system, complete the following steps:

1. In the *Properties* section of the *Project* ribbon, click the **Change Working Time** button.

Module 04

Information: You can also determine which alternate enterprise calendars are available for use with your new project by clicking the **View Calendars** button in the custom *Definition* ribbon.

Microsoft Project displays the *Change Working Time* dialog shown in Figure 4 - 22. Notice that the enterprise *Standard* calendar contains company holidays through the year 2023, indicated by the names of the holidays and their dates in the *Exceptions* grid in the bottom half of the dialog.

Warning: When using Microsoft Project while connected to Project Web App, the *Change Working Time* dialog **does not** allow you to create new calendars or to edit existing calendars. Even though the name of the dialog is *Change Working Time*, you **cannot change working time** in this dialog. Only your application administrator can create or edit enterprise calendars. If there is no enterprise calendar in the system that meets your scheduling needs, reach out to your organization's application administrator for assistance.

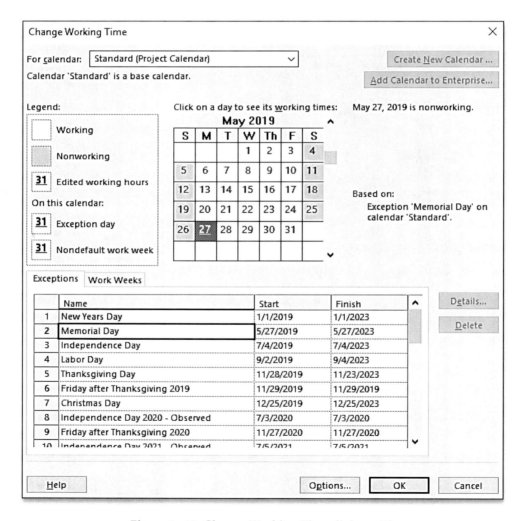

Figure 4 - 22: Change Working Time dialog with
enterprise Standard calendar displayed

2. Click the **For calendar** pick list at the top of the *Change Working Time* dialog and select any calendar other than the enterprise *Standard* calendar. For example, Figure 4 - 23 shows the *Canada Work Schedule* enterprise calendar, which contains Canadian national holidays through the year 2023.

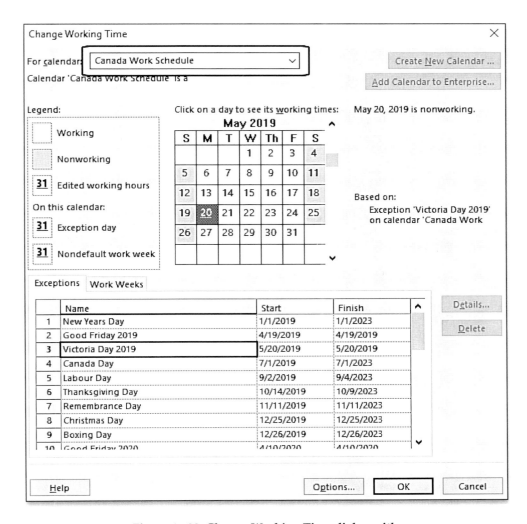

Figure 4 - 23: Change Working Time dialog with
Canada Work Schedule calendar displayed

3. After you determine which enterprise calendar you want to use to schedule your project, close the *Change Working Time* dialog by clicking either the **OK** button or the **Cancel** button.

If you need to specify an alternate calendar other than the enterprise *Standard* calendar, you must select the alternate calendar *twice* in your enterprise project: first as the **Project calendar** and the second time as the **Non-working time calendar**. Microsoft Project uses the *Project calendar* to set the initial schedule of all tasks in your project prior to assigning resources to tasks. Think of the *Project calendar* as the master schedule for the project. Microsoft Project uses the *Non-working time calendar* for a single purpose: to display nonworking time from the *Project calendar* (such as weekends and holidays) as gray shaded vertical bands in the *Gantt Chart* pane.

If you need to specify an alternate calendar other than the enterprise *Standard* calendar as the *Project calendar*, complete the following steps:

1. In the *Properties* section of the *Project* ribbon, click the **Project Information** button.

Module 04

 Information: You can also specify the *Project calendar* for your new project by clicking the **Set Project Calendar** button in the custom *Definition* ribbon.

2. In the *Project Information* dialog, click the **Calendar** pick list and select the alternate calendar. For example, if most of the project work will be performed by a specialized team in Canada, I would select the *Canada Work Schedule* calendar, such as shown in Figure 4 - 24.

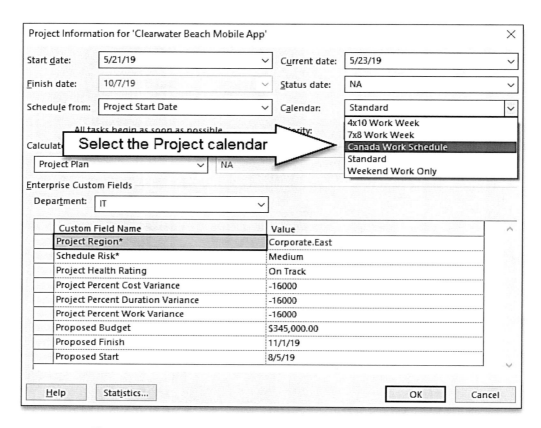

Figure 4 - 24: Project Information dialog – set the Project calendar

3. Click the **OK** button.

If you specified an alternate calendar other than the enterprise *Standard* calendar as the *Project calendar*, Microsoft Project ***does not*** automatically set this alternate calendar as the *Non-working time calendar*. Instead, you must manually specify the ***same*** alternate calendar as the *Non-working time calendar*. Remember that the *Non-working time calendar* displays the nonworking time from *the Project calendar* (such as weekends and holidays) as gray shaded vertical bands in the *Gantt Chart* pane. To set the *Non-working time calendar*, complete the following steps:

1. Double-click anywhere in the **Timescale** bar and then click the **Non-working time** tab in the *Timescale* dialog.

Creating an Enterprise Project

Information: You can also specify the *Non-working time calendar* for your new project by clicking the **Set Nonworking Time Calendar** button in the custom *Definition* ribbon.

2. On the *Non-working time* page of the *Timescale* dialog, click the **Calendar** pick list and select the *same alternate calendar* you specified previously as the *Project calendar*. For example, notice in Figure 4 - 25 that I am selecting the *Canada Work Schedule* enterprise calendar. Notice also that Microsoft Project indicates that this calendar is the *Project calendar* for the project.

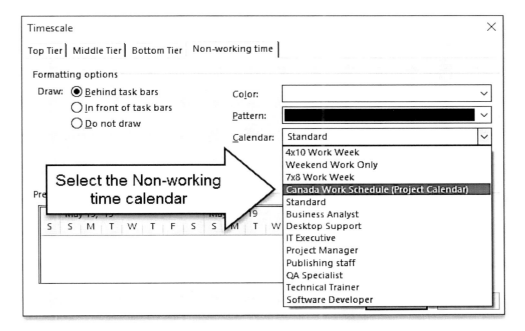

Figure 4 - 25: Timescale dialog, Non-working time page

3. Click the **OK** button when finished.

Warning: You ***must complete both sets of steps*** to set an alternate working schedule for your project. If you set the *Project calendar* but fail to select the *Non-working time calendar*, Microsoft Project schedules each task correctly, but you cannot confirm the schedule because you cannot see the holidays as gray shaded bands in the *Gantt Chart* pane.

Set Options Unique to the Project

Microsoft Project allows you to specify three types of options settings in the *Project Options* dialog, which are as follows:

- Application options that control how the software looks and works
- Options specific to any project currently open
- Options for all new projects created from a blank project

To specify all three types of options settings, click the **File** tab and then click the **Options** tab in the *Backstage*.

Information: You can also specify options settings for your new project by clicking the **Set Project Options** button in the custom *Definition* ribbon.

Microsoft Project displays the *General* page of the *Project Options* dialog shown in Figure 4 - 26.

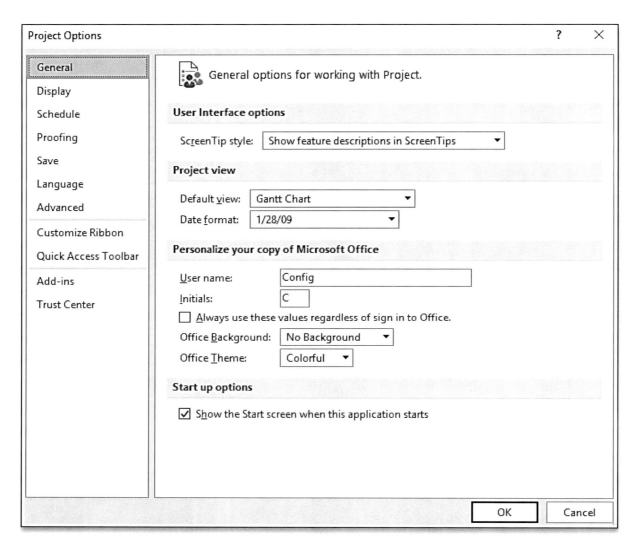

Figure 4 - 26: Project Options dialog

Notice in Figure 4 - 26 shown previously that the *Project Options* dialog includes tabs for eleven pages of options. In this module, I do not discuss all of the options in this dialog, but emphasize important options settings recommended as a best practice by Projility. The following tables document the recommended options settings.

General page

Option	Recommended Setting
Date format	1/28/09
Show the Start screen when this application starts	Deselected

Table 4 - 1: Recommended options for the General page

Schedule page

Option	Recommended Setting
New tasks created	Auto Scheduled
New tasks are effort driven	Selected
Show that scheduled tasks have estimated durations	Deselected (only if you estimate *Work* rather than *Duration*)
New scheduled tasks have estimated durations	Deselected (only if you estimate *Work* rather than *Duration*)

Table 4 - 2: Recommended options for the Schedule page

Advanced page

Option	Recommended Setting
Automatically add new resources and tasks	Deselected
Minutes	m
Hours	h
Days	d
Weeks	w
Months	mo
Years	y
Show project summary task	Selected

Table 4 - 3: Recommended options for the Advanced page

Module 04

The options settings recommended for the *General* page, as shown in Table 4 - 1, are application options that impact every project you open. You only need to set these options once and then Microsoft Project maintains these options settings each time you launch the software.

The options settings recommended for the *Schedule* and *Advanced* pages, as shown in Table 4 - 2 and Table 4 - 3, are project-specific. This means you can change them for an individual project if that project has unique scheduling needs. When finished, click the **OK** button to close the *Project Options* dialog.

 Best Practice: Projility recommends as a best practice that whoever creates enterprise project templates in your organization should specify the options settings on the *Schedule* and *Advanced* pages of the *Project Options* dialog ***in every enterprise project template***. Doing so maintains consistency across all enterprise projects and eliminates the need for your project managers to manually specify these settings for each new project.

If your organization does not have enterprise project templates, this means you need to create every new project from the *Blank Project* template. In this case, you can set the options on the *Schedule* and *Advanced* pages for all new projects created from this template so that you only need to set these options once. To set the options for all new blank projects, complete the following steps:

1. Click the **File** tab and then click the **Options** tab in the *Backstage*.

2. In the *Project Options* dialog, click the **Schedule** tab.

3. Click the **Scheduling options for this project** pick list and select the **All New Projects** item, as shown in Figure 4 - 27.

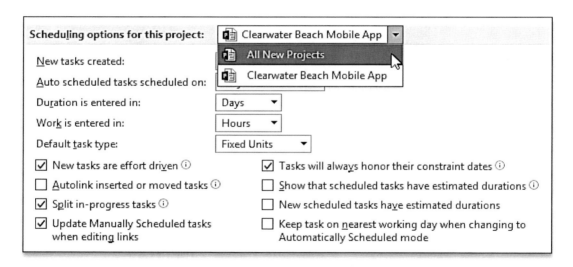

Figure 4 - 27: Set the All New Projects
option for Schedule options

4. Specify the options settings documented previously in Table 4 - 2.

5. Click the **Advanced** tab in the *Project Options* dialog.

6. Click the **General options for this project** pick list and select the **All New Projects** item.

7. Specify the first options setting documented previously in Table 4 - 3.

8. Click the **Display options for this project** pick list and select the **All New Projects** item.
9. Specify the other options setting documented previously in Table 4 - 3.
10. Click the **OK** button.

Save the Project

The final step in the definition process is to save your enterprise project according to your organization's naming convention for enterprise projects. If you created your project previously from the *Project Center* page in Project Web App, simply click the **File** tab and then click the **Save** tab to save the latest changes to your enterprise project.

 Information: You can also save the changes to your new project by clicking the **Save** button in the custom *Definition* ribbon.

If you created your project in Microsoft Project from an enterprise template or from the *Blank Project* template, click the **File** tab, click the **Save** tab in the *Backstage*, and then click the **Save** button on the *Save As* page. Microsoft Project displays the *Save to Project Web App* dialog shown in Figure 4 - 28.

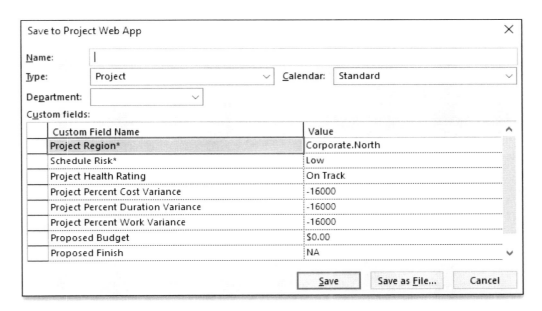

Figure 4 - 28: Save to Project Web App dialog

In the *Save to Project Web App* dialog, enter a name for the project in the **Name** field using your organization's naming convention. Click the **Department** pick list and select a department. Enter or select a value in each of the required and optional fields displayed in the *Custom fields* section of the dialog. Click the **Save** button to save your new enterprise project to Project Web App.

Module 04

Warning: When entering the name of your enterprise project, use only alphanumeric characters, spaces, and the underscore (_) character. **Do not** use any other special characters, such as ampersands (&), pound signs (#), dashes (-), or commas. Using special characters in the name of your enterprise project can lead to a corrupted project.

Hands On Exercise

Exercise 4 - 3

Properly define your new enterprise project using the six-step process.

Important Note: Complete the following set of steps if you created your class project from the *Project Center* page in Project Web App.

1. Launch Microsoft Project and connect to Project Web App using the *Login* dialog.
2. Click the **Open** button on the Quick Access Toolbar.
3. In the *Open* dialog, double-click the **Show me the list of all projects** link at the top of the project list.
4. Select the name of your class project (the project name that begins with **Training**, followed by your first name and last name) and then click the **Open** button.
5. In the *READ-ONLY* gold banner at the top of the project, click the **Check Out** button

Important Note: Complete the following additional set of steps regardless of which method you used to create your class project.

6. Click the **Definition** tab to display the *Definition* ribbon.
7. Click the **Set Project Start Date** button in the *Definition* ribbon.
8. Set the **Start date** value to *Monday of the current week*, specify any values needed for the fields in the *Enterprise Custom Fields* section of the dialog, and then click the **OK** button.
9. Click the **Set Project Properties** button in the *Definition* ribbon.
10. In the *Properties* dialog for your class project, enter the following information:
    - Title = Develop ERP User Training
    - Author = your name

98

- Manager = your name
- Company = your company name
- Comments = Create a user training class for our company's new ERP software

11. In the *Properties* dialog for your class project, click the **OK** button.
12. Select the **Display Project Summary Task** checkbox in the *Definition* ribbon.
13. Widen the *Task Name* column and then drag the split bar to the right edge of the *Duration* column.
14. Click the **Set Project Calendar** button in the *Definition* ribbon.
15. In the *Project Information* dialog, click the **Calendar** pick list and select the enterprise calendar used by your organization (if different from the *Standard* calendar), and then click the **OK** button.
16. Click the **Set Nonworking Time Calendar** button in the *Definition* ribbon.
17. In the *Timescale* dialog, click the **Calendar** pick list, select the same calendar you selected in step #14 above, and then click the **OK** button.
18. Click the **Save** button in the *Definition* ribbon.
19. If you created your class project in Microsoft Project using the steps in Exercise 4 - 2, enter a name for your project in the *Save to Project Web App* dialog using the following naming convention, and then click the **Save** button:

    Training YourFirstName YourLastName Develop ERP Class

Creating Resource Engagements Using Generic Resources

When you receive preliminary approval to proceed with the new proposed project, your organization may need you to create Resource Engagements in the project using Generic resources to forecast the high-level resource demand for the project. Once you submit the Resource Engagements, leaders in your organization can respond as follows:

- Resource managers can see the resource demand for the project and determine if there is sufficient resource capacity to meet the staffing needs of the project.
- If there is sufficient resource capacity to staff the project, executives can approve the project to move to the next step in the SharePoint workflow process.
- Once the project is approved to move to the planning stage, resource managers can substitute a human resource for each of the Generic resources in the project and provide you with your approved project team.

As you can see from the list of items above, your organization can gain tremendous benefits by using Resource Engagements with Generic resources in each new proposed project. I document the exact process for creating Resource Engagements in the following topical sections.

Setting Resource Utilization to Use Resource Engagements

Before you create Resource Engagements in your enterprise project, you need to tell Microsoft Project and Project Web App to calculate resource utilization using the Resource Engagements in the project by completing the following steps:

1. Click the **Project** tab to display the *Project* ribbon.
2. In the *Properties* section of the *Project* ribbon, click the **Project Information** button.
3. In the *Project Information* dialog, click the **Calculate Resource Utilization** from pick list and select the **Resource Engagements** item, as shown in Figure 4 - 29.

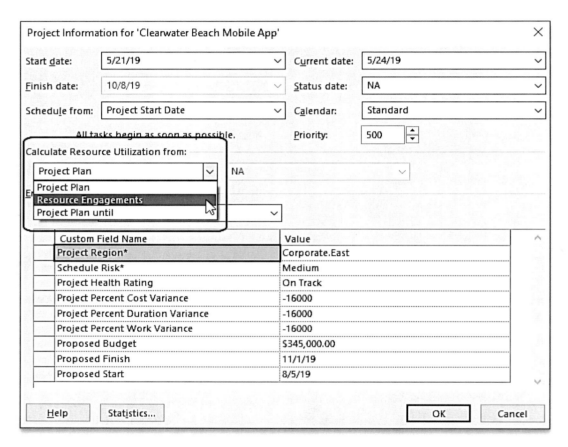

Figure 4 - 29: Calculate resource utilization from Resource Engagements

4. Click the **OK** button.
5. Click the **Save** button in your Quick Access Toolbar.

Building the Project Team with Generic Resources

Before you can create Resource Engagements, you also need to build your project team with Generic resources that represent the roles needed to staff the project. You should use Generic resources in your project team because the project is still in the proposed state, and at this point the exact human resources that will work in the project are most likely undefined. To build your project team using Generic resources, complete the following steps:

Creating an Enterprise Project

1. Click the **Resource** tab to display the *Resource* ribbon.
2. In the *Insert* section of the *Resource* ribbon, click the **Add Resources** pick list button and select the **Build Team from Enterprise** item, as shown in Figure 4 - 30.

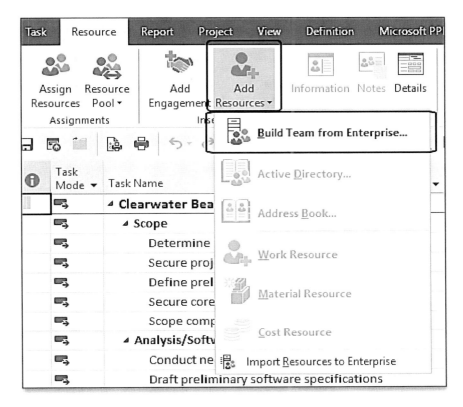

Figure 4 - 30: Select the Build Team from Enterprise item

 Information: If your organization's Enterprise Resource Pool contains 1,000 or more resources, Microsoft Project displays a "pre-filter" dialog initially. In this dialog, you can build your own custom filter to reduce the list to less than 1,000 resources, but the system does not require you to do so. If you do want to see every resource in the Enterprise Resource Pool, simply click the **OK** button in the "pre-filter" dialog to continue the process of building your project team with Generic resources.

Microsoft Project displays the *Build Team* dialog shown in Figure 4 - 31. This dialog displays all of the resources in your organization's Enterprise Resource Pool, which can include Work resources, Cost resources, and Material resources. Microsoft Project organizes the *Build Team* dialog as follows:

- The *Filter enterprise resources* section at the top of the dialog contains the *Existing filters* pick list that displays the default resource filters included with Microsoft Project, along with any enterprise custom filters created by your application administrator. The *Existing filters* pick list contains several useful filters for locating only Work, Cost, or Material resources, but it does not contain a filter to locate Generic resources.

101

- The *Customize filters (optional)* section in the middle of the dialog, when expanded, allows you to create an "ad hoc" filter to reduce the number of resources shown in the dialog. This section also allows you to show resource availability between the *Start* and *Finish* dates of your project, or between a custom set of dates you manually select.

- In the lower left corner of the *Build Team* dialog, you see the list of all resources in your organization's Enterprise Resource Pool. Immediately above the list of resources is the *Group by* pick list, which you can use to organize the resources into groups based on a field you select in the pick list.

- In the lower right corner of the *Build Team* dialog, you see the list of all resources currently on your project team. When you display the *Build Team* dialog for a new project created from an enterprise template, your project team may be blank, or it may consist of a team of Generic resources included as a part of the enterprise template. Notice in the lower right corner of the *Build Team* dialog shown in Figure 4 - 31 that the project team is currently empty. This means I need to manually add Generic resources to the project team.

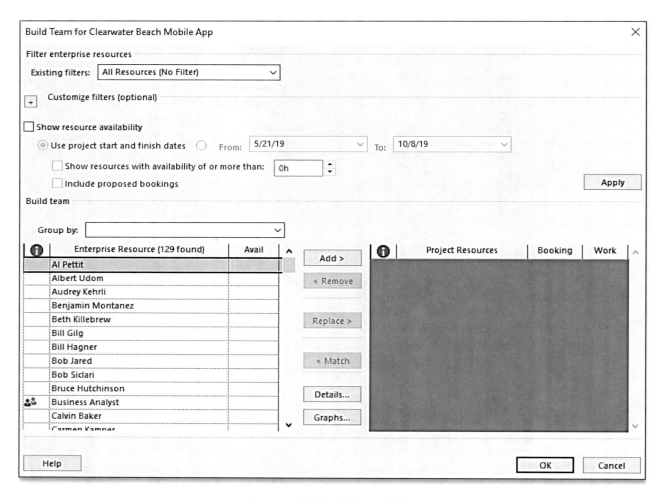

Figure 4 - 31: Build Team dialog

To build your project team, the fastest way to locate your Generic resources is to click the **Group by** pick list and select the **Generic** field. Collapse the *No* group and you can see that the *Yes* group contains all of your organization's Generic resources, such as shown in Figure 4 - 32.

Creating an Enterprise Project

Information: The Generic resources shown in the *Build Team* dialog in Figure 4 - 32 are resources added to our organization's Enterprise Resource Pool by our application administrator. In your own organization's Enterprise Resource Pool, you may find no Generic resources at all or an entirely different list of Generic resources than those shown in the figure.

Best Practice: If your organization's Enterprise Resource Pool does not contain any Generic resources, Projility recommends as a best practice that your application administrator add Generic resources representing the roles for resources that work in projects. To do this, Projility also recommends that the application administrator work with your organization's resource managers to determine which roles need to be modeled with Generic resources.

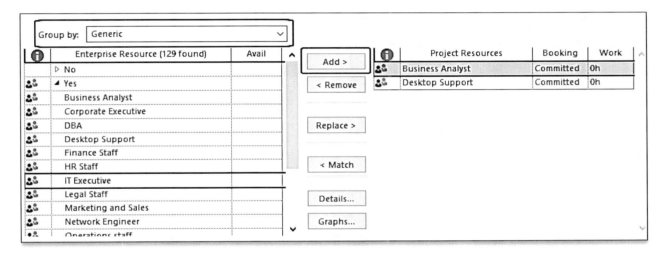

Figure 4 - 32: Group by the Generic field

To add Generic resources to your project team, select a Generic resource from the list on the left side of the dialog and then click the **Add** button to add the selected resource to the project team. Repeat this process until your project team includes all of the roles needed to create Resource Engagements. Click the **OK** button to close the *Build Team* dialog and then click the **Save** button in your Quick Access Toolbar to save the changes to your enterprise project.

Creating the Resource Engagements

After building your project team with Generic resources, you are now ready to create Resource Engagements in your enterprise project. To begin the process of creating Resource Engagements, click the **View** tab to display the *View* ribbon. In the *Resource Views* section of the *View* ribbon, click the **Resource Plan** button shown in Figure 4 - 33.

Module 04

Figure 4 - 33: Click the Resource Plan button

Microsoft Project displays the default *Resource Plan* view shown in Figure 4 - 34, which shows the Generic resources you added to your project team from the *Build Team* dialog. This view includes the default *Resource Plan* table on the left side of the view and a timephased grid on the right side of the view. The *Resource Plan* table includes the *Proposed Start, Proposed Finish, Proposed Max Units,* and *Engagement Status* columns. The timephased grid displays two rows of data for each resource and ultimately for each Resource Engagement as well: the *Proposed Max Units* and the *Committed Max Units* rows.

		Name	Proposed Start	Proposed Finish	Proposed Max Units	Engagement Status	Details	Qtr 2, 2019 Apr	May	Jun
1		Business Analyst	NA	NA			Prop. Max Units			
							Com. Max Units			
2		Desktop Support	NA	NA			Prop. Max Units			
							Com. Max Units			
3		IT Executive	NA	NA			Prop. Max Units			
							Com. Max Units			
4		Project Manager	NA	NA			Prop. Max Units			
							Com. Max Units			
5		Publishing staff	NA	NA			Prop. Max Units			
							Com. Max Units			
6		QA Specialist	NA	NA			Prop. Max Units			
							Com. Max Units			
7		Software Developer	NA	NA			Prop. Max Units			
							Com. Max Units			
8		Technical Trainer	NA	NA			Prop. Max Units			
							Com. Max Units			
							Prop. Max Units			
							Com. Max Units			

Figure 4 - 34: Resource Plan view

When you apply the *Resource Plan* view, Microsoft Project activates the *Engagements* ribbon shown in Figure 4 - 35. This ribbon is only available when you apply the *Resource Engagements* view, and contains most of the tools you need to use to create and work with Resource Engagements.

Creating an Enterprise Project

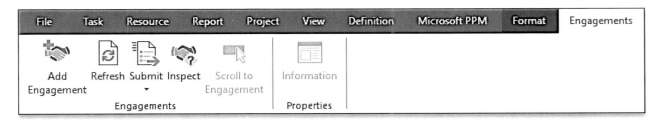

Figure 4 - 35: Engagements ribbon

To create a new Resource Engagement, complete the following steps:

1. Click the **Engagements** tab to display the *Engagements* ribbon.
2. Select the name of a Generic resource.
3. In the *Engagements* section of the *Engagements* ribbon, click the **Add Engagement** button.

Microsoft Project displays the *Engagement Information* dialog shown in Figure 4 - 36. Notice that the *Resource* field contains the name of the selected Generic resource.

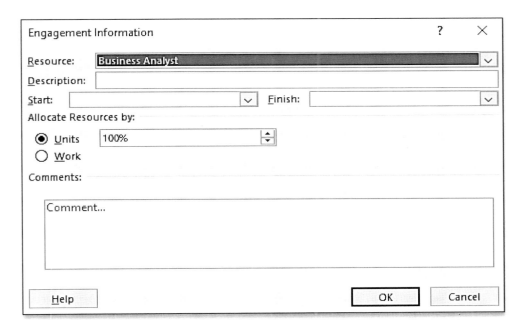

Figure 4 - 36: Engagement Information dialog - blank

4. In the **Description** field, enter an accurate description of the work to be performed by the resource.

Best Practice: If your organization requires resource managers to replace Generic resources with human resources for each Resource Engagement, Projility strongly recommends that you provide an accurate description of the resource's work in the *Description* field. The resource manager can use this description information to select the most qualified available resource for your project.

105

5. In the **Start** field, enter the date the resource begins work in the project, and in the **Finish** field, enter the date the resource finishes work in the project.

6. In the **Units** field, select or manually enter a value that represents the percentage of the resource's working time to the project work, where *100%* represents full-time work (an average of 8 hours/day) and *50%* represents half-time work (an average of 4 hours/day).

> **Information**: In the *Allocate Resources by* section of the *Engagement Information* dialog, you can select the *Work* option and then enter a *Work* estimate measured in hours, rather than as a percentage value using the *Units* field. If you select the *Work* value, you must enter the total number of hours of work performed by the resource over the time span between the *Start* and *Finish* dates you enter.

7. In the **Comments** field, enter an optional comment for the resource manager. For example, you might request a specific human resource for the project.

Figure 4 - 37 shows the completed *Engagement Information* dialog. Notice the following about the information I entered in the dialog:

- In the *Description* field, I provided a description of the work for the Business Analyst to help the resource manager with selecting the best analyst for this project.

- The dates I entered in the *Start* and *Finish* fields indicate I am requesting the Business Analyst for a specific role only during the month of October.

- I entered a *50%* value in the *Units* field to indicate that the Business Analyst will need to work half-time on this role in the project.

- In the *Comments* field, I am requesting Marilyn Ray as the Business Analyst for this project, in the hopes that her manger will assign her if she is available.

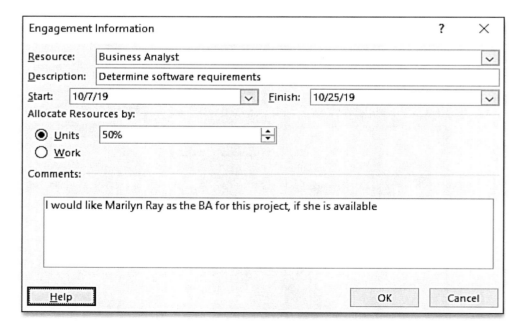

Figure 4 - 37: Engagement Information dialog – completed

Creating an Enterprise Project

8. Click the **OK** button to create the Resource Engagement for the Generic resource.

Information: To edit an existing Resource Engagement, double-click the engagement or click the *Information* button in the *Properties* section of the *Engagements* ribbon for the selected engagement.

9. Continue creating Resource Engagements for each of the Generic resources in your project.
10. Save the changes to your enterprise project.

When creating Resource Engagements, there are three scenarios of which you should be aware:

- You need the resource several different times during the life of the project, but not continuously.
- You need multiple resources with the same role for the project, such as three Network Engineers, for example.
- The *Units* value changes over the time span of the Resource Engagement, such as *50%* the first two months and *100%* the last two months, for example.

When you need the resource at several different times during the life of the project, but not continuously, the best way to model this need is to create a separate Resource Engagement for each time period of utilization. For example, notice in Figure 4 - 38 that I need a Business Analyst to work half-time in October to determine software requirements and I also need a Business Analyst to work 75% of full-time in January to develop use cases for testing. Breaking the work for the Business Analyst into two Resource Engagements allows the resource manager to assign two different analysts to staff these two engagements, if needed.

		Name	Proposed Start	Proposed Finish	Proposed Max Units
1		▲ Business Analyst	10/7/19	1/31/20	75%
		Determine software requirements	*10/7/19*	*10/25/19*	*50%*
		Develop use cases for testing	*1/6/20*	*1/31/20*	*75%*
2		Desktop Support	NA	NA	

Figure 4 - 38: Two different Resource Engagements
needed for the Business Analyst resource

When you need multiple resources with the same role in the project, the best way to model this need is to create multiple Resource Engagements for the Generic resource. For example, notice in Figure 4 - 39 that I need two full-time Software Developers from the *Proposed Start* date of the project to the *Proposed Finish* date of the project. Creating two Resource Engagements for the Software Developer resource allows the resource manager to assign a different developer to each engagement.

Module 04

		Name	Proposed Start	Proposed Finish	Proposed Max Units
7	👥	▲ Software Developer	10/7/19	2/28/20	200%
	✱	*Develop the software*	10/7/19	2/28/20	100%
	✱	*Develop the software*	10/7/19	2/28/20	100%
8	👥	▲ Technical Trainer	2/3/20	2/28/20	100%

Figure 4 - 39: Two Resource Engagements for the
Software Developer resource

 Information: If your organization does not require resource managers to replace Generic resources with human resources in Resource Engagements, the other way to model the need for two full-time Software Developers would be to create a single Resource Engagement with a *Units* value of *200%*.

If *Units* value changes over the time span of the Resource Engagement, the best way to model this is to manually contour (edit) the *Proposed Max. Units* values in the timephased grid for the selected Resource Engagement. For example, notice in Figure 4 - 40 that I do not need the QA Specialist to test the software at a *Units* value of *50%* for the entire life of the project. Instead, I only need the QA Specialist to work 10% of full-time for the first month, 25% of full-time for the second month, and then 50% of full-time for the remainder of the project.

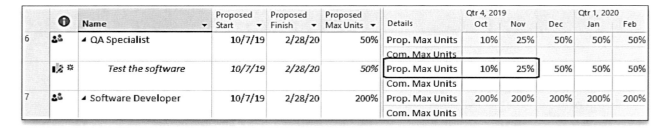

Figure 4 - 40: Manually contour the Proposed Max. Units values

Submitting the Resource Engagements

After creating at least one Resource Engagement for each of the Generic resources in your project team, submit the Resource Engagements and publish the project by completing the following steps:

1. Save the latest changes to your enterprise project.

2. In the *Engagements* section of the *Engagements* ribbon, click the **Submit** pick list button and select the **Submit All Engagements** item, as shown in Figure 4 - 41. When completed, Microsoft Project displays a *Submit Completed Successfully* confirmation message at the right end of the *Status* bar at the bottom of the application screen.

Creating an Enterprise Project

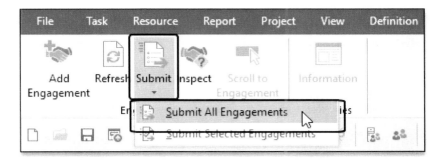

Figure 4 - 41: Submit All Engagements

 Warning: Microsoft Project does not allow you to submit Resource Engagements unless you published the enterprise project at least one time. If you created the project from the *Project Center* page in Project Web App, the system published the project initially during the creation process. If you created the project using an enterprise template in Microsoft Project, however, this means you have not yet published the project, which causes Microsoft Project to disable ("gray out") the *Submit* pick list button. To publish the project initially, click the **File** tab and then click the **Publish** button on the *Info* page of the *Backstage.* In the *Publish Project* dialog, click the **Publish** button. When the Publish job completes, Microsoft Project enables the *Submit* pick list button so that you can submit your pending Resource Engagements.

3. Click the **File** tab and then click the **Publish** button on the *Info* page of the *Backstage,* or click the **Publish** button in your Quick Access Toolbar to publish the latest changes to your enterprise project.

4. Close and check in your enterprise project.

 Hands On Exercise

Exercise 4 - 4

Create Resource Engagements for your enterprise project.

1. Click the **Project** tab to display the *Project* ribbon.

2. In the *Properties* section of the *Project* ribbon, click the **Project Information** button.

3. In the *Project Information* dialog, click the **Calculate Resource Utilization from** pick list and select the **Resource Engagements** item.

4. Click the **OK** button.

5. Click the **Resource** tab to display the *Resource* ribbon.

6. In the *Insert* section of the *Resource* ribbon, click the **Add Resources** pick list button and select the **Build Team from Enterprise** item.

7. In the *Build Team* dialog, click the **Group by** pick list and select the **Generic** field.

109

Module 04

8. Collapse the **No** group so that you can see all of the Generic resources in the *Yes* group.

9. In the list of resources on the left side of the dialog, select *four* Generic resources and then click the **Add** button to add them to your project team.

10. Click the **OK** button to close the *Build Team* dialog.

11. Click the **View** tab to display the *View* ribbon.

12. In the *Resource Views* section of the *View* ribbon, click the **Resource Plan** button.

13. Click the **Engagements** tab to display the *Engagements* ribbon.

14. Select the name of a Generic resource.

15. In the *Engagements* section of the *Engagements* ribbon, click the **Add Engagement** button to display the *Engagement Information* dialog.

16. In the **Description** field, enter a description of the work to be performed by the resource.

17. In the **Start** field, enter the *Start* date you previously entered for the project in Exercise 4 - 3..

18. In the **Finish** field, enter a date that is *approximately three months after* the *Start* date.

19. In the **Units** field, select or manually enter a value that represents the percentage of the resource's working time to the engagement.

20. In the **Comments** field, enter a comment for the resource manager.

21. Click the **OK** button to create the Resource Engagement for the selected Generic resource.

22. Create at least one Resource Engagement for each of the other three Generic resources in your class project with the following requirements:

    - Enter same *Start* and *Finish* dates entered in steps #17-18 for each of the Resource Engagements.
    - Enter a unique *Units* value for each of the Resource Engagements.
    - Manually contour the *Proposed Max Units* value for each month in the timephased grid for one of the Resource Engagements.

23. Save the changes to your class project.

24. In the *Engagements* section of the *Engagements* ribbon, click the **Submit** pick list button and select the **Submit All Engagements** item.

> **Information**: If the *Submit* pick list button is disabled, you must publish your project before you can submit your Resource Engagements. Click the **File** tab and then click the **Publish** button on the *Info* page of the *Backstage*. In the *Publish Project* dialog, click the **Publish** button. When the Publish job completes, Microsoft Project enables the *Submit* pick list button so that you can submit your pending Resource Engagements

25. Click the **File** tab and then click the **Publish** button on the **Info** page of the *Backstage*.

110

Module 05

Task Planning

Learning Objectives

After completing this module, you will be able to:

- Understand the task planning process
- Use basic task planning skills to plan the task list
- Use the correct task dependencies to model the task sequence
- Use constraints, deadline dates, and task calendars to constrain the task schedule
- Enter duration estimates

Inside Module 05

Understanding the Task Planning Process	**113**
Understanding the Task Mode Setting	**113**
Using Basic Task Planning Skills	**114**
Entering and Editing Tasks	*114*
Moving Tasks	*114*
Inserting Tasks	*115*
Deleting Tasks	*116*
Creating the Work Breakdown Structure (WBS)	*118*
Creating Milestones	*121*
Adding Task Notes	*124*
Applying Cell Background Formatting	*125*
Using Task Dependencies	**128**
Understanding Task Dependencies	*128*
Setting and Removing Task Dependencies	*130*
Using Lag Time with Dependencies	*132*
Using Lead Time with Dependencies	*133*
Setting Task Constraints and Deadline Dates	**137**
Setting Constraints	*137*
Understanding Flexible and Inflexible Constraints	*139*

Understanding Planning Wizard Messages about Constraints..140
Setting Deadline Dates..141
Understanding Missed Constraints and Deadline Dates..143
Applying Task Calendars 144
Estimating Task Durations 147

Understanding the Task Planning Process

After you define your project, the planning process begins. The first step in the planning process is **task planning**. If you do not create your project from an enterprise template, then you must manually complete a series of steps in Microsoft Project during the task planning process. Even if you do create your project from an enterprise template, there are still optional task planning steps that you must complete, such as deleting unneeded tasks or inserting additional tasks not included in the template.

When you create a task list manually, you must thoughtfully analyze the activities required to complete the project. Depending on the size of your project, this may mean lots of typing! You can use either a "top down" or "bottom up" approach to create the initial task list. The "top down" approach begins by listing the major phases of the project, as well as the project deliverables under each phase. Under each deliverable you list the activities necessary to produce the deliverable. To help you with the process of "top down" planning, Microsoft Project includes a feature that allows you to insert "top down" summary tasks. I discuss this new feature later in this module.

The "bottom up" approach works in the opposite direction. Using this approach, you list all of the activities in the project and then organize the activities into phase and deliverable summary sections. You can be effective in creating the task list for the project using either of these approaches. Whenever you create a new project manually, it is a good idea to apply the following top down methodology:

1. Create the task list.
2. Create summary tasks to organize the Work Breakdown Structure (WBS) of the project.
3. Create project milestones.
4. Set task dependencies and document unusual task dependencies with a task note.
5. Set task constraints and deadline dates, and document all task constraints with a task note.
6. Set task calendars for any task with an alternate working schedule and document the task calendars with a task note.
7. Estimate task effort or durations according to your preferred or required methodologies.

In successive topical sections in this module, I discuss each of the steps in the preceding task planning methodology.

Understanding the Task Mode Setting

The **Task Mode** setting in Microsoft Project offers two types of tasks for to use in your projects: **Auto Scheduled** tasks and **Manually Scheduled** tasks. An *Auto Scheduled* task is a task whose schedule is controlled by Microsoft Project, and whose schedule is updated automatically every time you make a scheduling change in your project. A *Manually Scheduled* task is a task whose schedule you must control manually. Every time you make a scheduling change in your project, Microsoft Project ***does not*** update the schedule of *Manually Scheduled* tasks. Instead, the software relies on you to manually reschedule the tasks as needed.

The default *Task Mode* setting for every new blank project is *Manually Scheduled* tasks, unless you changed this setting as documented previously in Module 04. Because of the very restrictive behavior of *Manually Scheduled* tasks, I recommend that you ***do not use them*** in your enterprise projects. Instead, always use *Auto Scheduled* tasks to allow Microsoft Project to control the schedule of your project and to update the schedule automatically every time you make a scheduling change in your project. Doing so guarantees you the most accurate schedule possible in your enterprise projects.

Best Practice: To prevent project managers from using *Manually Scheduled* tasks in their enterprise projects, Projility recommends that your organization's application administrator "lock down" this option. In the *Task Mode Settings* section of the *Additional Server Settings* page in Project Web App, the app admin should select the **Automatically Scheduled** option and leave the **Users can override default in Project Professional** checkbox ***deselected***. With just a few small permission changes, all project managers must use only *Auto Scheduled* tasks in every enterprise project.

Using Basic Task Planning Skills

You should possess a variety of basic task skills to use Microsoft Project effectively. I discuss each of these basic task planning skills in the following subtopical sections.

Entering and Editing Tasks

Entering project tasks in Microsoft Project is very similar to entering data in a Microsoft Excel spreadsheet. To enter a new task, complete the following steps:

1. In any blank row, select the blank cell in the *Task Name* column of the task sheet.
2. Manually type the task name.
3. Press the **Enter** key or **Down-Arrow** key on your computer keyboard.

To edit the name of an existing task, select the task and then use any of the following methods:

- Double-click the task and edit the name in the *Task Information* dialog.
- Retype the task name.
- Press the **F2** function key on your computer keyboard and edit the task name.
- Select the name of the task and then click anywhere in the cell to enable in-cell editing.

Moving Tasks

During the task planning process, you may create the task list with tasks in the wrong order. To move a task, complete the following steps:

1. Click the task *ID* number (row header) on the far left end of the task and then release the mouse button. Microsoft Project changes the mouse pointer to a four-headed arrow, indicating that you can drag the task to a new location in the schedule.
2. Click and hold the task *ID* number to "grab" the task.
3. Move the mouse pointer up or down on the screen to move the task.

As you move the mouse pointer, you see a gray I-beam bar to indicate that you are moving the task, as shown in Figure 5 - 1.

Task Planning

	Task Mode	Task Name	Duration	Start	Finish
0	⇉	⊿ New Project	1 d	6/1/20	6/1/20
1	⇉	Build	1 d	6/1/20	6/1/20
2	⇉	Test	1 d	6/1/20	6/1/20
3	⇉	Design	1 d	6/1/20	6/1/20
4	⇉	Implement	1 d	6/1/20	6/1/20

Figure 5 - 1: Moving a task

4. Drag the task until you position the gray I-beam indicator where you want to place the task.
5. Release the mouse button to complete the move and "drop" the task in its new location.

You can move a single task, or you can move a group of tasks using this technique. When you move a summary task, all of its subtasks move with it automatically.

Warning: Moving tasks by using cut and paste of entire task rows can lead to a corrupted enterprise project. Duplicating tasks by using copy and paste of entire task rows can also lead to a corrupted enterprise project. When you cut and paste or copy and paste entire task rows, the system creates duplicate GUIDs in the Project Online database, which inevitably leads to a corrupted enterprise project. If you need to duplicate tasks, the safe way to accomplish this is to select only the *Task Name* and *Duration* data for the tasks, copy the data to the Windows Clipboard, insert new blank rows, and then paste the data in the blank rows.

Inserting Tasks

While entering a task list, you may discover that you omitted one or more tasks. To insert new tasks in your project plan, select any cell in the row where you want to insert the new task and then use one of the following methods:

- Press the **Insert** key on your computer keyboard.
- In the *Insert* section of the *Task* ribbon, click the **Task** button.
- Right-click in the row and then select the **Insert Task** item on the shortcut menu.

Microsoft Project adds a new blank task row automatically above the selected task. If you add the new task using the **Insert** key on your keyboard, the system simply adds a new blank row. If you use the **Task** button on the *Task* ribbon or the **Insert Task** item on the shortcut menu, the system adds a new task named *<New Task>* with a default *Duration* value of *1 day*. After using any of these three methods, you must still enter the name of the new task.

To add multiple new tasks simultaneously, select as many rows as the number of new tasks you would like to add, and use one of the preceding methods. For example, notice in Figure 5 - 2 that I want to add two new tasks before the Implement task, so I select the Implement task and the following blank task as the location of the two new tasks.

115

ID		Task Mode	Task Name	Duration	Start	Finish
0			⊿ New Project	1 d	6/1/20	6/1/20
1			Design	1 d	6/1/20	6/1/20
2			Build	1 d	6/1/20	6/1/20
3			Test	1 d	6/1/20	6/1/20
4			Implement	1 d	6/1/20	6/1/20

Figure 5 - 2: Preparing to insert two new tasks after the Test task

When I press the Insert key on my computer keyboard, notice in Figure 5 - 3 that Microsoft Project inserts two new blank rows before the Implement task.

ID		Task Mode	Task Name	Duration	Start	Finish
0			⊿ New Project	1 d	6/1/20	6/1/20
1			Design	1 d	6/1/20	6/1/20
2			Build	1 d	6/1/20	6/1/20
3			Test	1 d	6/1/20	6/1/20
4						
5						
6			Implement	1 d	6/1/20	6/1/20

Figure 5 - 3: Two new tasks inserted before the Implement task

Deleting Tasks

While entering a task list, you may find that you no longer need one or more tasks in the project plan. This is especially true if you created your project from an enterprise template. To delete a task, complete the following steps, select the ID number of the task you want to delete and use one of the following methods:

- Click the *ID* number (row header) of the task you want to delete, and then press the **Delete** key on your computer keyboard.

- Right-click in the selected task and then select the **Delete Task** item on the shortcut menu.

If you select only the name of the task in the *Task Name* column (rather than selecting the task *ID* number) and then press the **Delete** key on your computer keyboard, the software displays a Smart Tag to the left of the cell. When you float your mouse pointer over the Smart Tag, the system displays a pick list. Click the Smart Tag pick list and then select whether to clear the contents of only the task name or to delete the entire task. Figure 5 - 4 shows the Smart Tag for deleting a task after selecting only the name of the task, rather than the task ID number.

Task Planning

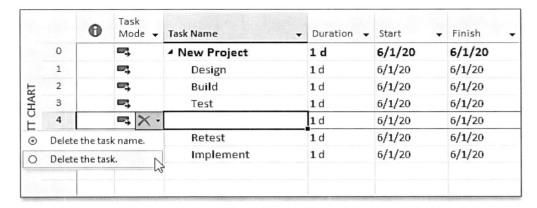

Figure 5 - 4: Smart Tag displayed after deleting a task

 Hands On Exercise

Exercise 5 - 1

Use basic task planning skills to define the task list in a project.

1. Click the **Task** tab to display the *Task* ribbon.

2. In the *View* section of the *Task* ribbon, click the **Gantt Chart** pick list button and select the **Gantt Chart** view.

3. Right-click on task ID #10, the *Create Pre-Class Skills Assessment* task, and select the **Delete Task** item on the shortcut menu.

4. Click the *Task Name* cell in the first blank row at the bottom of the task list and then manually enter a task named **Student Sample Files**.

5. Click the hold the row header for task ID #13, the *Student Sample Files* tab, then drag and drop the task between task IDs #4 and #5.

6. Select task IDs #6-8, from the *Instructor Files* task to the *Create Instructor Demo* Files task.

7. Press the **Insert** key on your computer keyboard to add three new blank rows.

8. Beginning with task ID #6, enter the names of the following tasks in the three blank rows:

    - *Create Sample Files for Chapter 1*
    - *Create Sample Files for Chapter 2*
    - *Create Sample Files for Chapter 3*

9. Save the changes to your enterprise project.

Module 05

Creating the Work Breakdown Structure (WBS)

The Work Breakdown Structure (WBS) divides the project tasks into meaningful and logical components. The WBS consists of summary tasks representing major aspects of the project, such as phase and deliverable sections, along with subtasks in each summary section. Figure 5 - 5 shows a simple generic Work Breakdown Structure comprised of a single phase section with two deliverable sections, and with four subtasks in each deliverable section.

		Task Mode	Task Name	Duration	Start	Finish
0			**⊿ Project Summary Task**	1 d	3/2/20	3/2/20
1			⊿ PHASE I	1 d	3/2/20	3/2/20
2			⊿ Deliverable 1	1 d	3/2/20	3/2/20
3			Design	1 d	3/2/20	3/2/20
4			Build	1 d	3/2/20	3/2/20
5			Test	1 d	3/2/20	3/2/20
6			Implement	1 d	3/2/20	3/2/20
7			⊿ Deliverable 2	1 d	3/2/20	3/2/20
8			Design	1 d	3/2/20	3/2/20
9			Build	1 d	3/2/20	3/2/20
10			Test	1 d	3/2/20	3/2/20
11			Implement	1 d	3/2/20	3/2/20

Figure 5 - 5: Work Breakdown Structure

To create a Work Breakdown Structure in a project, you must create a series of summary tasks and subtasks. The purpose of summary tasks is to summarize or "roll up" the data contained in the subtasks. Microsoft Project offers you several ways to create summary tasks and subtasks. To use the first method, complete the following steps:

1. Type the name of a summary task, along with the names of each of its subtasks.

2. Select the tasks that you want to convert to subtasks of the summary task.

3. In in the *Schedule* section of the *Task* ribbon, click the **Indent** button.

In the project shown in Figure 5 - 6, notice that I selected the Design task through the Implement task to prepare to convert them to subtasks of the PHASE I summary task.

Task Planning

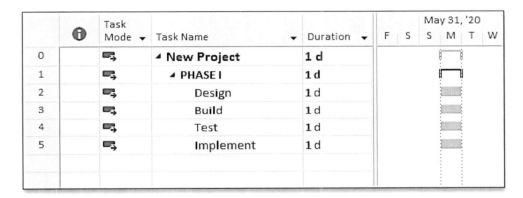

Figure 5 - 6: Prepare to make PHASE I a summary task

After clicking the **Indent** button in the *Schedule* section of the *Task* ribbon, Microsoft Project indents the four selected tasks to convert them to subtasks of the PHASE I summary task. Figure 5 - 7 shows the result of this procedure.

		Task Mode	Task Name	Duration	May 31, '20 F S S M T W
0			▲ **New Project**	1 d	
1			▲ **PHASE I**	1 d	
2			Design	1 d	
3			Build	1 d	
4			Test	1 d	
5			Implement	1 d	

Figure 5 - 7: PHASE I summary task with four subtasks

Notice in Figure 5 - 7 how the software displays summary tasks and subtasks:

- It formats PHASE I with bold text.
- It shows an outline indicator (black triangle symbol) to the left of the PHASE I task name.
- It changes the Gantt bar shape for PHASE I to a black square bracket.
- It indents the Design task through the Implement task one level to the right of the PHASE I summary task.

 Warning: Promoting a task to a summary task causes the system to roll up summary information from its subtasks. This changes the behavior of the task, but it does not change any underlying data that you previously entered. If you demote the summary task back to a normal task, any previously entered data reappears. Because this can cause surprising schedule changes, Projility recommends that you create all of your summary tasks from new tasks rather than repurposing tasks in which you previously entered work or duration values.

The second method for creating summary tasks in Microsoft Project is to use the **Insert Summary Task** feature. This feature allows you to directly insert summary tasks into your project, which makes "top down" task planning much easier. To use this feature, select the row where you want to insert the summary task, then click the **Insert**

119

Summary Task button in the *Insert* section of the *Task* ribbon. The software inserts a new unnamed summary task along with one unnamed subtask, as shown in Figure 5 - 8.

		Task Mode	Task Name	Duration	Start	Finish
0			▲ New Project	1 d	6/1/20	6/1/20
1			▲ <New Summary Task>	1 d	6/1/20	6/1/20
2			<New Task>	1 d	6/1/20	6/1/20

Figure 5 - 8: Newly Inserted Summary Task

After inserting the new summary task and subtask pair, you should edit the name of the summary task, replacing the default value with the name of the phase or deliverable section it represents. Similarly, you eventually need to edit the name of the subtask and add additional subtasks as needed. Alternately, you can leave the name of the subtask with its original *<New Task>* name as a placeholder for a future subtask until you are ready to add detail tasks to the summary section.

If you insert a summary task below another summary task or a subtask, Microsoft Project automatically indents the new summary task at the same level of indenture as the task immediately preceding it. This is the default behavior of the tool, and you cannot change it. For example, Figure 5 - 9 shows a new summary task and subtask pair inserted after the Design task. Notice that the system indented the new summary task at the same level as the Design task preceding it. To resolve the indenting situation shown in Figure 5 - 9, select the new summary task and then click the **Outdent Task** button in the *Schedule* section of the *Task* ribbon.

		Task Mode	Task Name	Duration	Start	Finish
0			▲ New Project	1 d	6/1/20	6/1/20
1			▲ PHASE I	1 d	6/1/20	6/1/20
2			Design	1 d	6/1/20	6/1/20
3			▲ <New Summary Task>	1 d	6/1/20	6/1/20
4			<New Task>	1 d	6/1/20	6/1/20

Figure 5 - 9: New summary task indented at same level as the Design task preceding it

In addition to inserting summary tasks during "top down" task planning, Microsoft Project also makes it easier to insert a summary task for a selected group of subtasks. For example, consider the set of four tasks shown in Figure 5 - 10. I want to show that each of these four selected tasks is a subtask of a new Phase I section in the project.

Task Planning

		Task Mode ▼	Task Name ▼	Duration ▼	Start ▼	Finish ▼
0		⇶	▲ New Project	1 d	6/1/20	6/1/20
1		⇶	Design	1 d	6/1/20	6/1/20
2		⇶	Build	1 d	6/1/20	6/1/20
3		⇶	Test	1 d	6/1/20	6/1/20
4		⇶	Implement	1 d	6/1/20	6/1/20

Figure 5 - 10: Four tasks ready for inclusion as
subtasks of a new Phase I summary task

To make these four selected tasks become subtasks of a new Phase I section of the project, leave the four tasks selected and then click the **Insert Summary Task** button in the *Insert* section of the *Task* ribbon. Microsoft Project automatically inserts a new unnamed summary task and indents the four tasks as subtasks of the summary task, as shown in Figure 5 - 11. You can then rename the new summary task as desired.

		Task Mode ▼	Task Name ▼	Duration ▼	Start ▼	Finish ▼
0		⇶	▲ New Project	1 d	6/1/20	6/1/20
1		⇶	▲ <New Summary Task>	1 d	6/1/20	6/1/20
2		⇶	Design	1 d	6/1/20	6/1/20
3		⇶	Build	1 d	6/1/20	6/1/20
4		⇶	Test	1 d	6/1/20	6/1/20
5		⇶	Implement	1 d	6/1/20	6/1/20

Figure 5 - 11: Four tasks inserted as subtasks
below the new unnamed summary task

Creating Milestones

In project management terms, we define a milestone as "a significant point in time" in a project. Most typically, a milestone represents the start or finish of something important. You may use a milestone to indicate the beginning point of a project, as well as the completion point for a phase, a deliverable, or even the entire project. Most projects contain multiple milestones.

Best Practice: Projility recommends as a best practice that you add a milestone to every phase and deliverable section of your enterprise projects. In addition, Projility also recommends that you include a final milestone at the end of each project to denote the completion date of the entire project.

Microsoft Project offers you several ways to create milestone tasks. One method for creating a milestone is to insert a new task and then change its **Duration** value to **0 days**. The software automatically converts the new task to a milestone task. Notice in Figure 5 - 12 that the Gantt Chart symbol for a milestone is a black diamond with the *Finish* date of the milestone displayed to the right of the symbol.

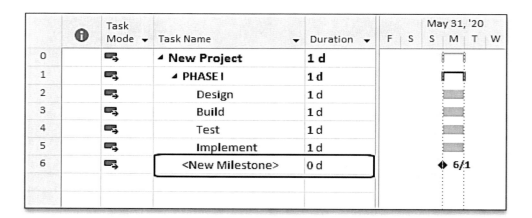

Figure 5 - 12: New milestone task created

The other method for creating a milestone is to click the **Insert Milestone** button in the *Insert* section of the *Tasks* ribbon. Microsoft Project inserts a new unnamed task with a *Duration* value of *0 days*, such as shown in Figure 5 - 13. Rename the new milestone task as needed.

Figure 5 - 13: Insert new milestone task

Best Practice: In Figure 5 - 12 and Figure 5 - 13 shown previously, notice that I outdented the new milestone task so that its indenture level is the same as the PHASE I summary task. For ease of high-level reporting, Projility recommends that you outdent milestone tasks at the same level of indent as the summary tasks they represent. With this structure in place, when you collapse *Outline Level 1* summary tasks (the project phases), you can see each phase along with its corresponding milestone. In addition, the milestone symbol in the *Gantt Chart* pane displays the *Finish* date of the milestone, which represents the *Finish* date of the phase as well.

Hands On Exercise

Exercise 5 - 2

Create summary tasks and milestones in your enterprise project.

1. Select task IDs #2-4, from the *Write Chapter 1* task to the *Write Chapter 3* task.
2. In the *Schedule* section of the *Task* ribbon, click the **Indent Task** button.
3. Select task IDs #6-8, from the *Create Sample Files for Chapter 1* task to the *Create Sample Files for Chapter 3* task.
4. In the *Schedule* section of the *Task* ribbon, click the **Indent Task** button.
5. Select task IDs #10-12, from the *Create Instructor Training Outline* task to the *Create Instructor PowerPoint* task.
6. In the *Schedule* section of the *Task* ribbon, click the **Indent Task** button.
7. Select task IDs #14-16, from the *Create Student Learning Aids* task to the *Create Class Completion Certificate* task.
8. In the *Schedule* section of the *Task* ribbon, click the **Indent Task** button.
9. Widen the *Task Name* column, as needed, and then drag the split bar to the right edge of the *Duration* column.
10. Select task ID #5, the *Student Sample Files* summary task, and then press the **Insert** key on your computer keyboard.
11. In the new blank line, enter a task named **Course Manual Complete** and change its **Duration** value to **0 days**.
12. Select task ID #5, the new *Course Manual Complete* milestone task, then click the **Outdent Task** button in the *Schedule* section of the *Task* ribbon.
13. Select task ID #10, the *Instructor Materials* summary task, and then press the **Insert** key on your computer keyboard.
14. In the new blank line, enter a task named **Student Sample Files Complete** and change its **Duration** value to **0 days**.
15. Select task ID #10, the new *Student Sample Files Complete* milestone task, then click the **Outdent Task** button in the *Schedule* section of the *Task* ribbon.
16. Select task ID #15, the *Student Handouts* summary task, and then press the **Insert** key on your computer keyboard.
17. In the new blank line, enter a task named **Instructor Materials Complete** and change its **Duration** value to **0 days**.

Module 05

18. Select task ID #15, the new *Instructor Materials Complete* milestone task, then click the **Outdent Task** button in the *Schedule* section of the *Task* ribbon.

19. In the first blank line at the bottom of the project, enter two new tasks named **Student Handouts Complete** and **Project Complete**, and then change their **Duration** values to **0 days**.

20. Select the two new milestone tasks at the bottom of the project and then click the **Outdent Task** button in the *Schedule* section of the *Task* ribbon.

21. Save the latest changes to your enterprise project.

Adding Task Notes

Task notes are an important part of project documentation and are essential to understanding the historical information about any project. You can add notes to tasks at any time during the life of the project, from planning through closure. To add a note to a task, use any of the following methods:

- Double-click the task and then click the **Notes** tab in the *Task Information* dialog.
- Select the task and then click the **Task Notes** button in the *Properties* section of the *Task* ribbon.
- Right-click on the task and then select the **Notes** item on the shortcut menu.

Using any of the preceding methods, the software displays the *Notes* page of the *Task Information* dialog. Type the text of the note in the *Notes* section, such shown in Figure 5 - 14. Click the **OK** button when finished.

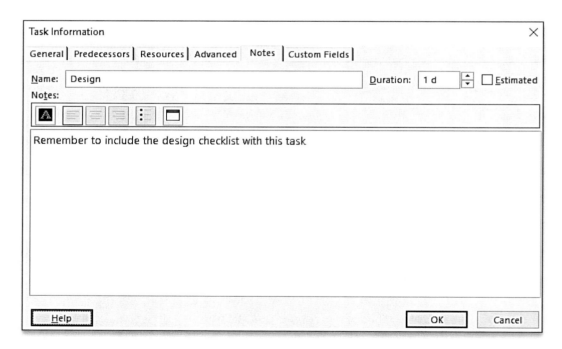

Figure 5 - 14: Task Information dialog, Notes page

After you add a note to a task, Microsoft Project displays a yellow note indicator in the *Indicators* column to the left of the task. You can read the text of the note by floating your mouse pointer over the note indicator, such as shown in Figure 5 - 15.

124

Task Planning

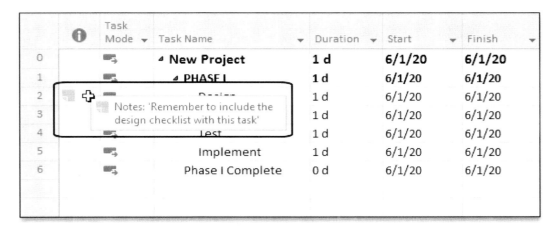

Figure 5 - 15: Notes indicator with screen tip text displayed

Applying Cell Background Formatting

To easily identify tasks of interest, you can use the cell background formatting feature in Microsoft Project. This feature is similar to cell background formatting in Microsoft Excel. To apply cell background formatting for one or more tasks, complete the following steps:

1. Select the *ID* numbers (row headers) of the tasks you want to format.

2. In the *Font* section of the *Task* ribbon, click the **Background Color** pick list button.

3. In the *Background Color* menu, select a color in either the *Theme Colors* or *Standard Colors* section, as shown in Figure 5 - 16.

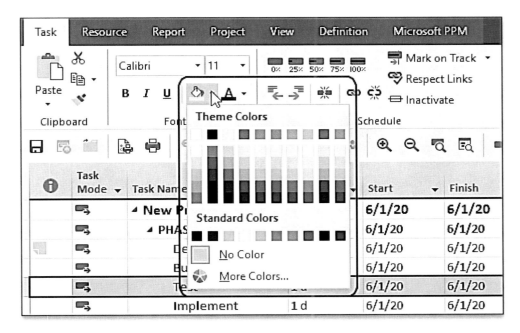

Figure 5 - 16: Background Color pick list

125

Module 05

If you do not see your desired color in the *Background Color* menu, click the **More Colors** item at the bottom of the menu. Microsoft Project displays the *Colors* dialog. This dialog contains two pages: the *Standard* page shown in Figure 5 - 17 and the *Custom* page shown in Figure 5 - 18.

On the *Standard* page of the *Colors* dialog, select a color from the **Colors** palette, and then click the **OK** button. On the *Custom* page of the *Colors* dialog, click anywhere in the **Colors** palette to select the initial color, and then drag the **Shading** slider to the right of the *Colors* palette to "dial in" the exact color you want. Click the **OK** button when finished.

Warning: You can only apply cell background formatting to the default views included with the Microsoft Project desktop application. These default views include task views such as the *Gantt Chart*, *Tracking Gantt*, or *Task Usage* views. When you apply cell background formatting to a default view, the software keeps the formatting with the view in your enterprise project so that you see the formatting every time you open the project.

If you attempt to apply cell background formatting to an enterprise view created by your application administrator, on the other hand, Microsoft Project only allows you to apply the formatting **temporarily**. The next time you open your enterprise project, the cell background formatting **disappears** from the view because only the application administrator can permanently edit enterprise views. Therefore, to avoid frustration and wasted time, limit your use of cell background formatting to only the default task views included with Microsoft Project.

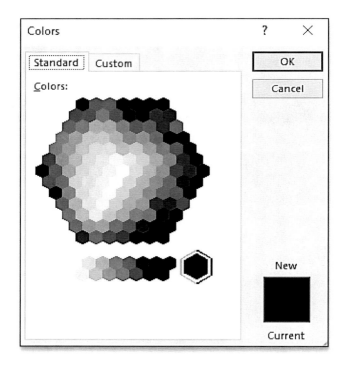

Figure 5 - 17: Colors dialog – Standard page

126

Task Planning

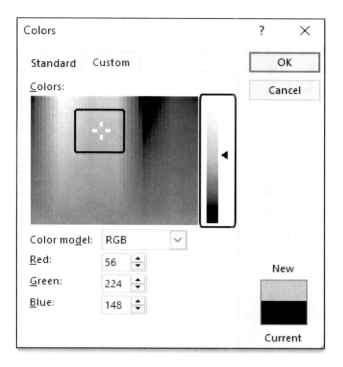

Figure 5 - 18: Colors dialog – Custom page

Figure 5 - 19 shows background cell formatting for the *Test* task in a project. To format this task, I selected the *Light Green* color in the *Standard Colors* section of the *Background Color* menu.

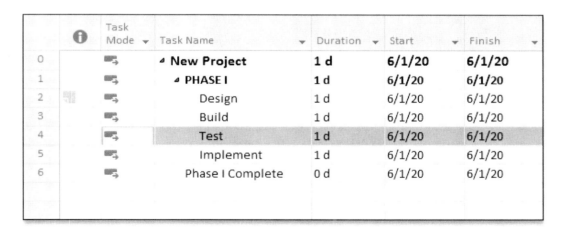

Figure 5 - 19: Cell background formatting applied to the Test task

127

Hands On Exercise

Exercise 5 - 3

Add a note and apply cell background formatting to a task in your enterprise project.

1. Double click task ID #4, the *Write Chapter 3* task, and then click the **Notes** tab in the *Task Information* dialog.

2. Manually enter the following note on the *Notes* page:

   Meet with consultant to determine the content for this chapter.

3. Click the **OK** button to close the *Task Information* dialog.

4. Float your mouse pointer over the indicator shown in the *Indicators* column for the selected task and then read the body of the note in the tooltip.

5. Click the *ID* number (row header) for task ID #12, the *Create Instructor Training Outline* task, to select the entire row.

6. In the *Font* section of the *Task* ribbon, click the **Background Color** pick list button.

7. In the *Theme Colors* section of the *Background Color* menu, select the **Orange, Lighter 60%** color.

8. Select any other task in the project and then notice how task ID #12 "stands out" because of the cell background formatting you applied to this task.

9. Save the latest changes to your enterprise project.

Using Task Dependencies

After completing the initial steps in the task planning process, your next step is to determine the sequence in which tasks occur. The sequence planning process requires you to determine and set task dependencies between the tasks in your project.

Understanding Task Dependencies

When you set a task dependency between two tasks, Microsoft Project designates one task as the **Predecessor** task and designates the other task as the **Successor** task in the dependency relationship. Many people confuse the meaning of these two terms, wrongly assuming that the Predecessor is the task that "comes before" before the Successor task. A better way to think of these two terms is as follows:

- The Predecessor is the "driving task" that **drives** the schedule of the Successor task.
- The Successor is the "driven task" whose schedule **is driven** by the Predecessor task.

Microsoft offers you the following four default task dependency types, with examples of each of them shown in Figure 5 - 20:

- Finish-to-Start (abbreviated as FS)
- Start-to-Start (SS)
- Finish-to-Finish (FF)
- Start-to-Finish (SF)

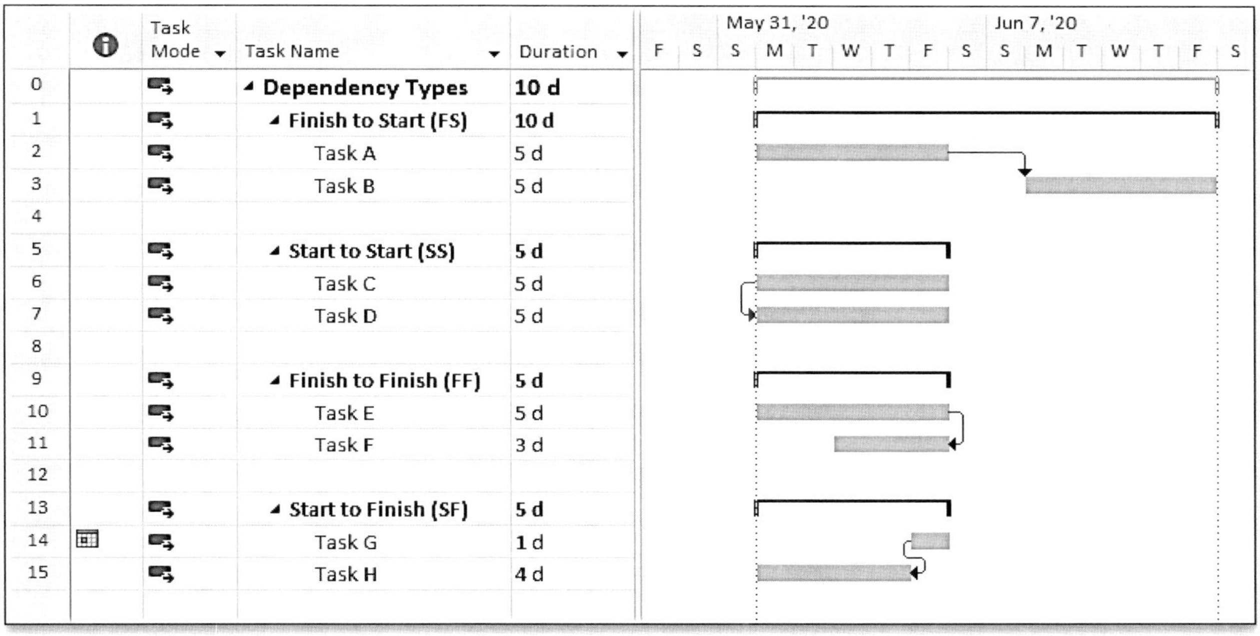

		Task Mode	Task Name	Duration
0			▲ Dependency Types	10 d
1			▲ Finish to Start (FS)	10 d
2			Task A	5 d
3			Task B	5 d
4				
5			▲ Start to Start (SS)	5 d
6			Task C	5 d
7			Task D	5 d
8				
9			▲ Finish to Finish (FF)	5 d
10			Task E	5 d
11			Task F	3 d
12				
13			▲ Start to Finish (SF)	5 d
14			Task G	1 d
15			Task H	4 d

Figure 5 - 20: Four default task dependency types

A **Finish-to-Start (FS)** dependency means that the *Finish* date of the Predecessor task drives the *Start* date of the Successor task. Figure 5 - 20 displays an FS dependency between Task A and Task B. In this task pair, the *Finish* date of Task A (the Predecessor) drives the *Start* date of Task B (the Successor).

A **Start-to-Start (SS)** dependency means that the *Start* date of the Predecessor task drives the *Start* date of the Successor task. Figure 5 - 20 displays an SS dependency between Task C and Task D. In this task pair, the *Start* date of Task C (the Predecessor) drives the *Start* date of Task D (the Successor). In practical terms, this means that these two tasks start at the same time.

A **Finish-to-Finish (FF)** dependency means that the *Finish* date of the Predecessor task drives the *Finish* date of the Successor task. . Figure 5 - 20 displays an FF dependency between Task E and Task F. In this task pair, the *Finish* date of Task E (the Predecessor) drives the *Finish* date of Task F (the Successor). In practical terms, this means that these two tasks finish at the same time.

A **Start-to-Finish (SF)** dependency means that the *Start* date of the Predecessor task drives the *Finish* date of the Successor task. Figure 5 - 20 displays an SF dependency between Task G and Task H. In this task pair, the Start date of Task G (the Predecessor) drives the *Finish* date of Task H (the Successor). Many people refer to the SF dependency as the "backwards" dependency.

Module 05

Setting and Removing Task Dependencies

To set a dependency in Microsoft Project, complete the following steps:

1. Select two or more tasks that are dependent on one another.

2. In the *Schedule* section of the *Task* ribbon, click the **Link the Selected Tasks** button.

When you complete these steps, Microsoft Project automatically sets a Finish-to-Start (FS) dependency on the selected tasks by default. To change the dependency type to one of the other three, continue with the following steps:

3. In the *Gantt Chart* pane, double-click the link line between two dependent tasks. The software displays the *Task Dependency* dialog shown in Figure 5 - 21.

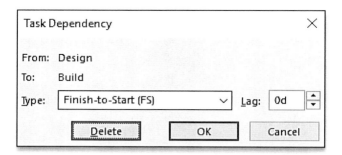

Figure 5 - 21: Task Dependency dialog

4. Click the **Type** pick list and select the desired dependency type.
5. Click the **OK** button.

 Information: Because link lines are very thin, it can be a bit of a challenge to double-click on a link line. The easiest target to double-click is the arrowhead at the end of the link line that points to the Successor task.

 Warning: Set task dependencies **only** on regular tasks and milestone tasks. **Do not** set task dependencies on summary tasks, as doing this can lead to circular reference errors any time in the future. Circular reference errors are very difficult to troubleshoot and resolve, so it is better to avoid them by not setting task dependencies on summary tasks.

 Warning: Be very careful when you use the **Control** key on your computer keyboard to select multiple tasks to link them. After selecting one task initially, when you press and hold the **Control** key and then select additional tasks, Microsoft Project assumes that the first task you selected is the Predecessor task and all other tasks you selected are Successor tasks. This means that when using the **Control** key to select and link multiple tasks, the order in which you select the tasks is very important in the dependency planning process!

To remove a dependency relationship between two or more tasks, complete the following steps:

1. Select the tasks from which you want to remove the dependencies.
2. In the *Schedule* section of the *Task* ribbon, click the **Unlink Tasks** button.

Hands On Exercise

Exercise 5 - 4

Review the basics of how to set task dependencies.

1. Open the **Dependency Planning 2019.mpp** sample file from your student sample files folder. The project file opens Read-Only.

The Design, Build, Test, and Implement tasks are a "chain of events" that must occur sequentially. Link them with a Finish-to-Start (FS) dependency.

2. Select the *Design* through *Implement* tasks and then click the **Link the Selected Tasks** button in the *Schedule* section of the *Task* ribbon.

The Install Windows task and the Install Electrical task must start at the same time, and the Install Windows task is the "driving event" between these two tasks. Set a Start-to-Start (SS) dependency on these two tasks.

3. Select the *Install Windows* and *Install Electrical* tasks and then click the **Link the Selected Tasks** button in the *Schedule* section of the *Task* ribbon.

4. In the *Gantt Chart* pane, double-click the link line between the Gantt bars for the *Install Windows* and *Install Electrical* tasks.

5. In the *Task Dependency* dialog, click the **Type** pick list and select the **Start-to-Start (SS)** item, and then click the **OK** button.

The Order Event Supplies task and the Order Promotional Materials task must finish at the same time, and the Order Event Supplies task is the "driving event" between these two tasks. Set a Finish-to-Finish (FF) dependency on these two tasks.

6. Select the *Order Event Supplies* and *Order Promotional Materials* tasks and then click the **Link the Selected Tasks** button in the *Schedule* section of the *Task* ribbon.

7. In the *Gantt Chart* pane, double-click the link line between the Gantt bars for the *Order Event Supplies* and *Order Promotional Materials* tasks.

8. In the *Task Dependency* dialog, click the **Type** pick list and select the **Finish-to-Finish (FF)** item, and then click the **OK** button.

You scheduled your PMP certification examination for 8:00 AM on Friday, April 9. You believe you need two days to study for the exam. The date and time of the exam is the "driving event" that determines when you need to finish studying. Set a Start-to-Finish (SF) dependency between the PMP Certification Exam task and the Study for the Exam task.

9. Select the *PMP Certification Exam* and *Study for the Exam* tasks and then click the **Link the Selected Tasks** button in the *Schedule* section of the *Task* ribbon.

10. In the *Gantt Chart* pane, double-click the link line between the Gantt bars for the *PMP Certification Exam* and *Study for the Exam* tasks.

11. In the *Task Dependency* dialog, click the **Type** pick list and select the **Start-to-Finish (SF)** item, and then click the **OK** button.

12. Leave the **Dependency Planning 2019.mpp** sample file open for the next exercise.

Using Lag Time with Dependencies

Lag time is a delay in either the *Start* date or the *Finish* date of a Successor task. You can use *Lag* time for a number of reasons, including situations such as the following:

- You need to plan for the delivery time delay between ordering equipment or supplies and receiving them in a Finish-to-Start (FS) dependency relationship.

- You require the completion of a portion (time or percentage) of the Predecessor task before the Successor task begins, such as when you want to show that the painters can start painting after a portion of the dry wall work is complete in a Start-to-Start (SS) dependency relationship.

You can enter *Lag* time as either a time value, such as *5 days*, or as a percentage of the duration of the Predecessor task, such as *50%*. To enter *Lag* time on a task dependency, complete the following steps:

1. Initially link the two dependent tasks with an FS dependency.

2. In the *Gantt Chart* pane, double-click the link line between two dependent tasks.

3. In the *Task Dependency* dialog, optionally change the **Type** value and then enter a *positive value* in the **Lag** field (either as a time unit or as a percentage).

4. Click the **OK** button.

In the *Task Dependency* dialog shown in Figure 5 - 22, notice that I added *2d* of *Lag* time to the FS dependency between the Design and Build tasks.

Task Planning

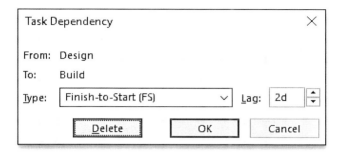

Figure 5 - 22: Task Dependency dialog
with 2d Lag time entered

Figure 5 - 23 shows two different dependencies with *Lag* time added. For the FS dependency between the Pour Foundation and Frame Walls tasks, I added *2 days* of *Lag* time. This dependency means that the Frame Walls task must start 2 days after the Pour Foundation task finishes. For the SS dependency between the Test for Bugs and Fix Bugs tasks, I added a *50%* *Lag* time. This means that the Test for Bugs task starts, and when it is 50% completed, the Fix Bugs task must start.

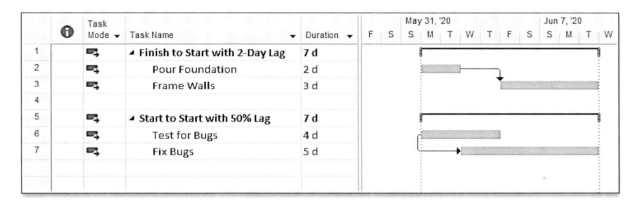

Figure 5 - 23: Lag time added to two different dependencies

Using Lead Time with Dependencies

Lead time is the opposite of *Lag* time. Most people use Lead time to create an overlap between two tasks linked with a Finish-to-Start (FS) dependency. To enter *Lead* time on a task dependency, complete the following steps:

1. Initially link the two dependent tasks with an FS dependency.

2. In the *Gantt Chart* pane, double-click the link line between two dependent tasks.

3. In the *Task Dependency* dialog, enter a ***negative value*** in the **Lag** field (either as a time unit, such as days, or as a percentage).

4. Click the **OK** button.

In the *Task Dependency* dialog shown in Figure 5 - 24, notice that I added *-2d* of *Lag* time (*2d* of *Lead* time) to the FS dependency between the Design and Build tasks.

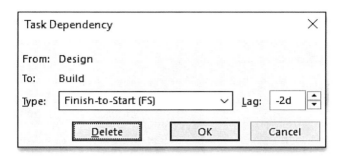

Figure 5 - 24: Task Dependency dialog
with 2d Lead time added

In an FS dependency, 2 days of *Lead* time means the Successor task can start 2 days **before** the *Finish* date of the Predecessor task. Adding 2 days of *Lead* time on an FS dependency creates an overlap between the dependent tasks as shown in Figure 5 - 25.

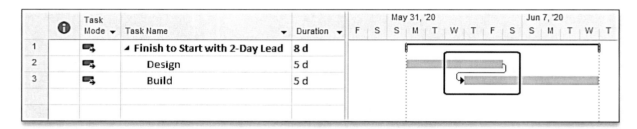

Figure 5 - 25: FS Dependency with 2 days of Lead time

Hands On Exercise

Exercise 5 - 5

Review the basics of how to set task dependencies with *Lag* time and *Lead* time.

1. Return to the **Dependency Planning 2019.mpp** sample file.

The Setup Servers task cannot begin until 5 days after the Order Servers task finishes. Set a Finish-to-Start (FS) dependency with 5 days of *Lag* time for these two tasks.

2. Select the *Order Servers* and *Setup Servers* tasks and then click the **Link the Selected Tasks** button in the *Schedule* section of the *Task* ribbon.

3. In the *Gantt Chart* pane, double-click the link line between the Gantt bars for the *Order Servers* and *Setup Servers* tasks.

4. In the *Task Dependency* dialog, enter **5d** in the **Lag** field and then click the **OK** button.

5. Float your mouse pointer over the link line between these two tasks to see the screen tip indicating that the FS dependency includes 5 days of *Lag* time.

The Decorate Rooms task cannot start until the Paint Rooms task is 50% completed. Set a Start-to-Start (SS) dependency with 50% *Lag* time for these two tasks.

6. Select the *Paint Rooms* and *Decorate Rooms* tasks and then click the **Link the Selected Tasks** button in the *Schedule* section of the *Task* ribbon.

7. I In the *Gantt Chart* pane, double-click the link line between the Gantt bars for the *Paint Rooms* and *Decorate Rooms* tasks.

8. In the *Task Dependency* dialog, click the **Type** pick list and select the **Start-to-Start (SS)** dependency.

9. In the *Task Dependency* dialog, manually type **50%** in the **Lag** field, and then click the **OK** button.

10. Float your mouse pointer over the link line between these two tasks to see the screen tip indicating that the SS dependency includes 50% *Lag* time.

The Create App Documentation task can start 2 days before the Create iPhone App task finishes. Set a Finish-to-Start (FS) dependency with 2 days of *Lead* time for these two tasks.

11. Select the *Create iPhone App* and *Create App Documentation* tasks and then click the **Link the Selected Tasks** button in the *Schedule* section of the *Task* ribbon.

12. In the *Gantt Chart* pane, double-click the link line between the Gantt bars for the *Create iPhone App* and *Create App Documentation* tasks.

13. In the *Task Dependency* dialog, enter **-2d** in the **Lag** field and then click the **OK** button.

14. Float your mouse pointer over the link line between these two tasks to see the screen tip indicating that the FS dependency includes 2 days of *Lead* time (-2 days of *Lag* time).

Your team must complete the Assemble Meeting Packets task 5 days before the Annual Shareholder Meeting task begins. Because the *Start* date of the Annual Shareholder Meeting task drives the *Finish* date of the Assemble Meeting Packets task, set a Start-to-Finish (SF) dependency between these two tasks with 5 days of *Lead* time.

15. Select the *Annual Shareholder Meeting* and *Assemble Meeting Packets* tasks, and then click the **Link the Selected Tasks** button in the *Schedule* section of the *Task* ribbon.

16. In the *Gantt Chart* pane, double-click the link line between the Gantt bars for the *Annual Shareholder Meeting* and *Assemble Meeting Packets* tasks.

17. In the *Task Dependency* dialog, click the **Type** pick list and select the **Start-to-Finish (SF)** dependency.

18. In the *Task Dependency* dialog, enter **-5d** in the **Lag** field, and then click the **OK** button.

19. Float your mouse pointer over the link line between these two tasks to see the screen tip indicating that the SF dependency includes 5 days of *Lead* time.

20. Close but ***do not*** save the **Dependency Planning 2019.mpp** sample file.

Exercise 5 - 6

Set task dependencies in your enterprise project.

1. Return to the enterprise project you created during this class, if necessary.
2. Select task IDs #2-5, from the *Write Chapter 1* task to the *Course Manual Complete* milestone task and link them with a Finish-to-Start (FS) dependency.
3. Select task IDs #7-10, from the *Create Sample Files for Chapter 1* task to the *Student Sample Files Complete* milestone task and link them with a Finish-to-Start (FS) dependency.
4. Select task IDs #12-15, from the *Create Instructor Training Outline* task to the *Instructor Materials Complete* milestone task and link them with a Finish-to-Start (FS) dependency.
5. Select task IDs #17-21, from the *Create Student Learning Aids* task to the *Project Complete* milestone task and link them with a Finish-to-Start (FS) dependency.
6. Select task ID #5, the *Course Manual Complete* milestone task, press and hold the **Control** key, and then select task ID #12, the *Create Instructor Training Outline* task.
7. Link the selected tasks with a Finish-to-Start (FS) dependency.
8. Select task ID #10, the *Student Sample Files Complete* milestone task, press and hold the **Control** key, and then select task ID #12, the *Create Instructor Training Outline* task.
9. Link the selected tasks with a Finish-to-Start (FS) dependency.
10. Select task ID #15, the *Instructor Materials Complete* milestone task, press and hold the **Control** key, and then select task ID #17, the *Create Student Learning Aids* task.
11. Link the selected tasks with a Finish-to-Start (FS) dependency.
12. Select task ID #2, the *Write Chapter 1* task, press and hold the **Control** key, and then select task ID #7, the *Create Sample Files for Chapter 1* task.
13. Link the selected tasks with a Start-to-Start (SS) dependency and with *4 days* of *Lag* time.
14. Select task ID #3, the *Write Chapter 2* task, press and hold the **Control** key, and then select task ID #8, the *Create Sample Files for Chapter 2* task.
15. Link the selected tasks with a Start-to-Start (SS) dependency and with *4 days* of *Lag* time.
16. Select task ID #4, the *Write Chapter 3* task, press and hold the **Control** key, and then select task ID #9, the *Create Sample Files for Chapter 3* task.
17. Link the selected tasks with a Start-to-Start (SS) dependency and with *4 days* of *Lag* time.
18. Save the changes to your enterprise project.

Task Planning

Setting Task Constraints and Deadline Dates

Microsoft Project offers you two way to constrain tasks in a project schedule:

- Constraints
- Deadlines

A **constraint** is a restriction that you set on the *Start* date or *Finish* date of a task. When you set a constraint on a task in Microsoft Project, you limit the software's ability to automatically reschedule the task when the schedule changes on predecessor tasks. Common reasons for using constraints include the following examples:

- **Contractual dates for task completion** when you have an obligation to complete a certain task in your project by a certain date.
- **Delivery dates for equipment and supplies** when a vendor guarantees delivery of equipment by a certain date.
- **Resource availability restrictions** when a resource cannot begin work on a task until after a certain date due to other project commitments.

A **deadline** is a way to set a "target date" for the completion of a task. Unlike constraints, deadlines do not limit the Microsoft Project scheduling engine, but the software does show an indicator if the *Finish* date of the task slips past the deadline date.

> **Warning**: A common mistake made by rookie users of Microsoft Project is manually typing a *Start* date and a *Finish* date for each task. This is a mistake because the schedule of each task should be governed by task dependency relationships and the *Duration* value for each task. In addition, when you manually type a *Start* date or a *Finish* date, Microsoft Project automatically sets a constraint on the task without asking your permission. If you manually enter a *Start* date for a task, the software adds a Start No Earlier Than (SNET) constraint on the task, using the date you entered as the constraint date. If you manually enter a *Finish* date on a task, the software adds a Finish No Earlier Than (FNET) constraint, using the date you entered as the constraint date. Because of this behavior, Projility recommends that you **never** manually type *Start* dates or *Finish* dates in your enterprise projects.

Setting Constraints

Microsoft Project offers you eight types of constraints that you can set on tasks, as needed. The available constraint types are as follows:

- **As Soon As Possible** – ASAP is the default constraint on every task when you schedule the project from a *Start* date. Microsoft Project schedules the task to start as soon as possible based on the project *Start* date and task dependencies.
- **As Late As Possible** – The software schedules the task to finish as late as possible without changing the *Finish* date of the project.
- **Finish No Earlier Than** – Microsoft Project does not allow the task to finish earlier than its constraint date, but does allow the task to finish later if needed.
- **Start No Earlier Than** – The software does not allow the task to start earlier than its constraint date, but does allow the task to finish later, if needed.

- **Finish No Later Than** – Microsoft Project does not allow the task to finish later than its constraint date, but does allow the task to finish earlier if needed.

- **Start No Later Than** – The software does not allow the task to start later than its constraint date, but does allow the task to start earlier if needed

- **Must Finish On** – Microsoft Project forces the task to finish on its constraint date, and does not allow the task to finish earlier or later than its constraint date.

- **Must Start On** – Microsoft Project forces the task to start on its constraint date, and does not allow the task to start earlier or later than its constraint date.

To set a constraint on a task, complete the following steps:

1. Double-click the task whose schedule you need to constrain.
2. In the *Task Information* dialog, click the **Advanced** tab. Figure 5 - 26 shows the *Advanced* page of the *Task Information* dialog for the Build task.
3. Click the **Constraint type** pick list and select a constraint.
4. Click the **Constraint date** pick list and select a date in the calendar date picker.

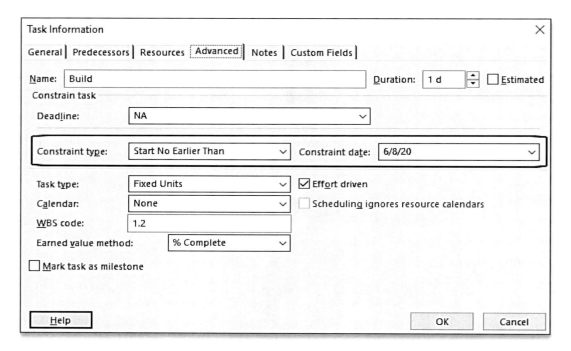

Figure 5 - 26: Task Information dialog – Advanced page

When setting a constraint on a task, it is wise to add a note to the task to document the reason for the constraint. Adding a note makes it easier for others to understand why you set the constraint originally, and you can use this information later to evaluate the historical data in your project. To add a constraint note to a task, continue with the following steps:

5. In the *Task Information* dialog, click the **Notes** tab.
6. Click anywhere in the **Notes** field and enter the body of your note.

A good "shorthand" method for documenting a constraint is to include the following information in the body of the note:

- Abbreviation or acronym for the constraint type, such as SNET for a Start No Earlier Than constraint
- Date of the constraint, such as 06/08/2020
- Reason for setting the constraint, such as contractual delivery date for supplies

Figure 5 - 27 shows the *Notes* page of the *Task Information* dialog documenting the reason for a constraint.

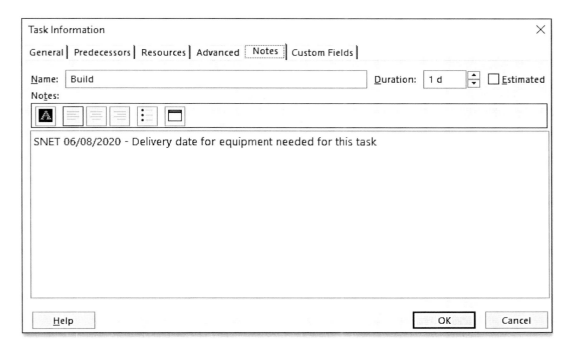

Figure 5 - 27: Task Information dialog, note documents a constraint

7. Click the **OK** button when finished setting the constraint and adding the note.

Understanding Flexible and Inflexible Constraints

In Microsoft, constraints are either **flexible** or **inflexible**. Some people like to refer to flexible constraints as "soft" constraints and inflexible constraints as "hard" constraints. A flexible constraint allows the software to reschedule the task as required, while an inflexible constraint limits or even stops the tool from using its normal scheduling behavior.

When you set a constraint, the software displays an indicator with a colored dot in the *Indicators* column to the left of the task name, and the color of the dot reveals whether the task is flexible or inflexible. If the indicator contains a red dot, the constraint is inflexible. If the indicator contains a blue dot, the constraint is flexible. When you schedule your project from a *Start* date, which is the default setting in Microsoft Project, the following constraints are **flexible** constraints:

- As Soon As Possible (no indicator displayed)
- As Late As Possible (no indicator displayed)

- Finish No Earlier Than
- Start No Earlier Than

When you schedule your project from a *Start* date, the following constraints are **inflexible** constraints:

- Finish No Later Than
- Must Finish On
- Must Start On
- Start No Later Than

Best Practice: Projility recommends as a best practice that you limit your use of constraints in your enterprise projects and use only flexible constraints when needed. Avoid using inflexible constraints, it at all possible, because they prevent Microsoft Project from using its normal scheduling behavior.

Understanding Planning Wizard Messages about Constraints

When you set a task constraint in a project scheduled from a *Start* date, Microsoft Project displays a *Planning Wizard* dialog when the resulting situation meets both of the following conditions:

- The constraint is *inflexible*, such as a Finish No Later Than constraint.
- The constrained task is a *successor task*, meaning that it has one or more predecessors.

Notice in the *Planning Wizard* dialog shown in Figure 5 - 28 that the software is warning how these two conditions can cause potential scheduling problems in the project, either now or any time in the future. The default response in this dialog is the first choice, which is to cancel setting the constraint. If you truly want to set an inflexible constraint and risk potential scheduling problems, you must select the third choice, **Continue**, and then click the **OK** button. The second option, by the way, makes no sense at all since selecting this option allows the software to change the constraint to a flexible constraint (Finish No Earlier Than), an outcome that defeats the purpose of setting the inflexible constraint in the first place!

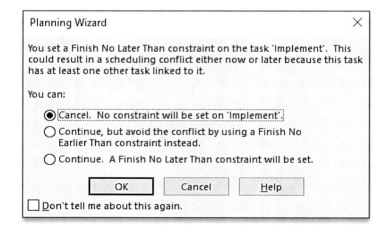

Figure 5 - 28: Planning Wizard message warns
about setting an inflexible constraint

Task Planning

 Best Practice: Projility recommends that you **do not** select the *Don't tell me about this again* checkbox to disable *Planning Wizard* messages about scheduling issues. These warning messages are a good way to confirm you selected the constraint you desire, especially if it is an inflexible constraint on a task that has predecessor tasks.

Setting Deadline Dates

In addition to constraints, Microsoft Project also allows you to set deadlines on tasks. Deadlines are similar to constraints, but do not limit the scheduling engine. When you set a *deadline* on a task, the software places a solid green arrow in the *Gantt Chart* pane on the same line as the task's Gantt bar. To set a deadline for any task, complete the following steps:

1. Double-click the task.

2. In the *Task Information* dialog, click the **Advanced** tab.

3. Click the **Deadline** pick list and select a date in the calendar date picker. Notice in Figure 5 - 29 that I set a deadline date of 6/9/20 on the Phase I Complete milestone task.

4. Click the **OK** button.

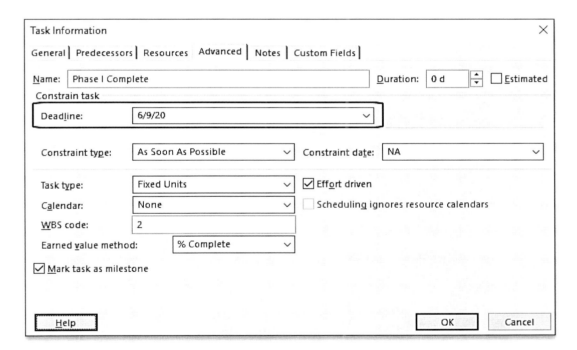

Figure 5 - 29: Set a deadline date in the Task Information dialog

Figure 5 - 30 shows how Microsoft Project displays the deadline symbol in the *Gantt Chart* pane. Notice that the software displays a solid green arrow to the right of the milestone symbol for the Phase I Complete milestone task.

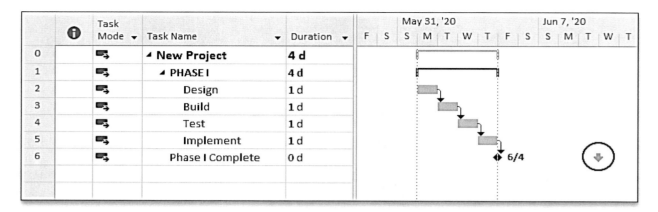

Figure 5 - 30: Deadline date indicator for the Phase I Complete milestone

Information: An organization with whom I consulted many years ago used a methodology that required the project manager to apply a deadline on the milestone task at the end of every phase and at the end of the project as well. This methodology required the project manager to negotiate the dates of each deadline with the customer, and then to apply the deadlines to the milestones in accordance with the negotiated dates. In addition, if a milestone slipped past its deadline date, the methodology required the project manager to meet with the customer and to renegotiate a new date for the missed deadline.

Hands On Exercise

Exercise 5 - 7

Set a constraint and a deadline in your enterprise project.

Because of a planned update to the ERP software that impacts the content of Chapter 3 in the training manual, the Write Chapter 3 task cannot begin until the Monday that is **5 weeks** after the *Start* date of the project.

1. Double-click task ID #4, the *Write Chapter 3* task.
2. In the *Task Information* dialog, click the **Advanced** tab.
3. Click the **Constraint Type** pick list and select the **Start No Earlier Than** item.
4. Click the **Constraint Date** pick list and select the date of the Monday that is **5 weeks** after the *Start* date of the project.
5. In the *Task Information* dialog, click the **Notes** tab.
6. In the *Notes* section of the *Notes* page, click at the end of the existing note text and then press the **Enter** key on your computer keyboard to add a new line of text.

Task Planning

7. In the *Notes* section of the *Notes* page, manually enter the following text:

    SNET (date of constraint) – ERP upgrade impacts content to Chapter 3

8. Click the **OK** button.

The project Sponsor requested that the project be completed by the Friday that is **12 weeks and 4 days** after the project *Start* date.

9. Double-click task ID #21, the *Project Complete* milestone task.
10. In the *Task Information* dialog, click the **Advanced** tab.
11. Click the **Deadline** pick list and select the date of the Friday that is **12 weeks and 4 days** after the project *Start* date.
12. Click the **OK** button.
13. Save the schedule changes to your enterprise project.

Understanding Missed Constraints and Deadline Dates

Microsoft Project gives you a limited warning when you miss a constraint date or a deadline date. When the *Start* or *Finish* date of a task slips past the constraint date for an inflexible constraint, such as when you miss a Finish No Later Than (FNLT) constraint, Microsoft Project displays the *Planning Wizard* dialog shown in Figure 5 - 31. In this dialog, the software warns you of the schedule conflict, and gives you an opportunity to cancel the action or to allow the schedule conflict to occur.

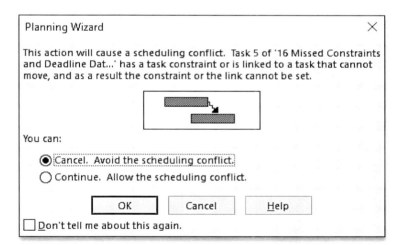

Figure 5 - 31: Planning Wizard message for
a missed inflexible constraint date

The default option in this *Planning Wizard* dialog is to cancel the action that caused the scheduling conflict. If you select the second option, *Continue*, Microsoft Project completes the scheduling change and allows the scheduling conflict, but it does not remove the constraint that is causing the scheduling conflict. Figure 5 - 32 shows the missed constraint date on the Phase I Complete milestone task. Notice how the link line between the Implement task and the Phase I Complete milestone task "wraps back" in time, an indication of the missed constraint date.

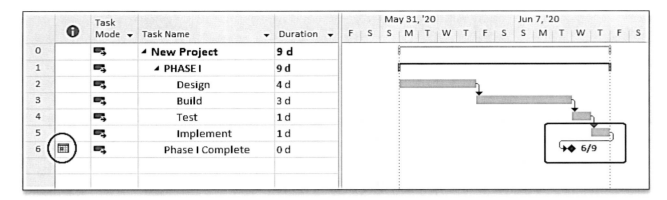

Figure 5 - 32: Missed constraint on the Phase I Complete milestone

When the *Finish* date of a task slips past its deadline date, the software does not display a *Planning Wizard* warning dialog. Instead, it displays a *Missed Deadline* indicator in the *Indicators* column. The indicator is a red diamond with a white exclamation point. For example, Figure 5 - 33 shows the missed deadline date on the Phase I Complete milestone task with a *Missed Deadline* indicator in the *Indicators* column to the left of the task name.

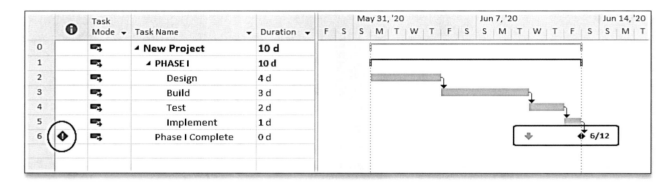

Figure 5 - 33: Missed deadline date on the Phase I Complete milestone

 Information: In addition to the behavior documented above, when a task misses an inflexible constraint date or a deadline date, Microsoft Project calculates a **negative Total Slack value** for the task and all of its direct predecessors. For any task whose *Total Slack* value is negative, the software also marks that task as a Critical task. If you applied Critical task formatting to any Gantt-based view, such as the *Gantt Chart* view, Microsoft Project displays a red Gantt bar for every task with a negative *Total Slack* value.

Applying Task Calendars

You must assign a **Task Calendar** when you want to manually override the current schedule of any task with a completely different schedule defined by a custom calendar. In Figure 5 - 34, notice that Microsoft Project schedules the Upgrade Servers task to start on a Thursday. However, this task cannot occur on a weekday because of potential disruptions to normal business processes, and this task must be scheduled over a weekend when most employees are gone for the weekend. To respond to this scheduling need, I need to apply a *Task* calendar to completely override the schedule of the Upgrade Servers task.

Task Planning

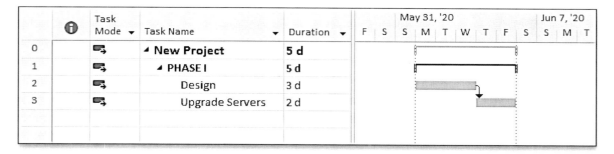

Figure 5 - 34: Upgrade Servers task starts on Thursday

 Information: The hierarchy of calendars in a Microsoft Project schedule is as follows:

- The *Project* calendar controls the initial schedule of every task in the project **before** you assign resources to the tasks.
- When you assign a resource to a task, the *Resource* calendar of the assigned resource controls the schedule of the task.
- To totally override the schedule of a task with a unique schedule defined by an enterprise calendar, apply a *Task* calendar to the task.

To apply a *Task* calendar to any task, complete the following steps:

1. Double-click the task whose schedule you want to override.

2. In the *Task Information* dialog, click the **Advanced** tab.

3. Click the **Calendar** pick list and select a custom calendar from the list.

For example, notice in Figure 5 - 35 that selected *Weekend Work Only* custom calendar on the *Calendar* pick list. This special custom calendar schedules work only on Saturdays and Sundays, with Monday through Friday marked as nonworking time.

4. Select the **Scheduling ignores resource calendars** checkbox.

5. Click the **OK** button.

 Information: The *Scheduling ignores resource calendars* option forces Microsoft Project to ignore the calendars of resources assigned to the task during the scheduling process. With this option selected, the software schedules assigned resources to work even when their calendars indicate that they are not available for work. If you leave this option deselected, then Microsoft Project schedules the task using the **common working time** between the *Project* calendar (set in the *Project Information* dialog) and the *Resource* calendars of the resources assigned to the task. Selecting this option is useful when you need to schedule resources to work on days that are nonworking time, such as on weekends or company holidays.

145

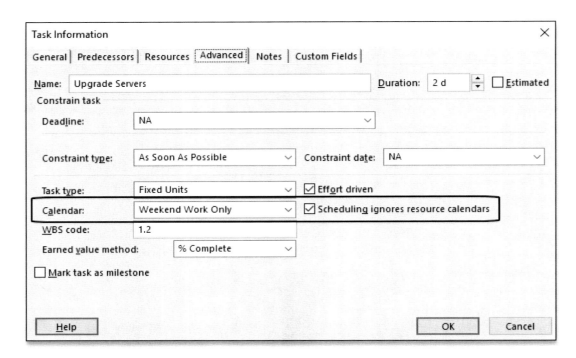

Figure 5 - 35: Select a task calendar in the Task Information dialog

Figure 5 - 36 shows the same project after assigning the *Weekend Work Only* custom calendar as a *Task* calendar to override the schedule of the Upgrade Servers task. Notice how the software rescheduled the task over a Saturday and Sunday. Notice also how Microsoft Project displays a special task calendar indicator in the *Indicators* column for the task.

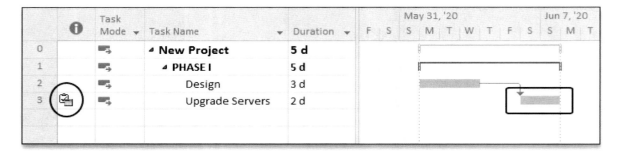

Figure 5 - 36: Upgrade Servers task scheduled on a weekend

Best Practice: When you assign a *Task* calendar to override the schedule of a task, Projility recommends that you always add a note to the task to document the reason for using the *task calendar*.

Task Planning

Information: Before you can apply a custom calendar as a *Task* calendar, the custom calendar must actually exist. If there is no custom calendar that meets your scheduling needs, reach out to your organization's Project Online application administrator to create the custom enterprise calendar for you. Remember the system does not allow you to create or edit enterprise calendars by default, as only the application administrator has permission to perform these actions.

Hands On Exercise

Exercise 5 - 8

Important Note: You should **only** work this exercise if your Project Online system includes a custom enterprise calendar that schedules work 7 days/week. If not, you may skip this exercise.

The project Sponsor requests that you schedule the resource assigned to the *Create Sample Files for Chapter 3* task to work 7 day/week to "make up" for the delay in starting work on Chapter 3. Therefore, override the schedule of this task by applying a task calendar.

1. Double-click task ID #9, the *Create Sample Files for Chapter 3* task.
2. In the *Task Information* dialog, click the **Advanced** tab.
3. Click the **Calendar** pick list and select a custom calendar that schedules work 7 days/week.
4. Select the **Scheduling ignores resource calendars** checkbox.
5. In the *Task Information* dialog, click the **Notes** tab.
6. In the *Notes* section of the *Notes* page, click at the end of the existing note text and then press the **Enter** key on your computer keyboard to add a new line of text.
7. In the *Notes* section of the *Notes* page, enter a note that describes why you applied the *task calendar* to this task.
8. Click the **OK** button.

Estimating Task Durations

After you create the task schedule, including setting task dependencies and constraints, you may need to estimate task durations, if required in your organization. According to Microsoft Project, duration is "the total span of active working time for a task." Another way to think of duration is the "window of opportunity" during which the team members work on the task.

147

Many novice users of Microsoft Project wrongly assume that duration and work are interchangeable terms in a project. In some cases, this may be true, but in most cases, duration and work are two entirely different numbers. Consider the following examples:

- A resource must perform 40 hours of work during a 10-day time period. The duration of this task is 10 days because it is the "window" during which the resource performs the work (40 hours).

- We allow an executive 5 days to approve a deliverable, but the executive will only perform 2 hours of actual work on the approval. The duration of this task is 5 days because this is the "window" during which the executive performs the work (2 hours).

Notice in the two preceding examples that the duration or "window of opportunity" does not consider the amount of work performed on the task. The duration is simply the time span during which team members perform the work, regardless of how much or how little work the task requires. There are several ways to determine a task duration estimate:

- Obtain the estimate from the team member who will actually perform the work on the task. This allows you to tap the skills, knowledge, and experience of the team member. This is a Project Management Institute (PMI) best practice, by the way.

- If you cannot obtain the duration estimate from a team member, then obtain an estimate from a team leader who has experience in this type of work.

- If you cannot obtain a duration estimate from a team leader, study your organization's repository of completed projects and obtain an estimate based on historical data for similar tasks.

- If you cannot use any of the previous methods, then make your best "guestimate" for the task, but then validate your duration estimate later against the actual completion data for the task.

To enter duration values, simply type your estimate in the **Duration** column for each task. You may enter the duration value using any time unit, including hours, days, weeks, months, etc. By default, however Microsoft Project formats *Duration* values in days.

> **Information**: If your organization uses Work-based estimates rather than Duration-based estimates, then you do not need to enter *Duration* values for the tasks in your Microsoft Project schedules. Instead, leave the *Duration* value at its default setting of *1 day*. During the assignment planning process when you assign resources to tasks using a *Units* value and a *Work* estimate in hours, the software calculates the task *Duration* value for you automatically based on the resource assignment values.

Module 06

Resource Planning

Learning Objectives

After completing this module, you will be able to:

- Understand the available types of enterprise resources
- Use the features of the Build Team dialog to build a project team
- Create local resources in an enterprise project

Inside Module 06

Understanding Enterprise Resources	**151**
Building a Project Team	**152**
Filtering Resources	*154*
Grouping Resources	*158*
Viewing Resource Information	*158*
Viewing Resource Availability	*159*
Adding Resources to Your Project Team	*162*
Matching and Replacing Resources	*162*
Using Proposed vs. Committed Booking	*163*
Using Local Resources in a Project Team	**165**

Resource Planning

Understanding Enterprise Resources

The **Enterprise Resource Pool** in Project Online and Project Server contains all of the enterprise resources needed for enterprise projects in your organization. In simple terms, resources are the people, equipment, and materials required to execute a project. In accounting terms, resources are the elements of project costs. Project Online defines enterprise resources in a variety of ways and organizes them in the Resource Organization Chart shown Figure 6 - 1.

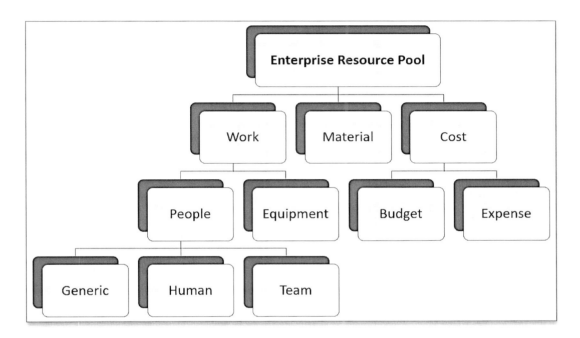

Figure 6 - 1: Resource Organization Chart

Project Online offers you three basic resource types: **Work**, **Material**, and **Cost**. You use *Work* resources to model people and equipment. You use *Material* resources to represent the supplies consumed during the life of the project. You use *Cost* resources to specify overall budgetary amounts for the project and to track non-labor expenses. *Work* resources affect both the schedule and the cost of the project, while *Material* and *Cost* resources affect only the project cost.

 Warning: Project Online has no built-in way to distinguish between *Work* resources that represent people and *Work* resources that represent equipment. Because of this, Projility recommends using an enterprise custom field to identify whether each *Work* resource represents a person or a piece of equipment. Using an enterprise custom field to distinguish this difference may be valuable to your organization since you may also use this information for reporting purposes, as well as for grouping and filtering Work resources in views.

Project Online organizes *Work* resources into the following three groups:

- Generic resources
- Non-Generic resources (aka Human resources)
- Team resources

151

Module 06

A **Generic** resource is a skill-based or placeholder resource, such as a SQL Server DBA. Generic resources allow you to specify the skills required for a task assignment before you know which Human resources are available to work on the task. You can use skill-set matching to replace Generic resources with available Human resources who possess the same skills. A **Non-Generic** or **Human** resource is a specific individual you can identify by name, such as Mickey Cobb, for example

A **Team** resource is a special type of *Work* resource that represents a team of people. You should only use *Team* resources if your organization uses dedicated project teams and the team members do not change from project to project. When you assign a *Team* resource to a task and publish the project, any members of that team can self-assign the task, taking full ownership of the task, and can then report their progress on the task in Project Web App.

Building a Project Team

After you define a new enterprise project and complete the task planning process, you are ready to begin the resource planning process by building your project team with enterprise resources. The *Build Team* dialog in Microsoft Project provides you with tools for searching through the Enterprise Resource Pool to find the right resources for your project team. To build your project team, click the **Resource** tab to display the *Resource* ribbon. In the *Insert* section of the *Resource* ribbon, click the **Add Resources** pick list button, and then select the **Build Team from Enterprise** item as shown in Figure 6 - 2.

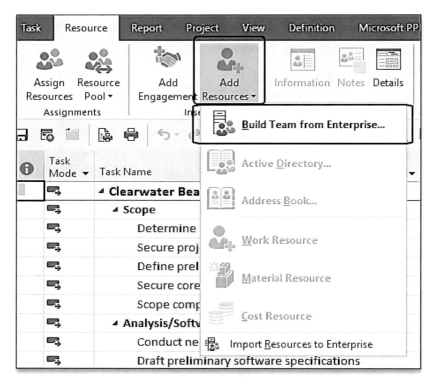

Figure 6 - 2: Select the Build Team from Enterprise item

Microsoft Project displays the *Build Team* dialog shown in Figure 6 - 3.

Resource Planning

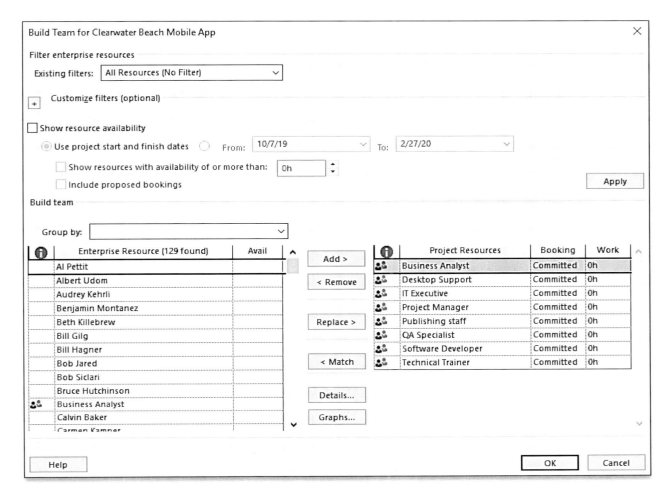

Figure 6 - 3: Build Team dialog

 Information: If your Enterprise Resource Pool contains more than 1,000 resources, Microsoft Project displays a pre-filter dialog to prompt you to filter the Enterprise Resource list. The system continues to prompt you to filter the list until the total number of resources displayed is less than 1,000.

The *Build Team* dialog shown previously in Figure 6 - 3 displays all of the resources in your organization's Enterprise Resource Pool, which can include *Work*, *Cost*, and *Material* resources. Microsoft Project organizes the *Build Team* dialog as follows:

- The *Filter enterprise resources* section at the top of the dialog contains the *Existing filters* pick list that displays the default resource filters included with Microsoft Project, along with any enterprise custom filters created by your application administrator.

- The *Customize filters (optional)* section in the middle of the dialog, when expanded, allows you to create an "ad hoc" filter to reduce the number of resources shown in the dialog. This section also allows you to show resource availability between the *Start* and *Finish* dates of your project, or between a custom set of dates you manually select.

- In the lower left corner of the *Build Team* dialog, you see the list of all resources in your organization's Enterprise Resource Pool. Immediately above the list of resources is the *Group by* pick list, which you can use to organize the resources into groups based on a field you select in the pick list.

- In the lower right corner of the *Build Team* dialog, you see the list of all resources currently on your project team. Notice in the lower right corner of the *Build Team* dialog shown previously in Figure 6 - 3 that the project team includes the Generic resources I added to the team so that I could forecast the resource demand for my project by creating Resource Engagements.

The default permissions in Project Online allow members of the Project Managers security group to see all resources in the Enterprise Resource Pool. Using the features in the *Build Team* dialog, you can search through your Enterprise Resource Pool to identify resources with the skills and availability to do the job.

Information: When applicable, the *Build Team* dialog may display indicators to the left of the names in your project team, such as overallocation indicators and note indicators. Float your mouse pointer over any indicator to display the information for that indicator.

Filtering Resources

Because the default permissions in Project Online allow you to see all of the resources in the Enterprise Resource Pool, the *Build Team* dialog offers three methods to filter the list of resources:

- Use existing filter
- Create ad hoc custom filter
- Filter for availability within a date range filter

To use an existing filter, click the **Existing filters** pick list at the top of the dialog and select a filter. The *Existing filters* pick list contains standard filters, enterprise filters, and your own personal filters. When you select a filter, Microsoft Project applies it immediately and restricts the resources shown in the *Enterprise Resource* list on the lower left corner of the dialog.

Information: Your application administrator can create and save custom enterprise resource filters in the Enterprise Global file, which makes these custom filters available to all users. You see these custom filters in the *Existing filters* pick list.

To create your own ad hoc custom filter, you must first expand the *Customize filters (optional)* section. By default, Microsoft Project displays the *Build Team* dialog with this section collapsed each time display the dialog. Click the **Expand (+)** button to the left of the *Customize Filters (optional)* section to expand the section, as shown in Figure 6 - 4.

Resource Planning

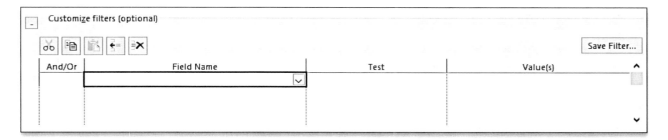

Figure 6 - 4: Customize filters (optional) section expanded

To create your own ad hoc custom filter, enter your filter criteria in the data grid shown in the *Customize filters (optional)* section of the dialog. The data grid consists of four columns: the *And/Or*, *Field Name*, *Test*, and *Value(s)* columns. Create your ad hoc custom filter by completing the following steps:

1. In the *Field Name* column, click the pick list button and select a field on which to filter from the list of standard and custom fields.

2. In the *Test* column, click the pick list button and select the comparison test for your filter criteria.

3. In the *Value(s)* column, select or enter the value for which to test. If the field you selected in the *Field Name* column contains a lookup table of values, the *Value(s)* pick list displays only the list of values available in the lookup table.

4. If you want to specify multiple filtering criteria, click the pick list in the *And/Or* column on the second row, select your Boolean condition, and then create the second set of filtering criteria.

5. Click the **Apply** button.

Notice in *Customize filters (optional)* section shown in Figure 6 - 5 that I created a custom filter to locate resources whose *Corporate Role* value is *Software Developer* and whose *IT Skill* value is *C Sharp*. After clicking the *Apply* button, this filter displayed only six resources in the Enterprise Resource list (not shown).

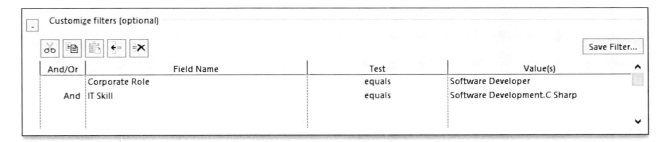

Figure 6 - 5: Custom filter with multiple filter criteria

After creating and applying your ad hoc custom filter, Microsoft Project allows you to save it for future use by clicking the **Save Filter** button. The system displays the *Save Filter* dialog. Enter an original name for your filter in the dialog, such as shown in Figure 6 - 6 and then click the **OK** button to save it.

155

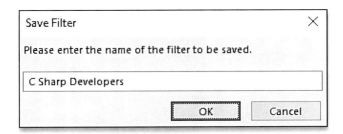

Figure 6 - 6: Save Filter dialog

 Warning: When you save your ad hoc custom filter using the *Save Filter* dialog, Microsoft Project saves the filter ***in the active project only***. To save your custom filter for use in all of your current and future projects, close the *Build Team* dialog, click the **File** tab, and then click the **Info** tab in the *Backstage*. At the bottom of the *Info* page, click the **Organizer** pick list button and select the **Organizer** item. In the *Organizer* dialog, click the **Filters** tab and then select the **Resource** option to display only Resource filters. In the list on the right side of the *Organizer* dialog, select your custom filter and then click the **<< Copy button** to copy the filter to your Global.mpt file. Click the **Close** button when finished.

In addition to filtering resources using the *Existing filters* pick list and creating custom filters in the *Custom filters (optional)* section of the *Build Team* dialog, you can also restrict the *Enterprise Resource* list by filtering for availability in a specific date range. To use availability filtering, select the **Show resource availability** checkbox in the *Build Team* dialog. The system activates the options in this section of the dialog, as shown in Figure 6 - 7.

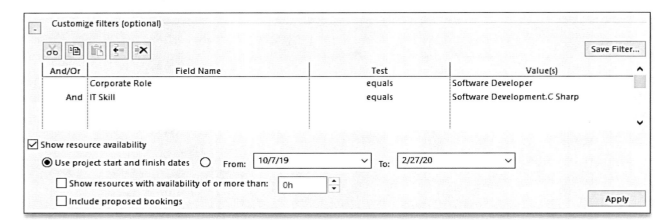

Figure 6 - 7: Show resource availability options

Select the **Use project start and finish dates** option to use the scheduled *Start* date and calculated *Finish* date of the project. Alternately, you can select the **From** option and enter a specific date range in the **From** and **To** fields. If you select the *From* option, the system sets the *From* date to the scheduled *Start* date of the project, and sets the *To* date to the current calculated *Finish* date of the project. After selecting the *From* option, can change either date to filter during a specific date range.

After selecting the desired date range, click the **Apply** button. The system calculates the availability for each resource shown in the *Enterprise Resource* list on the lower left corner of the dialog and then displays this information in the *Avail* column, as shown in Figure 6 - 8.

Resource Planning

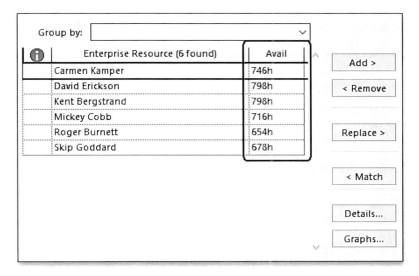

Figure 6 - 8: Availability calculation applied
in the Build Team dialog

The system calculates the availability for each resource using the formula **Availability = Capacity – Work** for the specified date range. Notice in the *Enterprise Resource* list shown previously in Figure 6 - 8 that David Erickson has 798 hours of availability during the time span from 10/07/19 to 02/27/20. Notice also that Roger Burnett has only 654 hours of availability during the same time span. If you see a resource with 0 hours of availability, this means the resource is completely booked on other projects during the specific time span of your project, and is not available to work on your project.

 Warning: Even if a resource has 0 hours of availability to work on your project, Microsoft Project allows you to add the resource to your project team and to assign the resource to tasks in your project. Doing so is unwise as this action creates cross-project overallocations for the resource.

To continue filtering by availability for a specific time period, select the **Show resources with availability of or more than** option and then enter the number of hours in the corresponding field. Click the **Apply** button to show only those resources with the minimum amount of availability for your project.

When you filter for availability over a specific time period, Project Online allows you to determine whether the system considers *Proposed* bookings in the filtering process. A *Proposed* booking indicates a tentative resource commitment to a project, while a *Committed* booking indicates a firm commitment to a project. To include *Proposed* bookings, select the **Include proposed bookings** option and then click the **Apply** button.

 Information: To remove all filtering from the list of enterprise resources, click the **Existing filters** pick list and select the **All Resources (No Filter)** item.

157

Grouping Resources

The *Group by* pick list provides another way of refining your resource selections by applying grouping to the *Enterprise Resource* list in the lower left corner of the *Build Team* dialog. Click the **Group by** pick list and then select any resource field available in the system, including both standard and custom fields. When you select a field in the *Group by* pick list, Microsoft Project applies grouping using the resource values in the field to the resources in the *Enterprise Resource* list, with each group expanded to show all members of the group. As you review your resources groups, you can collapse or expand each group as needed.

Figure 6 - 9 shows the list of resources after I selected the *RBS* field in the *Group by* pick list. By the way, the *RBS* field defines the "pseudo org chart" for each resource in the Enterprise Resource Pool. Notice that Larry McKain is the manager of the Desktop team while Bob Siclari, Kate Wittkowski, LaVon Catron, and Susan Tartaglia all report to Larry McKain, as defined by their hierarchical position below him in the *RBS* grouping.

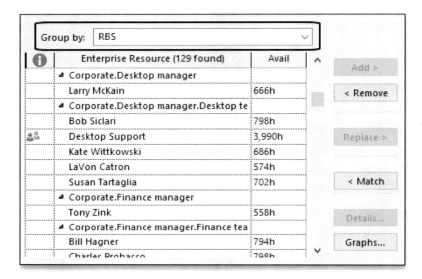

Figure 6 - 9: RBS to Enterprise Resource list

 Information: To remove grouping from the *Enterprise Resource* list, click the **Group by** pick list and select the blank row at the top of the pick list.

Viewing Resource Information

Select any resource in the list of resources in the *Enterprise Resource* section of the *Build Team* dialog, and then click the **Details** button. Microsoft Project displays the *Resource Information* dialog for the selected resource, such as shown in Figure 6 - 10 for a resource named Audrey Kehrli.

In the *Resource Information* dialog, examine the information shown for the selected resource on the *General*, *Costs*, *Notes*, or *Custom Fields* page. In particular, study the *Custom Fields* page of the dialog to see all the enterprise custom field values for the selected resource.

Information: Remember that only your organization's application administrator can edit enterprise resource information for any of the resources in your Enterprise Resource Pool. As a project manager, you can view the resource information, but you cannot edit it.

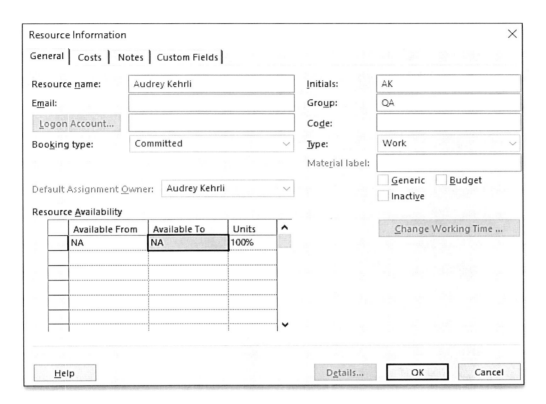

Figure 6 - 10: Resource Information dialog, General page

Viewing Resource Availability

During the process of building your project team, you probably want to know which resources are available to work in your project, based on their task assignments in other projects. The *Build Team* dialog gives you direct access to resource availability information using resource graphs. Select one or more resources in the list of resources on the right or left side of the dialog, and then click the **Graphs** button. The system launches your default web browser, logs into Project Web App, navigates to the *Capacity Planning* page with the *Resource Utilization* view applied, and displays the availability information for the selected resources as shown in Figure 6 - 11. For example, notice that the *Capacity Planning* page shows availability information for Mickey Cobb and Roger Burnett.

 Information: By default, the *Capacity Planning* page shows a two-week "look ahead" from the current date for the resources you selected in the *Build Team* dialog. For purposes of clarity, I modified the settings on this page to show a 9-month "look ahead" from the current date, as shown in Figure 6 - 11.

The *Capacity Planning* page displays a graph in the top half of the page and a *Details* grid in the bottom half. The graph is a combination line chart/stacked column chart. The line chart portion of the graph displays the *Capacity* data for all the selected resources. Project Online calculates the *Capacity* for each resource using the schedule shown on their calendar, multiplied by their *Max. Units* value. The stacked column chart shows the total amount of work per time period for each selected resource.

The left side of the *Details* grid lists each selected resource, along with rows that show their *Capacity* and *Availability*, along with rows that displays the projects in which they are assigned tasks . The right side of the grid shows the timephased *Capacity*, *Availability*, and work for each resource by time period. In Figure 6 - 11, notice that the right side of the *Details* grid is timephased by months.

Availability is the difference between the resource's *Capacity* and their assigned work. When *Availability* is a positive number, this indicates the resource is available to perform at least a certain number of hours of work in your project. When *Availability* is a negative number, this indicates the resource is overallocated and is not available to work in your project.

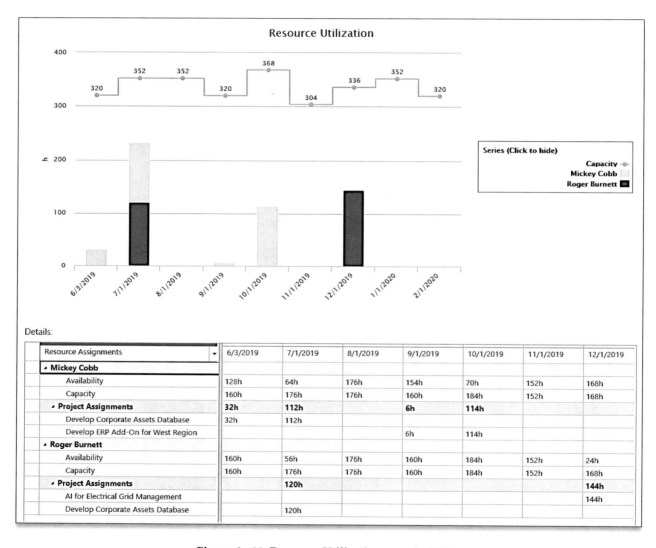

Figure 6 - 11: Resource Utilization page in PWA

To manipulate the data shown in the *Capacity Planning* page, complete the following steps:

1. In the *Filters* section of the *Availability* ribbon shown in Figure 6 - 12, click the **Timescale** pick list and select a timescale value, such as months for example.

Resource Planning

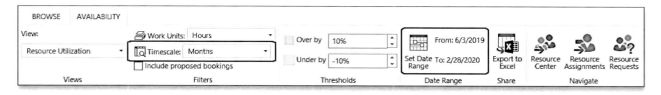

Figure 6 - 12: Availability ribbon

2. In the *Date Range* section of the *Availability* ribbon, click the **Set Date Range** button. Project Web App displays the *Set Date Range* dialog shown in Figure 6 - 13.

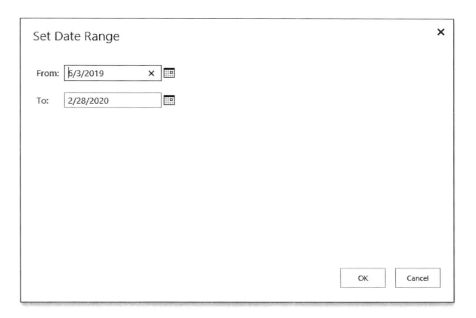

Figure 6 - 13: Set Date Range dialog

3. In the *Set Date Range* dialog, select a date in the **From** and **To** fields, and then click the **OK** button.

4. In the *Series* section to the right of the graph, click the names of individual resources to hide their assigned work in the graph so you can see the assigned work for the other resources. For example, I can click the name of Roger Burnett to hide his assigned work and only view the assigned work for Mickey Cobb.

 Warning: When you hide the names of resources in the *Series* section, the graph **does not** update the *Capacity* line chart to show the total *Capacity* for only the unhidden resources. The *Capacity* line chart always shows the total *Capacity* for all of the resources selected in the *Build Team* dialog, even when you hide some of them in the *Series* section of the page. Because of this default behavior in the *Capacity Planning* page, the *Capacity* line chart has limited usefulness when viewing and analyzing resource availability.

When you finish analyzing resource availability information on the *Capacity Planning* page in PWA, close your web browser to return to the *Build Team* dialog in Microsoft Project.

Adding Resources to Your Project Team

To add resources to your project team, select one or more resources from the *Enterprise Resource* list on the left side of the *Build Team* dialog and then click the **Add** button. The system adds the selected resources to the *Project Resources* list on the right side of the dialog. To remove resources from your project team, select one or more resources from the *Project Resources* list on the right and then click the **Remove** button.

Matching and Replacing Resources

If your application administrator configured your Project Online system to support the process of matching Human resources with Generic resources, you can use the *Match* button in the *Build Team* dialog. To match Human resources with a Generic resource, add the Generic resource to the *Project Resources* list on the right (if necessary) and then click the **Match** button. The system matches resources based on resource attributes such as skills or role. For example, notice in Figure 6 - 14 that I matched the QA Specialist generic resource with five Human resources who have this role in the organization.

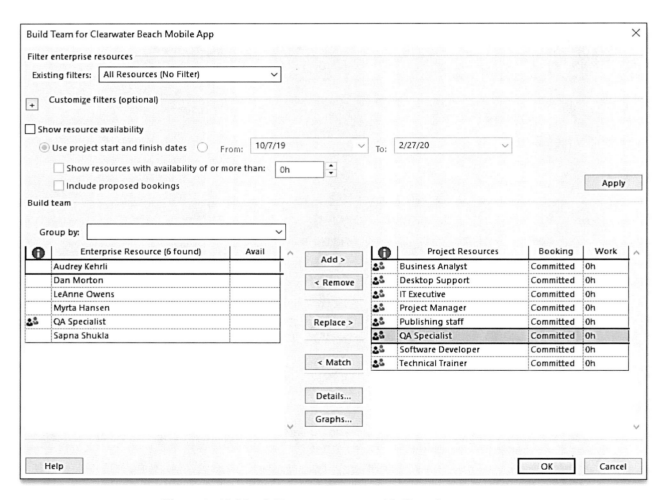

Figure 6 - 14: Match Human resources with Generic resources

When you use the *Match* feature in the *Build Team* dialog, Microsoft Project creates and applies an ad hoc custom filter for the resources in the *Enterprise Resource* list using the matching attributes of the selected Generic resource.

Resource Planning

In our organization, the application administrator configured the *Corporate Role* field for matching Human resources with Generic resources. To view the custom filter that the *Match* button uses, expand the **Customize filters (optional)** section at the top of the *Build Team* dialog. Notice in the ad hoc filter shown in Figure 6 - 15 that the *Field Name* value is *Corporate Role*, the *Test* value is *contains*, and the value in the *Value(s)* column is *QA Specialist*.

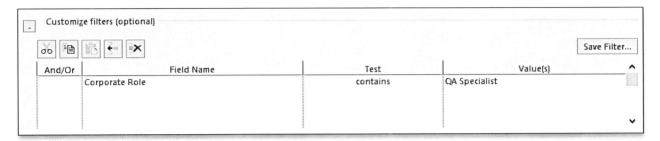

Figure 6 - 15: Customize filter data grid after a matching operation

 Information: In most cases, you use the *Match* button to match specific skills between Generic resources and Human resources. However, because the system shows all matches for a selected resource, you can use the matching feature to make human-to-human matches as well.

To replace a Generic resource in the *Project Resources* list with a Human resource from the *Enterprise Resources* list, select a resource in each list, and then click the **Replace** button. If you previously assigned the Generic resource to tasks in the project schedule, Microsoft Project replaces the Generic resource with the Human resource on every task assignment in the entire project schedule.

Using Proposed vs. Committed Booking

The *Booking* column in the *Project Resources* list on the right side of the *Build Team* dialog allows you to specify a booking type for each resource you add to your project team. You may book team members as either **Proposed** or **Committed**. As their names imply, a *Proposed* booking indicates a tentative commitment for the resource, while a *Committed* booking indicates a firm commitment. To change the *Booking* value for a project team member, click the **Booking** pick list for the resource and select the desired booking type.

When a project manager books a team member as a *Proposed* resource on a project, Project Online treats all task assignments for the *Proposed* resource as ***proposed assignments***. The consequences of setting the *Booking* value to *Proposed* for a project team member are as follows:

- Published task assignments do not display on their *Tasks* page or their *Timesheet* page in Project Web App. In other words, the proposed team member cannot see or report progress on their tasks in PWA.

- On the *Capacity Planning* page in PWA, you do not see projects assigned to a proposed resource by default. You can only see the projects assigned to the proposed resource by selecting the **Include proposed bookings** checkbox in the *Filters* section of the *Availability* ribbon at the top of the *Capacity Planning* page.

- On the *Resource Assignments* page in PWA, you do not see proposed assignments.

When you finish building your project team with enterprise resources, and confirmed that you set the *Booking* value to *Committed* for each team member, click the **OK** button to close the *Build Team* dialog. Be sure to save the latest changes to your Microsoft Project schedule.

Hands On Exercise

Exercise 6 - 1

Build your project team in your enterprise project using the features in the Build Team dialog.

1. Click the **Resource** tab to display the *Resource* ribbon.

2. In the *Insert* section of the *Resource* ribbon, click the **Add Resources** pick list button and select the **Build Team from Enterprise** item.

Information: You should only perform steps #3-6 below if an enterprise custom field exists in your Project Online system that identifies resource attributes such as role or skill. If such a field does not exist, simply skip steps #3-6 and continue with step #7.

3. In the *Build Team* dialog, click the **Group by** pick list and select an enterprise custom field that groups the *Enterprise Resource* list by a resource attribute, such as role or skill.

4. Scroll through the list of resources in the *Enterprise Resource* list on the left side of the dialog and study the resource groups.

5. Click the **Group by** pick list and select the blank item at the top of the list to remove the grouping applied to the *Enterprise Resource* list.

Information: You should only perform steps #6-7 below if your application administrator set up your Project Online system to allow matching between Generic resources and Human resources. If your system is not set up to provide these matching capabilities, simply skip steps #4-5 and continue with step #6.

6. In the list of Generic resources displayed in the *Project Resources* list on the right side of the dialog, select a Generic resource that represents ***your role or skill in the organization***, and then click the **Match** button.

7. Expand the *Customize filters (optional)* section and examine the custom ad hoc filter created and applied by Microsoft Project.

8. Collapse the *Customize filters (optional)* section.

9. In the *Enterprise Resource* list on the left side of the dialog, select ***your own name***.

10. In the list of Generic resources displayed in the *Project Resources* list on the right side of the dialog, confirm that the Generic resource representing ***your*** role or skill in the organization is still selected (if not, reselect it), and then click the **Replace** button.

11. Click the **Existing filters** pick list at the top of the dialog and select the **All Resources (No Filter)** item to display all resources.

Resource Planning

12. In the *Enterprise Resource* list on the left side of the dialog, select the name of a Human resource who typically serves as a team member in your own projects.

13. In the list of Generic resources displayed in the *Project Resources* list on the right side of the dialog, select a Generic resource and then click the **Replace** button.

14. Repeat steps #12-13 above, replacing each remaining Generic resource with a Human resource that is a typical team member of your own projects.

Your project team displayed in the *Project Resources* list on the right side of the dialog should now include four Human resources and no Generic resources.

15. In the *Enterprise Resource* list on the left side of the dialog, select the name of one more Human resource who typically serves as a team member in your own projects.

16. In the middle of the *Build Team* dialog, click the **Add** button to add the resource to the project team.

17. Set the **Booking** value to **Proposed** for each of your fellow project team members, but leave the **Book** value set to **Committed** *only for you*.

Information: Setting the *Booking* value to *Proposed* for your fellow project team members means that Project Web App will not send them e-mail messages about your new enterprise project, nor will they see their tasks from your project on their *Tasks* page or *Timesheet* page in PWA. Because your Booking value is Committed, you will receive e-mail messages about your new enterprise project, and you will see your tasks from your project on your *Tasks* page or *Timesheet* page in PWA.

18. Click the **OK** button to close the *Build Team* dialog.

19. Save the latest changes to your enterprise project.

Using Local Resources in a Project Team

A **local resource** is any project resource not listed in the Enterprise Resource Pool. You can use local resources in an enterprise project to represent resources such as consultants or contractors, who are not employees of your organization and who will not report time or task progress in the Project Web App. To add a local resource to an enterprise project, complete these steps:

1. Click the **Task** tab to display the *Task* ribbon.

2. In the *View* section of the *Task* ribbon, click the **Gantt Chart** pick list button and select the **Resource Sheet** view.

3. In the first blank row at the bottom of the resource list, type the name of the local resource and press the **Enter** key on your computer keyboard.

4. Enter general information for the local resource, as needed, such as information in the *Initials, Group, Max. Units, Std. Rate, Ovt. Rate,* and *Base Calendar* columns.

Microsoft Project displays the new local resource, as shown in Figure 6 - 16. Notice the special indicator displayed in the *Indicator* column for the local resource. If you float your mouse pointer over this indicator, the software

165

displays a *Local Resource* message in the tooltip, confirming that the resource is truly a local resource and not an enterprise resource.

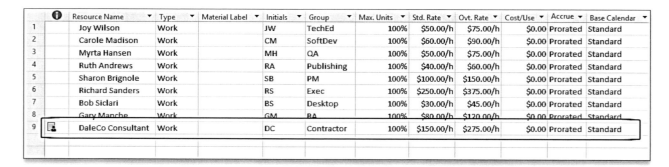

Figure 6 - 16: Resource Sheet view, new resource

 Information: Before you create a local resource in one of your enterprise projects, confirm that doing so complies with your organization's policies for staffing projects. Some organizations do not allow the use of local resources in enterprise projects because the assignment information for the local resource appears only minimally in Project Online reports.

 Hands On Exercise

Exercise 6 - 2

Add a local resource to the project team in your enterprise project.

1. In the *View* section of the *Resource* ribbon, click the **Team Planner** pick list button and select the **Resource Sheet** view.

2. Select the first blank line at the end of the resource list and then type the name **Irondale Partners** in the **Resource Name** field.

3. Widen the *Resource Name* column, if needed.

4. Manually type **IP** in the **Initials** field and then type **Contractor** in the **Group** field.

5. Widen the *Group* column, if needed.

6. Leave the **Max. Units** value set to **100%**.

7. Manually type **$100** in the **Std. Rate** field and type **$150** in the **Ovt. Rate** field.

8. Save the latest changes to your enterprise project.

Module 07

Assignment Planning

Learning Objectives

After completing this module, you will be able to:

- Assign resources to tasks using two different methods
- Understand the Task Types and Effort Driven features
- Assign Budget and Expense Cost resources to tasks
- Assign Material resources to tasks
- Understand and level resource overallocations
- Create Resource Engagements for human resources

Inside Module 07

Assigning Work Resources to Tasks **169**
Assigning Resources Using the Task Entry View *169*
Assigning Resources Using the Assign Resources Dialog *174*
Understanding Other Factors in Assignment Planning **177**
Understanding the Duration Equation *178*
Understanding Task Types *179*
Understanding Effort Driven Scheduling *183*
Assigning Cost Resources **189**
Adding Cost Resources to the Project Team *189*
Assigning a Budget Cost Resource *189*
Assigning an Expense Cost Resource *193*
Recommendations for Using Cost Resources *195*
Assigning Material Resources to Tasks **197**
Understanding Resource Overallocation **199**
Locating and Analyzing Resource Overallocations *200*
Leveling Overallocated Resources **204**
Using a Leveling Methodology *204*
Setting Leveling Options *205*

Leveling an Overallocated Resource *208*
Viewing Leveling Results *208*
Clearing Leveling Results *210*
Setting Task Priority Numbers *210*
Creating Resource Engagements for Human Resources **213**

Assigning Work Resources to Tasks

After you build your project team with enterprise resources, you are ready to assign your team members to tasks in your enterprise project. Microsoft Project offers two powerful tools for assigning resources to tasks, which are:

- **Task Entry** view
- **Assign Resources** dialog

Each method has its own advantages and disadvantages. I present an in-depth treatment of how to use each of these methods for assignment planning and analysis, and I provide additional information on how to analyze and level resource overallocations.

Warning: During the resource assignment process, do not assign resources to summary tasks, as this greatly increases the work hours and costs for your project. Instead, if you need to show a resource as the responsible person for a summary section of the project, use the built-in task *Contact* field instead.

Assigning Resources Using the Task Entry View

The *Task Entry* view is the most powerful way to assign resources to tasks because it gives you total control over all of the elements in the Microsoft Project scheduling engine. Using *the Task Entry* view, you can do all of the following in a single location:

- Assign multiple resources simultaneously, and specify different *Units* and *Work* values for each resource.
- Enter the *Duration* value of the task.
- Set the *Task Type* for the task to determine whether the software fixes or "locks" the *Units*, *Work*, or *Duration* value for the task.
- Specify the *Effort Driven* status of the task to determine what happens when you add or remove resources on the task.

Information: The biggest disadvantage of using the *Task Entry* view is that you cannot assign resources to multiple tasks simultaneously. If you need to assign the same resource to multiple tasks simultaneously, you must use the *Assign Resources* dialog instead.

To apply the *Task Entry view*, complete the following steps:

1. Click the **Task** tab to display the *Task* ribbon.
2. In the *View* section of the *Task* ribbon, click the **Gantt Chart** pick list button and select the **Gantt Chart** view (if you do not have the *Gantt Chart* view displayed already).
3. Click the **View** tab to display the *View* ribbon.
4. In the *Split View* section of the *View* ribbon, select the **Details** checkbox. Microsoft Project displays the *Task Entry* view, shown in Figure 7 - 1.

Module 07

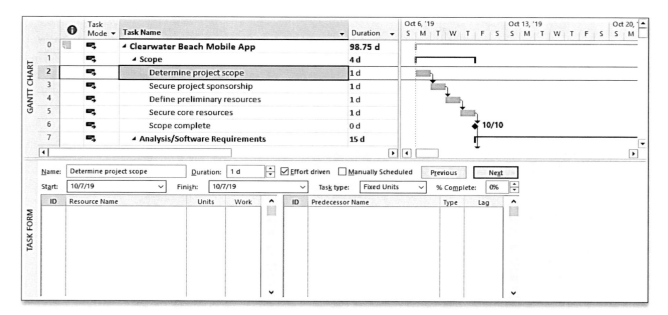

Figure 7 - 1: Task Entry view

Information: The fastest way to display the *Task Entry* view is to right-click anywhere in the *Gantt Chart* pane and to select the **Show Split** item in the shortcut menu. To close the *Task Form* pane and return to the *Gantt Chart* view, right-click anywhere in the *Gantt Chart* pane or the *Task Form* pane and **deselect** the **Show Split** item in the shortcut menu.

Information: The *Task Form* pane allows you to apply multiple sets of details (fields and columns) to the view. To change the set of details displayed in the *Task Form* view, right-click anywhere in the view. Microsoft Project displays a shortcut menu with a list of details you can apply. The default set of details is *Resources & Predecessors*. Apply the *Work* details if you need to enter planned *Overtime Work* for a resource. Apply the *Schedule* details so that you can specify when each resource begins their work on the task, assuming the task has multiple resources assigned. If you use any set of details other than the default, always reapply the *Resources & Predecessors* details when you finish.

The *Task Entry* view is a combination view consisting of two other views, each displayed in a separate pane. The *Task Entry* view includes the *Gantt Chart* view in the top pane and the *Task Form* view in the bottom pane. To assign a resource to a task using the *Task Entry* view, take the following steps:

1. Select a single task in the *Gantt Chart* pane.

2. In the *Task Form* pane, select the first blank cell in the **Resource Name** column and then click the pick list button in that cell. The software displays the list of your project team members, displayed alphabetically, as shown in Figure 7 - 2.

3. In the pick list of team members shown in the *Resource Name* column, select a resource.

Assignment Planning

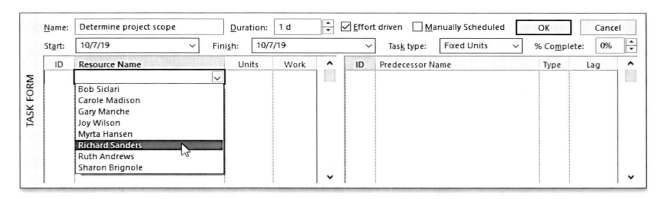

Figure 7 - 2: Task Entry view, select a resource

4. In the **Units** cell for the selected resource, select or enter a **Units** value. The Units value you select should indicate the resource's commitment to the task, where a Units value of 100% represents full-time work (8 hours per day) and 50% indicates half-time work (4 hours per day).

 Information: In the *Units* column, you can use the spin control feature to select a value, but Microsoft Project displays *Units* values only in 50% increments (0%, 50%, 100%, etc.). You cannot use the spin control to select *Units* values such as 25% or 75%. Instead, you must manually type the *Units* value you need.

5. In the **Work** cell for the selected resource, enter the estimated work in hours for the resource.

6. Repeat steps #2-5 for each additional resource you want to assign to the task.

7. In the *Task Form* pane, click the **OK** button.

 Warning: *Do not* click the *OK* button in the *Task Form* pane until you finish selecting all of the resources you want to assign to the task, and you finish setting both the *Units* and *Work* values for each resource.

Microsoft Project assigns the resource(s) to the task and then calculates the *Duration* value based on the *Units* and *Work* values you entered for each resource. Notice in Figure 7 - 3 that I assigned a resource named Richard Sanders to the task at a *Units* value of *50%* and a *Work* value of *16 hours*. Based on these two numbers, the software calculated a *Duration* value of *4 days* for the task.

Notice also in Figure 7 - 3 that the software changed the *OK* and *Cancel* buttons to the *Previous* and *Next* buttons. Using the *Previous* and *Next* buttons, you can navigate easily from task to task during the assignment process.

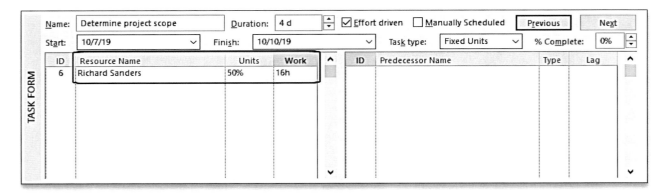

Figure 7 - 3: Task Entry view, Duration calculated
after assigning a resource with Units and Work

8. Click the **Next** button to select the next task in the project and to continue assigning resources to tasks.

To assign resources to a task for which you already entered an estimated *Duration* value previously, complete the following steps:

1. In the *Gantt Chart* pane, select the task with the estimated *Duration* value.
2. In the *Task Form* pane, select the name of a resource from the list in the **Resource Name** column.
3. Enter a **Units** value for the resource in the **Units** column.
4. *Do not* enter a **Work** value.
5. Repeat steps #2-4 for each additional resource you want to assign to the task.
6. Click the **OK** button.

Microsoft Project calculates the *Work* value for each resource assigned to the task. Notice in Figure 7 - 4 that I assigned three resources to work full-time (*Units* value of *100%*) on the task that already had a *Duration* value of *5 days*, so the software calculated *40 hours* of *Work* for each resource.

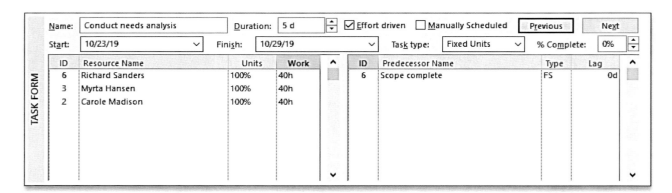

Figure 7 - 4: Task Entry view, Work calculated after
assigning resources with Units and Duration

Hands On Exercise

Exercise 7 - 1

Assign resources to tasks in your enterprise project using the *Task Entry* view.

1. Click the **Task** tab to display the *Task* ribbon.

2. In the *View* section of the *Task* ribbon, click the **Gantt Chart** pick list button and select the **Gantt Chart** view.

3. Right-click in the *Gantt Chart* pane and select the **Show Split** item on the shortcut menu.

4. Select task ID #2, the *Write Chapter 1* task.

5. In the *Task Form* pane, assign *yourself* to the task at a **Units** value of **100%** and a **Work** value of **80 hours**.

6. In the *Task Form* pane, click the **OK** button and then click the **Next** button to select task ID #3, the *Write Chapter 2* task.

7. In the *Task Form* pane, assign *yourself* to the task at a **Units** value of **100%** and a **Work** value of **80 hours**.

8. In the *Task Form* pane, click the **OK** button and then click the **Next** button to select task ID #4, the *Write Chapter 3* task.

9. In the *Task Form* pane, assign *yourself* to the task at a **Units** value of **100%** and a **Work** value of **80 hours**.

10. In the *Task Form* pane, click the **OK** button.

11. In the *Gantt Chart* pane, select task ID #7, the *Create Sample Files for Chapter 1* task.

12. In the *Task Form* pane, assign *your first team member* to the task at a **Units** value of **50%** and a **Work** value of **40 hours**.

13. In the *Task Form* pane, click the **OK** button and then click the **Next** button to select task ID #8, the *Create Sample Files for Chapter 2* task.

14. In the *Task Form* pane, assign *your first team member* to the task at a **Units** value of **50%** and a **Work** value of **40 hours**.

15. In the *Task Form* pane, click the **OK** button and then click the **Next** button to select task ID #9, the *Create Sample Files for Chapter 3* task.

16. In the *Task Form* pane, assign *your first team member* to the task at a **Units** value of **50%** and a **Work** value of **40 hours**.

17. In the *Task Form* pane, click the **OK** button.

Warning: If you see an overallocation indicator in the *Indicators* column for the Write Chapter 3 task, and you have a 7x8 calendar assigned as a *task calendar* for the task, the task calendar is the cause of the overallocation. This is an overallocation that you cannot resolve because there is a conflict between the working time shown on the assigned resource's calendar (5 days/week) and the task calendar (7 days/week).

18. In the *Gantt Chart* pane, select task ID #12, the *Create Instructor Training Outline* task.

19. In the *Task Form* pane, assign **yourself** to the task at a **Units** value of **100%** and a **Work** value of **40 hours**.

20. In the *Task Form* pane, click the **OK** button and then click the **Next** button to select task ID #13, the *Create Instructor Demo Files* task.

21. In the *Task Form* pane, assign the following resources:

    - *Yourself* at a **Units** value of **100%** and a **Work** value of **24 hours**
    - *Your second team member* at a **Units** value of **100%** and a **Work** value of **24 hours**

22. In the *Task Form* pane, click the **OK** button and then click the **Next** button to select task ID #14, the *Create Instructor PowerPoint* task.

23. In the *Task Form* pane, assign **yourself** to the task at a **Units** value of **100%** and a **Work** value of **32 hours**.

24. In the *Task Form* pane, click the **OK** button.

25. Right-click in the *Gantt Chart* pane and *deselect* the **Show Split** item in the shortcut menu.

26. Save the latest changes to your enterprise project.

Assigning Resources Using the Assign Resources Dialog

The *Assign Resources* dialog is a second method you can use in the assignment process. The power of the *Assign Resources* dialog is that it allows you to select a block of tasks and then to assign one or more resources to all of the selected tasks simultaneously. This dialog is ideal for assigning resources to tasks in which you previously entered a *Duration* value. The *Assign Resources* dialog is also useful for replacing one resource with another resource on multiple tasks simultaneously.

Although this dialog offers you a simple interface to assign resources to tasks quickly, keep in mind that it does not have all of the options available in the *Task Entry* view. Using the *Assign Resources* dialog, you have no control over most of the attributes of the scheduling engine. This means you cannot specify the *Duration*, the *Work*, the *Task Type*, or the *Effort Driven* status for a task. To display the *Assign Resources* dialog, use one of the following methods:

- In the *Assignments* section of the *Resource* ribbon, click the **Assign Resources** button.
- Right-click any task and then select the **Assign Resources** item on the shortcut menu.

Information: If you customized your ribbon and Quick Access Toolbar using the directions in Exercise 2 – 1, you can also display the *Assign Resources* dialog by clicking the **Assign Resources** button in the Quick Access Toolbar.

Assignment Planning

Microsoft Project displays the *Assign Resources* dialog with your project team members sorted alphabetically, as shown in Figure 7 - 5.

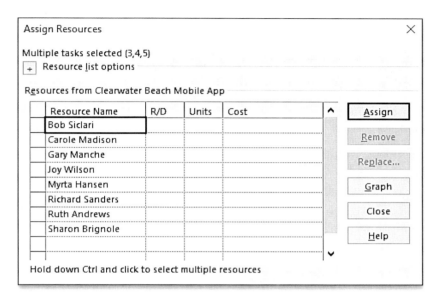

Figure 7 - 5: Assign Resources dialog

To assign a resource to tasks using the *Assign Resources* dialog, complete the following steps:

1. Select one or more tasks.

2. Select a single resource in the list of resources shown in the dialog.

3. Select or enter a **Units** value.

4. Click the **Assign** button.

The *Assign Resources* dialog indicates that you assigned the resource to the selected tasks by moving the assigned resource to the beginning of the list, and by adding a checkmark indicator to the left of the resource's name. Notice in Figure 7 - 6 that I assigned Ruth Andrews to work full time (*Units* value of *100%*) on the selected task. Microsoft Project calculates the value in the *Cost* column using the *Standard Rate* of the assigned resource multiplied by the calculated number of *Work* hours for the task. Notice also in Figure 7 - 6 that Ruth Andrews' assignment on the selected task costs the project $1600, as indicated by the value in the *Cost* column.

175

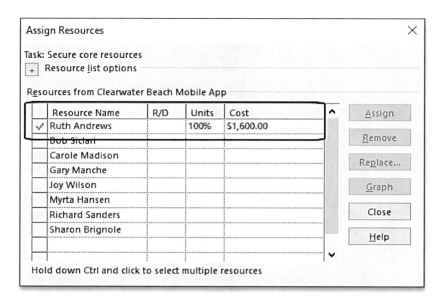

Figure 7 - 6: Assign Resources dialog with
one resource assigned to a task

To assign multiple resources to tasks using the *Assign Resources* dialog, complete the following steps:

1. Select one or more tasks.

2. In the *Assign Resources* dialog, select multiple resources using either the **Control** key or the **Shift** key on your computer keyboard.

3. *Do not* set a **Units** value for any resource.

4. Click the **Assign** button.

If you do not enter a *Units* value when assigning a resource in the *Assign Resources* dialog, Microsoft Project enters the *Max. Units* value for the resource from the *Resource Sheet* view of your project. This means that if the *Max. Units* value for a resource is *50%*, and you do not supply a *Units* value when assigning this resource, the *Assign Resources* dialog assigns the resource at a *Units* value of *50%* automatically. The exception to this rule is for resources that have a *Max. Units* value greater than *100%*, such as Generic resources. In this case, if you do not supply a *Units* value for the Generic resource, the *Assign Resources* dialog assigns the Generic resource at a *Units* value of *100%* automatically.

 Warning: Do not use the *Assign Resources* dialog to assign multiple resources to a task that requires using a different *Units* value for each resource. The software assigns the first resource at the *Units* value you select and then adds each of the other resources as helpers on the task using *Effort Driven* scheduling, decreasing the *Duration* of the task accordingly. This behavior is one of the major reasons why new users find Microsoft Project so frustrating! Instead, use the *Task Entry* view when you need to assign multiple resources with different *Units* values.

Assignment Planning

Hands On Exercise

Exercise 7 - 2

Assign resources to tasks in your enterprise project using the *Assign Resources* dialog.

1. Enter estimated *Duration* values on the following tasks as shown below:

   - Create Student Learning Aids   4 days
   - Create Survey Monkey Survey   2 days
   - Create Class Completion Certificate   1 day

2. Select task ID #17, the *Create Student Learning Aids* task.

3. Click the **Assign Resources** button in your Quick Access Toolbar.

4. In the *Assign Resources* dialog, select your **third and fourth team members** and then click the **Assign** button (**do not** enter a *Units* value for either resource).

Notice that Microsoft Project filled in the *Units* value for each resource using their *Max. Units* values from the *Resource Sheet* view of your project.

5. Select task ID #18, the *Create Survey Monkey Survey* task.

6. In the *Assign Resources* dialog, select *yourself*, enter a **Units** value of **75%**, and then click the **Assign** button.

7. Select task ID #19, the *Create Class Completion Certificate* task.

8. In the *Assign Resources* dialog, select your **fourth team** member, enter a **Units** value of **50%**, and then click the **Assign** button.

9. Click the **Close** button to close the *Assign Resources* dialog.

10. Save the latest changes to your enterprise project.

11. When the *Save* job completes, **close and check in** your enterprise project.

Understanding Other Factors in Assignment Planning

During the assignment planning process, three factors impact the assignment planning process in Microsoft Project. These factors are:

- **Duration Equation** – Microsoft Project uses the Duration Equation to calculate the *Duration*, *Work*, or *Units* value when you assign a resource to a task.

- **Task Type** – You use the *Task Type* option to fix or "lock" the *Duration*, *Work*, or *Units* value when assigning resources to a task.

- **Effort Driven Scheduling** – You use *Effort Driven* scheduling to reduce the *Duration* value on a task by adding one or more helpers to the task.

I discuss each of these factors individually as subtopics in this section of the module.

Understanding the Duration Equation

When you assign a resource to a task using the *Task Entry* view, and you enter a *Units* value and a *Work* value for the resource, Microsoft Project calculates the *Duration* value for the task automatically. How does the software calculate the *Duration* value? The software uses a formula known as the **Duration Equation**, written by default as follows:

$$\text{Duration} = \text{Work} \div (\text{Hours Per Day} \times \text{Units})$$

The default *Hours Per Day* value is *8 hours* per day in Microsoft Project. You can locate this value by clicking the **File** tab and then clicking the **Options** tab in the **Backstage**. In the *Project Options* dialog, click the **Schedule** tab. You find the *Hours per day* field in the *Calendar options for this project* section of the dialog, as shown in Figure 7 - 7.

To demonstrate how the Duration Equation works, I assign a resource to a task at a *Units* value *50%* and a *Work* value of *40 hours*. Using the Duration Equation, Microsoft Project calculates the *Duration* value as follows:

Duration = Work ÷ (Hours Per Day × Units)

Duration = 40 ÷ (8 × 50%)

Duration = 40 ÷ 4

Duration = 10 days

When you assign a resource to a task and enter *Duration* and *Units* values (rather than *Units* and *Work* values), Microsoft Project calculates the *Work* value. How does the software calculate the *Work* value? The software uses a modified version of the Duration Equation, rewritten to solve for the *Work* variable as follows:

$$\text{Work} = \text{Duration} \times \text{Hours Per Day} \times \text{Units}$$

For example, when I assign a resource with a *Units* value of *50%* to a task with a *Duration* value of *10 days*, Microsoft Project calculates *40 hours* of *Work* as follows:

Work = Duration × Hours Per Day × Units

Work = 10 × 8 × 50%

Work = 10 × 4

Work = 40 hours

When you select a task with an existing *Duration* value and you assign a resource to the task by entering a *Work* value, Microsoft Project calculates the *Units* value. Again, the software uses a modified version of the Duration Equation, except rewritten to solve for the *Units* variable as follows:

$$\text{Units} = \text{Work} \div (\text{Duration} \times \text{Hours Per Day})$$

For example, when I assign a resource at *40 hours* of *Work* on a task with an existing *Duration* value of *10 days*, Microsoft Project calculates a *Units* value of *50%* as follows:

Assignment Planning

Units = Work ÷ (Duration × Hours Per Day)

Units = 40 ÷ (10 × 8)

Units = 40 ÷ 80

Units = .5 or 50%

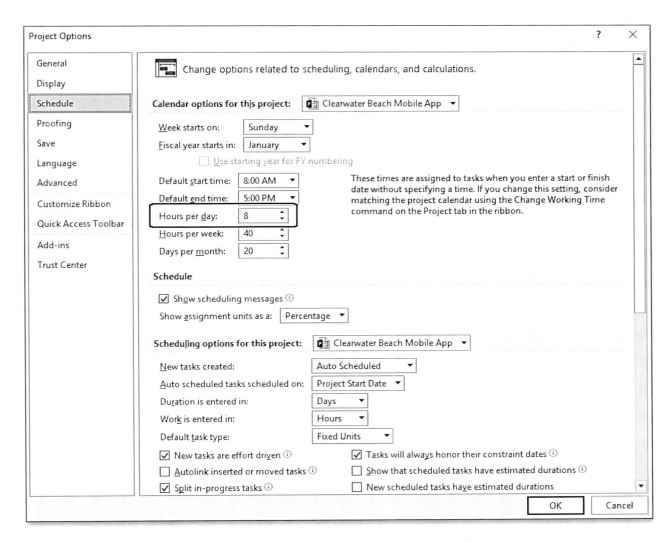

Figure 7 - 7: Project Options dialog, Schedule tab
shows the Hours per day setting

Understanding Task Types

You can specify the **Task Type** setting for each task using one of three types: *Fixed Units*, *Fixed Work*, or *Fixed Duration*. You can select only one *Task Type* setting for each task. The default *Task Type* setting for every task is *Fixed Units*, unless you specify otherwise in the *Project Options* dialog. To specify the *Task Type* setting for a task, select the task and then use one of the following methods:

- Apply the *Task Entry* view. In the *Task Form* pane, click the **Task Type** pick list and select the desired *Task Type* setting, and then click the **OK** button.

179

- Double-click the task and then click the **Advanced** tab in the *Task Information* dialog. Click the **Task Type** pick list and select the desired *Task Type* setting, and then click the **OK** button.

- In the *Properties* section of the *Task* ribbon, click the **Information** button, and then click the **Advanced** tab in the *Task Information* dialog. Click the **Task Type** pick list and select the desired *Task Type* setting, and then click the **OK** button.

- Right-click the task and then select the **Information** item on the shortcut menu. In the *Task Information* dialog, click the **Advanced** tab. Click the **Task Type** pick list and select the desired *Task Type* setting, and then click the **OK** button.

> **Information**: You can also specify the *Task Type* setting for multiple tasks simultaneously by selecting a group of tasks first. In the *Properties* section of the *Task* ribbon, click the **Information** button and then click the **Advanced** tab in the *Multiple Task Information* dialog. Click the **Task Type** pick list and select the desired *Task Type* setting for the selected tasks, and then click the **OK** button.

Specify the *Task Type* setting for tasks using the following information as your guide:

- **Fixed Units** – Microsoft Project locks the *Units* value for all resources assigned to a *Fixed Units* task. This is the default *Task Type* for every task, by the way. Use the *Fixed Units* setting when a resource has a known availability to perform work on tasks in your project. For example, you assign a resource to work on a task at a *Units* value of 50% because the resource also works half time on the Help Desk. Use the *Fixed Units* task type on this task to guarantee that the software does not recalculate the *Units* value if you change either the *Work* or *Duration* values on the task.

- **Fixed Work** – Microsoft Project locks the *Work* value for all resources assigned to a *Fixed Work* task. Use the *Fixed Work* setting when you are certain about the number of hours needed to complete a task. For example, you hire a consultant to work on a project task, and the work is set at 40 hours by contract. Use the *Fixed Work* task type on this task to guarantee that Microsoft Project does not recalculate the *Work* value if you change either the *Units* or *Duration* values on the task.

- **Fixed Duration** – Microsoft Project locks the *Duration* value on a *Fixed Duration* task. Use the *Fixed Duration* setting when you are certain of the *Duration* value for a task, such as when you have a known "window of opportunity" to complete the task. For example, you have a task called Shareholder Conference and the conference lasts 3 days. Use the *Fixed Duration* task type on this task to guarantee that the software does not recalculate the *Duration* value of *3 days* if you change either the *Units* or *Work* values on the task.

The *Task Type* setting you select fixes or "locks" one of the three variables in the Duration Equation for the selected task. When you change one of the two non-fixed variables, Microsoft Project calculates the other non-fixed variable automatically. Table 7 - 1 documents the behavior of all three *Task Types* when you change the non-fixed variable, and when you change the fixed variable as well.

Assignment Planning

Task Type	Fixed Value	You Change	Recalculated Value
Fixed Units	Units	Work	Duration
Fixed Units	Units	Duration	Work
Fixed Units	Units	Units	Duration
Fixed Work	Work	Units	Duration
Fixed Work	Work	Duration	Units
Fixed Work	Work	Work	Duration
Fixed Duration	Duration	Units	Work
Fixed Duration	Duration	Work	Units
Fixed Duration	Duration	Duration	Work

Table 7 - 1: Task Type behavior

Information: The release of Microsoft Project 2010 introduced some new functionality in the software that affects *Task Type* behavior. When a change made to either the *Work* value or the *Duration* value leads to the recalculation of the *Units* value, Microsoft Project no longer displays the new *Units* value in the *Task Form* pane. Instead, the software stores the recalculated value in the new *Peak* field. Because the *Peak* field is an *Assignment* field, you can display it only in the *Task Usage* view or the *Resource Usage* view. To compare the original *Units* value, stored in a field named *Assignment Units*, with the new value stored in the *Peak* field, insert both of these fields in the *Task Usage* view and then study the differences.

When you change a non-fixed variable for any task type Microsoft Project automatically recalculates the other non-fixed variable. When you change the fixed variable, however, the software invokes one of the programming decisions implemented by the software development team many years ago. For example, which variable should the software recalculate when you change the *Units* value on a *Fixed Units* task, or you change the *Work* value on a *Fixed Work* task, or you change the *Duration* value on a *Fixed Duration* task? We refer to the decisions made by the software development team as their programming biases. These programming biases are as follows:

- If you change the *Units* variable on a *Fixed Units* task, Microsoft Project **always** recalculates the *Duration* variable.

- If you change the *Work* variable on a *Fixed Work* task, Microsoft Project **always** recalculates the *Duration* variable.

- If you change the *Duration* variable on a *Fixed Duration* task, Microsoft Project **always** recalculates the *Work* variable.

As you can see, when you change the fixed variable, Microsoft Project has a bias to calculate changes in *Duration* rather than to *Work* or *Units*. If the software cannot change *Duration*, it has a bias to calculate changes in *Work* rather than *Units*.

Module 07

 Hands On Exercise

Exercise 7 - 3

Learn more about Task Types by changing the variables in the Duration Equation for tasks with different Task Types.

1. Open the **Understanding Task Types 2019.mpp** sample file from your student sample files folder. The project file opens Read-Only.
2. Click the **Task** tab to display the *Task* ribbon.
3. In the *View* section of the *Task* ribbon, click the **Gantt Chart** pick list button and select the **Task Usage** view.
4. Right-click anywhere in the timephased grid on the right side of the *Task Usage* view and then select the **Show Split** item on the shortcut menu.
5. In the *Task Usage* pane, select the *Fixed Units 1* task.
6. Click the **Scroll to Task** button in your Quick Access Toolbar to scroll the *Work* hours into view.
7. In the *Task Form* pane, change the **Work** value to **48h** and then click the **OK** button.
8. Click the **Next** button in the *Task Form* pane to select the *Fixed Units 2* task.
9. In the *Task Form* pane, change the **Duration** value to **8d** and then click the **OK** button.

For these *Fixed Units* tasks, notice that when you change the *Work* variable, Microsoft Project recalculates the *Duration* variable. When you change the *Duration* variable, the software recalculates the *Work* variable.

10. Click the **Next** button in the *Task Form* pane to select the *Fixed Units 3* task.
11. In the *Task Form* pane, change the **Units** value to **100%** and then click the **OK** button.

For this *Fixed Units* task, notice that when you change the *Units* variable, Microsoft Project recalculates the *Duration* variable. This behavior is the first of three programming biases I discussed in the previous topical section.

12. Click the **Next** button in the *Task Form* pane to select the *Fixed Work 1* task.
13. In the *Task Form* pane, change the **Units** value to **100%** and then click the **OK** button.

For this *Fixed Work* task, notice that when you change the *Units* variable, Microsoft Project recalculates the *Duration* variable.

14. Click the **Next** button in the *Task Form* pane to select the *Fixed Work 2* task.
15. In the *Task Form* pane, change the **Duration** value to **5d** and then click the **OK** button.

Assignment Planning

For this *Fixed Work* task, notice that when you change the *Duration* variable, Microsoft Project ***did not*** recalculate the *Units* variable as expected. Instead, the *Peak* field contains the new *Units* value. Notice that the *Peak* field shows the correct *Units* value of *100%*. Notice also in the timephased grid on the right side of the view that the software schedules the *Work* hours correctly at 8 hours per day based on the *Units* value of *100%* in the *Peak* field.

16. Click the **Next** button in the *Task Form* pane to select the *Fixed Work 3* task.
17. In the *Task Form* pane, change the **Work** value to **32h** and then click the **OK** button.

For this *Fixed Work* task, notice that when you change the *Work* variable, Microsoft Project recalculates the *Duration* variable. This behavior is the second of three programming biases I discussed in the previous topical section.

18. Click the **Next** button in the *Task Form* pane to select the *Fixed Duration 1* task.
19. In the *Task Form* pane, change the **Units** value to **100%** and then click the **OK** button.

For this *Fixed Duration* task, notice that when you change the *Units* variable, Microsoft Project recalculates the *Work* variable.

20. Click the **Next** button in the *Task Form* pane to select the *Fixed Duration 2* task.
21. In the *Task Form* pane, change the **Work** value to **60h** and then click the **OK** button.

For this *Fixed Duration* task, notice that when you change the *Work* variable, Microsoft Project did not recalculate the *Units* variable. Once again, notice that the *Peak* field contains the expected value of *75%*. Notice also in the timephased grid that the software schedules the *Work* hours correctly at 6 hours per day based on the *75%* value in the *Peak* field.

22. Click the **Next** button in the *Task Form* pane to select the *Fixed Duration 3* task.
23. In the *Task Form* pane, change the **Duration** value to **6d** and then click the **OK** button.

For this *Fixed Duration* task, notice that when you change the *Duration* variable, Microsoft Project recalculates the *Work* variable. This behavior is the final of three programming biases I discussed in the previous topical section.

24. Close the **Understanding Task Types 2019.mpp** sample file and ***do not*** save the changes.

Understanding Effort Driven Scheduling

In Microsoft Project, you can designate each task individually as either an **Effort Driven** task or a **non-Effort Driven** task. The *Effort Driven* status of any task determines how the software responds when you add or remove resources from a task to which you previously assigned one or more resources. The default setting for every task in Microsoft Project is *Effort Driven*.

To assign additional resources as helpers to a task using *Effort Driven* scheduling, complete the following steps:

1. Select a task to which you previously assigned at least one resource.
2. Apply the *Task Entry* view.

3. In the *Task Form* pane, select the **Effort driven** option, if not already selected.

4. Select one or more additional resources in the *Resource Name* column, and set a *Units* value for each additional resource.

5. *Do not* enter a *Work* value for any of the additional resources.

6. Click the **OK** button.

When you add a resource to an *Effort Driven* task, the software holds the *Remaining Work* value constant and allocates the *Remaining Work* proportionately to each resource based on each resource's *Units* value. When a task is unstarted, by the way, the *Remaining Work* value is equal to the *Work* value. Consider this first example of how Microsoft Project distributes the *Remaining Work* when the *Units* values **are the same** for each resource on an unstarted task:

- I initially assign Larry Barnes to the Design task at a *Units* value of *100%* and a *Work* value of *80 hours*. The software calculates a *Duration* value of *10 days*, as shown in Figure 7 - 8.

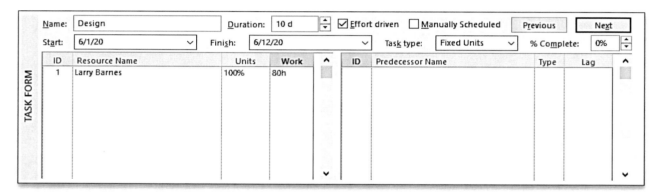

Figure 7 - 8: Original Duration of 10 days for the Design task

- Using *Effort Driven* scheduling, I later add Jerry King as a helper on the task at a *Units* value of *100%*. Microsoft Project reduces the *Duration* value to *5 days* and allocates the *80 hours* of *Work* proportionately between the two resources (40 hours each to Larry Barnes and to Jerry King) based on their *Units* values, as shown in Figure 7 - 9.

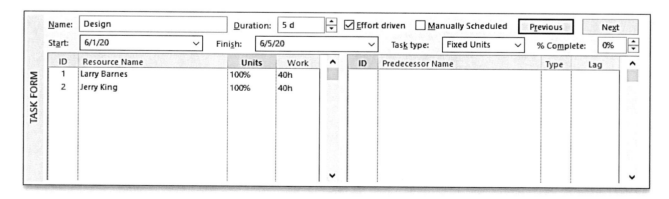

Figure 7 - 9: Using Effort Driven scheduling with identical Units values

How does Microsoft Project actually determine the proportionate split of the original 80 hours of work for the Design task? Because the *Units* values are identical for Larry Barnes and Jerry King, Microsoft Project applies a "50-50" proportional split, allocating 40 hours of work to Larry and 40 hours of Jerry.

Consider this second example of how Microsoft Project distributes the *Remaining Work* when the *Units* values **are different** for each resource assigned to an unstarted task:

- I initially assign Sue Uland to the Build task at a *Units* value of *100%* and a *Work* value of *120 hours*. The software calculates a *Duration* value of *15 days*, as shown in Figure 7 - 10.

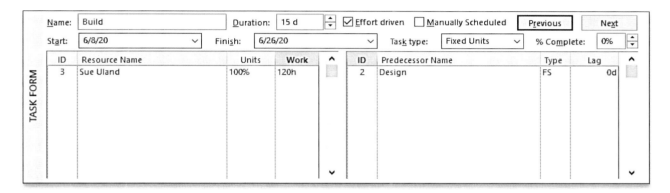

Figure 7 - 10: Original Duration value of 15 days for the Build task

- Using *Effort Driven* scheduling, I later add Myrta Hansen as a helper on the task at a *Units* value of *50%*. Microsoft Project reduces the *Duration* value to *10 days* and allocates the *120 hours* of *Remaining Work* proportionately between the two resources (80 hours to Sue Uland and 40 hours to Myrta Hansen) based on their *Units* values, as shown in Figure 7 - 11.

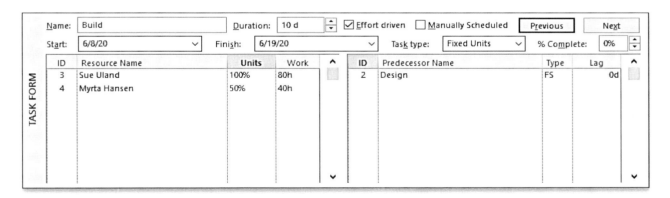

Figure 7 - 11: Effort Driven scheduling with different Units values

How does Microsoft Project actually determine the proportionate split of the original 80 hours of work? Sue Uland's *Units* value of *100%* is **two-thirds** of the total *Units* of *150%* for both resources (100/150 = 2/3), so the system allocates Sue Uland to two-thirds of the total *Work*, which is *80 hours* (120 x 2/3 = 80). Myrta Hansen's *Units* value of *50%* is one-third of the total *Units* of *150%* for both resources (50/150 = 1/3), so the system allocates Myrta Hansen to **one-third** of the total *Work*, which is *40 hours* (120 x 1/3 = 40).

Information: When you remove a resource from an *Effort Driven* task with multiple resources already assigned, Microsoft Project increases the *Duration* of the task and increases the *Work* value proportionately for each remaining resource. This behavior is due to *Effort Driven* scheduling, although most people do not realize this.

Remember that when you assign additional resources to an *Effort Driven* task that is unstarted, Microsoft Project reallocates the *Remaining Work* proportionately between all of the assigned resources. So how does the software respond when you add a helper to a task that already contains some completed work? Consider the following example:

- I assign Sarah Baker to the Test task at a *Units* value of *100%* and a *Work* value of *80* hours, and Microsoft Project calculates a *Duration* value of *10 days* for the task. Sarah completed *40 hours* of *Actual Work* on this task, which leaves *40 hours* of *Remaining Work* as shown in Figure 7 - 12.

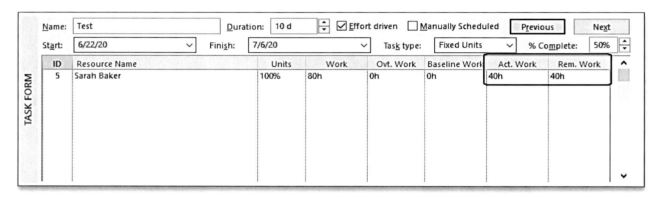

Figure 7 - 12: Test task has 40 hours of Actual Work

Using *Effort Driven* scheduling, I add Keith Doll to the task at a *Units* value of *100%*. Microsoft Project shortens the *Duration* to *7.5 days* and allocates the *40 hours* of *Remaining Work* evenly between the two resources (20 hours each to Sarah Baker and Keith Doll), as shown in Figure 7 - 13.

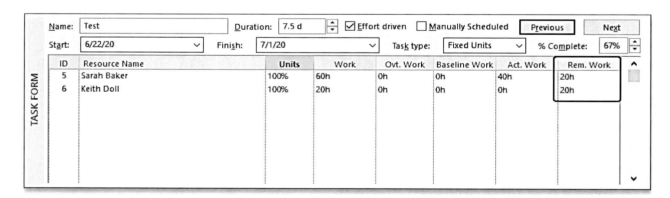

Figure 7 - 13: Using Effort Driven scheduling with an in-progress task

Figure 7 - 12 and Figure 7 - 13 show the *Task Form* view with the *Work* details applied. To view the *Work* details, right-click anywhere in the *Task Form* pane and then select the **Work** item on the shortcut menu. Notice in Figure 7 - 13 that the software holds the *40 hours* of *Remaining Work* constant, which gives Sarah Baker *60 hours* of total *Work*

Assignment Planning

(*40 hours* of *Actual Work* + *20 hours* of *Remaining Work*) and gives Keith Doll *20 hours* of total *Work* (*0 hours* of *Actual Work* + *20 hours* of *Remaining Work*).

Information: When you assign additional resources using *Effort Driven* scheduling for *Fixed Units* or *Fixed Work* tasks, Microsoft Project reduces the *Duration* of the task. When you assign additional resources using *Effort Driven* scheduling for a *Fixed Duration* task, however, Microsoft Project reduces the *Units* value for each assigned resource.

Hands On Exercise

Exercise 7 - 4

Learn about how Effort Driven scheduling works with each Task Type.

1. Open the **Understanding Effort Driven 2019.mpp** sample file from your student sample files folder. The project file opens Read-Only.

2. Right-click anywhere in the *Gantt Chart* pane and select the **Show Split** item on the shortcut menu to display the *Task Entry* view.

3. Select task ID #2, the *Design 1* task. This task is *Fixed Units* and *Effort Driven*.

4. In the *Task Form* pane, add **Linda Erickson** as a helper at a **Units** value of **100%**, and then click the **OK** button.

Notice how Microsoft Project reduces the *Duration* value to *5d* and splits the *Work* proportionately to *40h* for the two resources, based on their respective *Units* values.

5. In the *Task Form* pane, click the **Next** button to select the *Build 1* task. This task is *Fixed Work* and *Effort Driven*. By default, *Fixed Work* tasks **must** be *Effort Driven*.

6. In the *Task Form* pane, add **Jerry King** as a helper at a **Units** value of **50%**, and then click the **OK** button.

Notice how Microsoft Project reduces the *Duration* value to *10d* and splits the *Work* proportionately between the two resources, based on their respective *Units* values. The *Work* split is 2/3 for Gary Howard (80h) and 1/3 for Jerry King (40h), by the way.

7. In the *Task Form* pane, click the **Next** button to select the *Test 1* task. This task is *Fixed Duration* and *Effort Driven*.

8. In the *Task Form* pane, add **Mickey Cobb** as a helper at a *Units* value of **100%**, and then click the **OK** button.

Notice how Microsoft Project holds the *Duration* value constant at *10d* (because the *Task Type* of this task is *Fixed Duration*) and splits the *Work* proportionately at *40h* for each of the two resources, based on their respective *Units* values. This is the default behavior of *Fixed Duration* tasks that are *Effort Driven*.

9. In the *Task Form* pane, click the **Next** button to select the *Implement 1* task. This task is *Fixed Units* and *non-Effort Driven*.

10. In the *Task Form* pane, add **Terry Uland** as a helper at a **Units** value of **100%**, and then click the **OK** button.

Notice that the *Duration* was ***not cut*** in half and the *Work **was not*** split proportionately between the two resources, as you might have expected!

11. Scroll the *Gantt Chart* pane to the week of **April 11, 2021** and examine the tasks' Gantt bar.

Notice that the task is partially completed.

12. Right-click anywhere in the *Task Form* pane and select the **Work** item on the shortcut menu.

Because this task has *40 hours* of *Actual Work* completed by the original team member, Microsoft Project can only split the *Remaining Work* of *40 hours* between the original resource and the new helper. Notice how each resource now has *20 hours* of *Remaining Work* to perform on this task. Notice also that Susan Manche has *60h* of *Work* (*40h* of *Actual Work* and *20h* of *Remaining Work*), while Terry Uland has *20h* of *Work* (*0h* of *Actual Work* and *20h* of *Remaining Work*).

13. Right-click anywhere in the *Task Form* pane and select the **Resources & Predecessors** item on the shortcut menu.
14. In the *Gantt Chart* pane, select task ID #7, the *Design 2* task. This task is *Fixed Units* and *non-Effort Driven*.
15. In the *Task Form* pane, add **Linda Erickson** as a helper at a **Units** value of **100%**, and then click the **OK** button.

Notice how Microsoft Project holds the *Duration* value constant at *10d* and increases the total *Work* on the task to *160h*, with *80h* assigned to each resource.

16. In the *Task Form* pane, select task ID #9, the *Test 2* task. This task is *Fixed Duration* and *non-Effort Driven*.
17. In the *Task Form* pane, add **Mickey Cobb** as a helper at a **Units** value of **50%**, and then click the **OK** button.

Notice how Microsoft Project holds the *Duration* value constant at *10d* and increases the total *Work* on the task to *120h*, with *80h* assigned to Larry Barnes and *40h* assigned to Mickey Cobb, based on their respective *Units* values.

18. Close the **Understanding Effort Driven 2019.mpp** sample file and ***do not*** save the changes.

Assignment Planning

Assigning Cost Resources

If your Project Online system includes **Budget Cost** and **Expense Cost** resources, it is important that you know how to assign each type of *Cost* resource. Because Microsoft does not include any type of standard view or table to use for assigning *Cost* resources, I teach you how to customize the *Task Usage* view and then use it to assign each type of *Cost* resource.

Adding Cost Resources to the Project Team

Before you can assign *Cost* resources to your project, you must add them to your project team by completing the following steps:

1. Click the **Resource** tab to display the *Resource* ribbon.

2. In the *Insert* section of the *Resource* ribbon, click the **Add Resources** pick list button, and then select the **Build Team from Enterprise** item.

3. At the top of the *Build Team* dialog, click the **Existing filters** pick list and select the **Resources – Cost** filter.

4. In the *Enterprise Resource* list in the lower left corner of the dialog, select one or more *Budget Cost* and *Expense Cost* resources, and then click the **Add** button to add them to the project team.

5. Click the **OK** button to close the *Build Team* dialog.

Assigning a Budget Cost Resource

When you use a *Budget Cost* resource in a project, Microsoft Project allows you to assign the resource to only the Project Summary Task (Row 0 or Task 0). This allows you to set overall budgetary amounts for the entire project as a whole. To assign *Budget Cost* resources your project, complete the following steps:

1. Click the **Task** tab to display the *Task* ribbon.

2. In the *View* section of the *Task* ribbon, click the **Gantt Chart** pick list button and select the **Task Usage** view.

3. In your project, select the Project Summary Task (Row 0 or Task 0).

Information: If you do not see the Project Summary Task in your project, click the **Format** tab to display the **Format** ribbon. In the *Show/Hide* section of the ribbon, select the **Project Summary Task** option to display the Project Summary Task.

4. Click the **Resource** tab to display the *Resource* ribbon.

5. In the *Assignments* section of the *Resource* ribbon, click the **Assign Resources** button.

6. In the *Assign Resources* dialog, select your *Budget Cost* resources and click the **Assign** button.

7. Click the **Close** button to close the *Assign Resources* dialog. Notice in Figure 7 - 14 that I assigned three *Budget Cost* resources to the Project Summary Task so that I can enter the budgetary amounts for hardware, labor, and training.

189

	ⓘ	Task Mode	Task Name	Work
0		⇛	▲ **Clearwater Beach Mobile App**	**1,800 h**
			Hardware Budget	
			Labor Budget	
			Training Budget	
1		⇛	▲ Scope	96 h
2		⇛	▲ Determine project scope	32 h
			Sharon Brignole	*32 h*
3		⇛	▲ Secure project sponsorship	16 h
			Sharon Brignole	*16 h*

Figure 7 - 14: Budget Cost resource assigned
to the Project Summary Task

8. Click the **View** tab to display the *View* ribbon.

9. In the *Data* section of the *View* ribbon, click the **Tables** pick list button and select the **Cost** table.

10. Right-click on the **Fixed Cost** column header, select the **Insert Column** item on the shortcut menu, and then select the **Budget Cost** column.

11. In the **Budget Cost** column, enter the budgetary amount for each of your *Budget Cost* resources. Notice in Figure 7 - 19 that I entered budgetary amounts for hardware, labor, and training.

	Task Name	Budget Cost	Fixed Cost
0	▲ **Clearwater Beach Mobile App**	**$195,000.00**	**$0.00**
	Hardware Budget	$50,000.00	
	Labor Budget	$125,000.00	
	Training Budget	$20,000.00	
1	▲ Scope		$0.00
2	▲ Determine project scope		$0.00
	Sharon Brignole		
3	▲ Secure project sponsorship		$0.00
	Sharon Brignole		

Figure 7 - 15: Budgetary amounts entered
for each Budget Cost resource

12. Right-click anywhere in the timephased grid on the right side of the *Task Usage* view and then select the **Detail Styles** item on the shortcut menu, as shown in Figure 7 - 16. Microsoft Project displays the *Detail Styles* dialog.

Assignment Planning

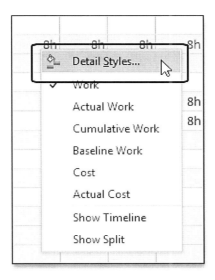

Figure 7 - 16: Select the Detail Styles
item on the shortcut menu

13. In the *Available fields* list on the left side of the *Detail Styles* dialog, select the **Budget Cost** field, and then click the **Show** button to add the field to the *Show these fields* list on the right side of the dialog. Figure 7 - 17 shows the *Detail Styles* dialog with the *Budget Cost* field added to the *Show these fields* list.

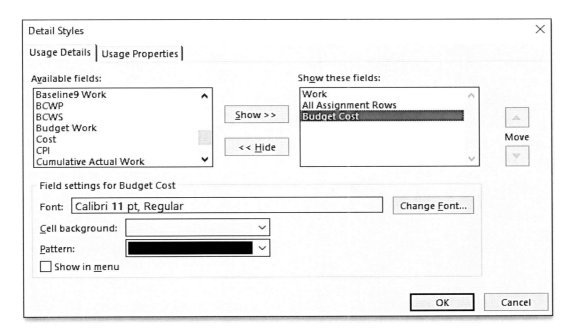

Figure 7 - 17: Detail Styles dialog,
add the Budget Cost field

14. Click the **OK** button to close the *Details Styles* dialog and to insert the *Budget Cost* field to the timephased grid.

15. In the timephased grid, double-click the right edge of the *Details* column header to "best fit" the column width. Figure 7 - 18 shows the customized *Task Usage* view with the *Budget Cost* column displayed in the *Cost* table on the left and the *Budget Cost* field displayed in the timephased grid on the right.

	Task Name	Budget Cost	Details	M	T	W	T	F
0	▲ **Clearwater Beach Mobile App**	$195,000.00	Work	8h	8h	8h	8h	8h
			Budget Cost	$1,826.70	$1,826.70	$1,826.70	$1,826.70	$1,826.70
	Hardware Budget	$50,000.00	Work					
			Budget Cost	$468.38	$468.38	$468.38	$468.38	$468.38
	Labor Budget	$125,000.00	Work					
			Budget Cost	$1,170.96	$1,170.96	$1,170.96	$1,170.96	$1,170.96
	Training Budget	$20,000.00	Work					
			Budget Cost	$187.35	$187.35	$187.35	$187.35	$187.35
1	▲ Scope		Work	8h	8h	8h	8h	8h
			Budget Cost					
2	▲ Determine project scope		Work	8h	8h	8h	8h	
			Budget Cost					
	Sharon Brignole		Work	8h	8h	8h	8h	
			Budget Cost					

Figure 7 - 18: Task Usage view ready for project budget entry

If your application administrator set the *Accrue At* field value to *Prorated* for each of your *Budget Cost* resources, Microsoft Project apportions the *Budget Cost* amounts evenly across the time span of the entire project. The *Accrue At* field is visible by default in the *Resource Sheet* view of your project. You can see the daily timephased *Budget Cost* amounts for each *Budget Cost* resource in the timephased grid shown previously in Figure 7 - 18. If necessary, you can reapportion the *Budget Cost* amounts in another manner, such as on a monthly basis, by following these additional steps:

1. Click the **View** tab to display the *View* ribbon.

2. In the *Zoom* section of the *View* ribbon, click the **Zoom** button to zoom the Timescale to the level of detail at which you wish to reapportion each *Budget Cost* amount.

Information: If you customized your ribbon and Quick Access Toolbar using the directions in Exercise 2 – 1, you can zoom the Timescale to the desired level of detail using the **Zoom In** or **Zoom Out** buttons in the Quick Access Toolbar.

Warning: If Microsoft Project displays pound signs (#####) instead of monetary values in any column, this indicates that the column is not wide enough to display the values. In the column header of any column displaying pound signs, click and hold the right gridline, and manually increase the column width until the software displays values in the column.

3. In the timephased grid, enter your anticipated *Budget Cost* values in the *Budget Cost* cells for each of your *Budget Cost* resources. In Figure 7 - 19, notice that I entered my *Budget Cost* values on a quarterly basis for each of my *Budget Cost* resources.

4. Save the latest changes to your enterprise project.

Assignment Planning

	Task Name	Budget Cost	Details	Q3	2020 Q4	Q1	Q2
0	▲ Clearwater Beach Mobile App	$195,000.00	Work		930h	870h	
			Budget Cost		$125,000.00	$70,000.00	
	Hardware Budget	$50,000.00	Work				
			Budget Cost		$40,000.00	$10,000.00	
	Labor Budget	$125,000.00	Work				
			Budget Cost		$75,000.00	$50,000.00	
	Training Budget	$20,000.00	Work				
			Budget Cost		$10,000.00	$10,000.00	
1	▲ Scope		Work		96h		
			Budget Cost				
2	▲ Determine project scope		Work		32h		
			Budget Cost				
	Sharon Brignole		Work		32h		
			Budget Cost				

Figure 7 - 19: Budget Cost information, Project Budget entry timephased grid

Assigning an Expense Cost Resource

After you enter your project's budgetary amounts using *Budget Cost* resources, you can also assign *Expense Cost* resources to your project so that you can track additional non-labor expenses. Microsoft Project allows you to assign *Expense Cost* resources to any type of task in the project, including summary tasks, subtasks, and milestone tasks. However, the software does not allow you to assign an *Expense Cost* resource to the Project Summary Task. To assign an *Expense Cost* resource to tasks in your project, complete the following steps:

1. Using the customized *Task Usage* view documented in the previous topical section, select any task in the project.

2. Click the **Resource** tab to display the *Resource* ribbon.

3. In the *Assignments* section of the *Resource* ribbon, click the **Assign Resources** button.

4. In the *Assign Resources* dialog, select your *Expense Cost* resource and then click the **Assign** button.

Figure 7 - 56 shows the *Assign Resources* dialog after assigning an *Expense Cost* resource named *Training Expense* to the selected task. I want to use this resource to capture the training expenditure on this task so that the cost is reportable in Project Online reports about project costs.

Module 07

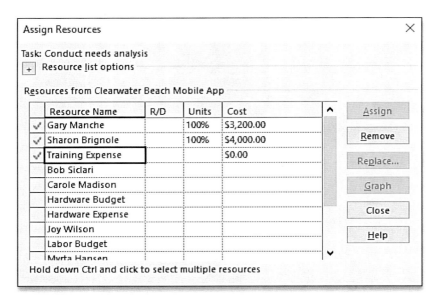

Figure 7 - 20: Assign Resources dialog after
assigning Expense Cost resource

5. In the **Cost** column of the *Assign Resources* dialog for the *Expense Cost* resource, manually enter the amount of anticipated expenditure and then press the **Enter** key on your computer keyboard. Figure 7 - 57 shows the *Assign Resources* dialog after entering my estimated expenditure for training expenses on the selected task. Notice that I anticipate a training expenditure of $5,000 for this task.

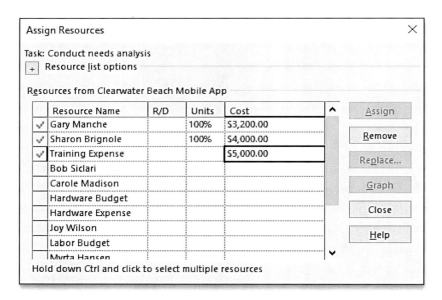

Figure 7 - 21: Assign Resources dialog after
assigning Expense Cost resource with Cost

6. Continue selecting tasks, assigning the *Expense Cost* resources, and entering anticipated cost expenditures until you finish.

7. Click the **Close** button to close the *Assign Resources* dialog.

Assignment Planning

8. Save the latest changes to your enterprise project.

Recommendations for Using Cost Resources

Cost resources do have some limitations, so follow Microsoft's recommendations for using Expense Cost resources effectively:

- Do not assign an *Expense Cost* resource on the same task as a *Work* resource if the resource reports actual progress from Project Web App.

- Do not assign a *Task calendar* using 24-hour elapsed days (edays) to a task assigned to an *Expense Cost* resource.

- Do not disable the *Actual costs are always calculated by Microsoft Project* option in your project. You find this option in the *Calculation options for this project* section of the *Advanced* page in the *Project Options* dialog.

- Avoid using the *Undo* feature if you edit the *Remaining Duration* field for a task assigned to an *Expense Cost* resource.

Hands On Exercise

Exercise 7 - 5

Assign a *Budget Cost* resource to your enterprise project and assign an *Expense Cost* resource to a task in your enterprise project.

Information: If your organization's Enterprise Resource Pool **does not** include *Cost* resources, you should skip this exercise entirely.

Add Cost Resources to Your Project Team

Add *Budget Cost* and *Expense Cost* resources to the project team in your enterprise project by completing the following steps:

1. Click the **Open** button in your Quick Access Toolbar.

2. In the *Open* dialog, select the enterprise project you created for this class, and then click the **Open** button.

3. In the *READ-ONLY* band at the top of the project schedule, click the **Check Out** button to check out the project for editing.

4. Click the **Resource** tab to display the *Resource* ribbon.

195

5. In the *Insert* section of the *Resource* ribbon, click the **Add Resources** pick list button, and then select the **Build Team from Enterprise** item.

6. At the top of the *Build Team* dialog, click the **Existing filters** pick list and select the **Resources – Cost** filter.

7. In the *Enterprise Resource* list in the lower left corner of the dialog, select one or more *Budget Cost* resources, and then click the **Add** button to add them to your project team.

8. In the *Enterprise Resource* list in the lower left corner of the dialog, select **only one** *Expense Cost* resource, and then click the **Add** button to add the resource to your project team.

9. Click the **OK** button to close the *Build Team* dialog.

Assign Budget Cost Resources to the Project Summary Task

Assign your Budget Cost resource(s) to the Project Summary Task of your enterprise project by completing the following steps:

1. Click the **Task** tab to display the *Task* ribbon.

2. In the *View* section of the *Task* ribbon, click the **Gantt Chart** pick list button and select the **Task Usage** view.

3. Widen the *Task Name* column, as needed, and then drag the split bar to the right edge of the *Work* column.

4. In your project, select the Project Summary Task (Row 0 or Task 0).

5. Click the **Assign Resources** button in your Quick Access Toolbar.

6. In the *Assign Resources* dialog, select one or more *Budget Cost* resources and then click the **Assign** button.

7. Click the **Close** button to close the *Assign Resources* dialog.

8. Click the **View** tab to display the *View* ribbon.

9. In the *Data* section of the *View* ribbon, click the **Tables** pick list button and select the **Cost** table.

10. Widen the *Task Name* column, as needed.

11. Right-click on the **Fixed Cost** column header, select the **Insert Column** item on the shortcut menu, and then select the **Budget Cost** column in the list of available task columns.

12. In the *Budget Cost* column, enter the budgetary amount for each of your *Budget Cost* resources, and then widen the *Budget Cost* column, if needed.

13. Right-click anywhere in the timephased grid on the right side of the *Task Usage* view and then select the **Detail Styles** item on the shortcut menu.

14. In the *Available fields* list on the left side of the *Detail Styles* dialog, select the **Budget Cost** field, and then click the **Show** button to add the field to the *Show these fields* list on the right side of the dialog.

15. Click the **OK** button to close the *Details Styles* dialog.

16. In the timephased grid, double-click the right edge of the *Details* column header to "best fit" the column width.

Assignment Planning

Assign Expense Cost Resources to Detailed Tasks

Assign *Expense Cost* resources to detailed tasks in your enterprise project by completing the following steps:

1. Using the customized *Task Usage* view, select task ID #8, the *Create Sample Files for Chapter 2* task.
2. Click the **Assign Resources** button in your Quick Access Toolbar.
3. In the *Assign Resources* dialog, select your *Expense Cost* resource and then click the **Assign** button.
4. In the *Cost* column of the *Assign Resources* dialog for the *Expense Cost* resource, manually enter the amount of anticipated expenditure and then press the **Enter** key on your computer keyboard.
5. Click the **Close** button to close the *Assign Resources* dialog.
6. Save the latest changes to your enterprise project.
7. When the *Save* job completes, **close and check in** your enterprise project.

Assigning Material Resources to Tasks

If your Project Online system includes **Material** resources, use the *Build Team* dialog to add *Material* resources to your project team, as needed. You use *Material* resources in a project to track project consumables. You can assign *Material* resources to tasks using either of two methods:

- Fixed consumption rate
- Variable consumption rate

You should assign a *Material* resource at a **fixed consumption rate** when the amount of the resource consumed does not depend upon the *Duration* of the task. For example, when you assign the *Material* resource named *Paper* at *25 Reams* to a *Project Administration* task, this action consumes 25 reams of paper regardless of whether the task takes 30 days or 12 months to complete.

You should assign a *Material* resource at a **variable consumption rate** when the amount of the resource consumed depends directly on the *Duration* of the task. For example, you assign the *Material* resource named *Gasoline* to the task *Excavate Site* at *100 gallons/day*. Using this method, you know that the excavation consumes 500 gallons of gasoline in 5 days or 2,000 gallons in 20 days.

When you assign a *Material* resource to a task, make sure that you use the correct assignment method. To assign a *Material* resource to a task, complete the following steps:

1. Apply the *Gantt Chart* view.
2. Click the **View** tab to display the *View* ribbon.
3. In the *Split View* section of the *View* ribbon, select the **Details** checkbox to apply the *Task Entry* view.
4. Select a task in the *Gantt Chart* pane.
5. In the *Resource Name* column of the *Task Form* pane, click the pick list and select a *Material* resource from the list of resources.
6. Enter the consumption for the *Material* resource in the **Units** column.

Information: To assign the *Material* resource using a fixed consumption rate, enter only a number in the *Units* field, such as **25** to show the consumption of the resource at 25 units. To assign the *Material* resource using a variable consumption rate, enter a number plus a slash symbol plus a time period in the *Units* field, such as **100/d** to show the consumption of the resource at 100 units per day.

7. Click the **OK** button.

Microsoft Project calculates the total consumption of the *Material* resource in the *Work* column of the *Task Form* pane. For example, Figure 7 - 22 shows the fixed consumption assignment of *25 reams* for the *Paper* resource assigned to the *Project Administration* task over a *Duration* of *30 days* for the task.

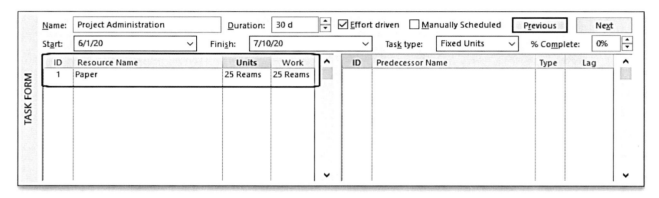

Figure 7 - 22: Material resource Paper assigned at a fixed consumption rate

Figure 7 - 23 shows the assignment after I change the *Duration* value of the task to *12 months*. Notice the number of reams of the *Paper* resource consumed ***does not change*** on this task because I assigned this *Material* resource using the fixed consumption rate method.

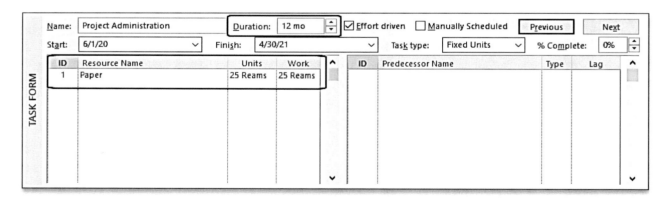

Figure 7 - 23: No change in Paper consumption at a fixed
consumption rate assignment when the Duration increases

Figure 7 - 24 shows the variable consumption rate of *100 gallons/day* applied to the *Gasoline* resource assignment on the *Excavate Site* task. Figure 7 - 25 demonstrates how the consumption amount changes after I increase the *Duration* value of the task from *5 days* to *20 days*. Notice that the amount of the *Gasoline* resource consumed changes in proportion to the *Duration* of the task, from *500 gallons* used in *5 days* to *2,000 gallons* used in *20 days*.

Assignment Planning

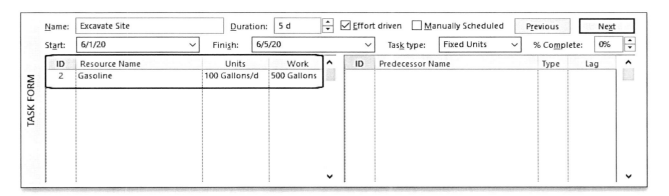

Figure 7 - 24: Material resource Gasoline assigned at a variable consumption rate

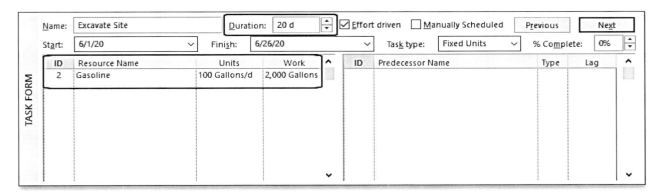

Figure 7 - 25: Consumption of the Gasoline resource changes as Duration changes

Understanding Resource Overallocation

During the resource assignment process, you may accidentally overallocate one or more resources in the project. A **resource overallocation** occurs when you assign more work to a resource than the resource can do during the working time available, resulting in a *Units* value that exceeds the *Max. Units* value for the resource. Each of the following situations can result in an overallocated resource:

- You assign a resource to work 32 hours in a single day.
- You assign a resource to work 160 hours in a single week.
- You assign a resource to work 30 minutes in a 15-minute time period.

Resource **leveling** is the process you use to resolve resource overallocations. The third bullet point above reveals an important truth about leveling overallocated resources:

Not all overallocations are worth leveling.

You should definitely level the overallocations I describe in the first two bulleted items above, because either situation would likely cause your project *Finish* date to slip. The third situation is not worth leveling, however, as the amount of time spent leveling this overallocation is not worth the effort, and likely will not impact your project.

199

Information: In addition to the three previous examples, you can also overallocate a resource by assigning the resource to a task using a *Units* value that exceeds the resource's *Max. Units* value. For example, you accidentally or intentionally overallocate a resource by assigning the resource at a *Units* value of *200%* when the *Max. Units* value for the resource is only *100%*. You cannot resolve this type of overallocation using the built-in leveling tool in Microsoft Project. Instead, you must manually resolve this type of overallocation by reducing the *Units* value for the resource's assignment on the task.

Locating and Analyzing Resource Overallocations

To begin the process of locating resource overallocations, apply any task view, such as the *Gantt Chart* view. Look for any task that displays a special overallocated resource indicator (red stick figure) in the *Indicators* column. In Figure 7 - 26, notice the overallocated resource indicators in the *Indicators* column for task IDs #13, 14, 15, and 17.

		Task Mode	Task Name	Duration
11			▲ Seek Input from Training Vendor	11 d
12			Define vendor deliverables - can they meet our needs?	1 d
13	👥		Obtain vendor commitment to training rollout schedule	3 d
14	👥		Review and customize training material	7 d
15	👥		Obtain approval for purchase orders to cover vendor invoices	2 d
16			▲ Perform End-User Skill Assessment	12 d
17	👥		Create a list of training recipients by department	2 d
18			Send skill assessment forms to department heads	2 d
19			Collect and compile results by department	5 d
20			Initiate end-user placement matrix	3 d
21			Finalize project budget	2 d

Figure 7 - 26: Overallocated resource indicators

To perform a preliminary analysis into the cause of the resource overallocations, complete the following steps:

1. Select a task that displays the overallocated resource indicator in the *Indicators* column.

2. Click the **Task** tab to display the *Task* ribbon.

3. In the *Tasks* section of the *Task* ribbon, click the **Inspect** pick list button and select the **Inspect Task** item.

On the left side of the project schedule, Microsoft Project displays the *Inspector* sidepane which reveals which resource is overallocated, along with the low-level cause of the resource overallocation. Notice in Figure 7 - 27 that the software indicates the overallocation for Renee Hensley is due to work on other tasks. Notice further that the *Inspector* sidepane ***does not reveal which other tasks*** are the source of the overallocation. To determine exactly which other tasks are the cause of the overallocation requires deeper analysis.

Assignment Planning

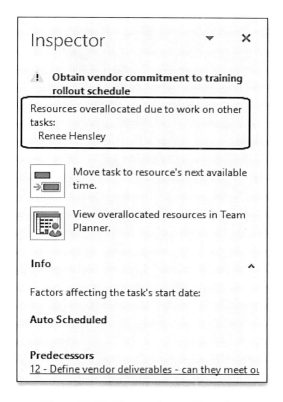

Figure 7 - 27: Cause of overallocation
for Renee Hensley

4. Continue selecting tasks that display the overallocated resource indicator in the *Indicators* column and analyzing the cause of the overallocation in the *Inspector* sidepane.

5. Click the **Close** button (**X** button) in the upper right corner of the *Inspector* sidepane.

 Warning: Even though the *Inspector* sidepane may offer you one or more ways to level the resource overallocations on the task, **do not** use any of the methods offered in the sidepane. Instead, Projility recommends that you perform further analysis into the cause of the overallocations and then use a different method for leveling the overallocations.

To perform a deeper analysis on the cause of your project's overallocated resources, use the *Resource Usage* view. To apply this view, click the **Resource** tab to display the *Resource* ribbon. In *View* section of the *Resource* ribbon, click the **Team Planner** pick list button and select the **Resource Usage** view.

Because Microsoft Project formats overallocated resources with the red font color, look for any resource names formatted in red and which display the overallocated resource indicator in the *Indicators* column. To determine the time periods during which resource is overallocated, select an overallocated resource. In the *Level* section of the *Resource* ribbon, click the **Next Overallocation** button. When you click this button, Microsoft Project scrolls the timephased grid and selects the start of the first resource overallocation, such as shown in Figure 7 - 28.

Module 07

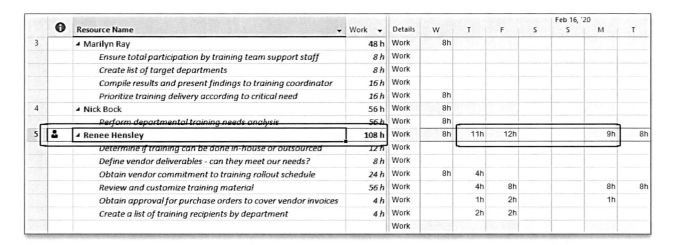

Figure 7 - 28: Resource Usage view shows first overallocation

In Figure 7 - 28, notice that I assigned Renee Hensley to work more than 8 hours each day in a 3-day period during the weeks of February 9 and February 16. I accidentally caused this overallocation when I assigned Renee to work full-time (*Units* value of *100%*) on two or more parallel tasks during each of those three days.

As you continue to click the **Next Overallocation** button, Microsoft Project selects the start of each successive overallocation. When the Microsoft Project cannot locate any more resource overallocations, it displays the dialog shown in Figure 7 - 29. Click the **OK** button to close the dialog.

Figure 7 - 29: No more resource overallocations

 Information: To analyze resource overallocations at an even deeper level, you can temporarily insert the *Max. Units*, *Assignment Units*, and *Peak* fields to the right of the *Resource Name* column. When the *Assignment Units* value exceeds the *Max. Units* value, this means that you must level the overallocation manually by adjusting the *Assignment Units* value so that it is less than or equal to the *Max. Units* value. When the *Peak* value exceeds the *Max. Units* value, you can use multiple methods to resolve the overallocation, including using the built-in leveling tool in Microsoft Project.

Assignment Planning

Hands On Exercise

Exercise 7 - 6

Locate and analyze resource overallocations.

1. Open the **Level Resource Overallocations 2019.mpp** sample file from your student sample files folder. The project file opens Read-Only.

2. Study the project and look for the overallocated resource indicators in the *Indicators* column for tasks with overallocated resources assigned.

3. Select task ID #4, the *Test 1* task.

4. Click the **Task** tab to display the *Task* ribbon.

5. In the *Tasks* section of the *Task* ribbon, click the **Inspect** pick list button and select the **Inspect Task** item.

In the *Inspector* sidepane, notice that Amy McKay is overallocated on this task because the project manager assigned her at a *Units* value of *100%* (shown in the sidepane as the *Peak Unit* value), which is greater than her *Max. Units* value of *50%*.

6. Individually click task IDs #8, 9, 13, and 14, and study the overallocation information displayed in the Inspector sidepane.

7. Click the **Close** button (**X** button) in the upper right corner of the *Inspector* sidepane to close the sidepane.

8. Click the **Resource** tab to display the *Resource* ribbon.

9. In the *View* section of the *Resource* ribbon, click the **Team Planner** pick list button and select the **Resource Usage** view.

Notice that this customized *Resource Usage* view contains the *Max. Units*, *Assignment Units*, and *Peak* columns to the right of the *Resource Name* column. Notice by the red font formatting that this project reveals three overallocated resources.

10. For each task assigned to an overallocated resource, compare the *Assignment Units* values with the *Max. Units* value for the resource.

11. In the *Level* section of the *Resource* ribbon, repeatedly click the **Next Overallocation** button and study the periods of overallocation for each overallocated resource.

12. Click the **OK** button in the dialog indicating no more overallocations.

13. Leave the **Level Resource Overallocations 2019.mpp** sample project open for the next exercise.

Leveling Overallocated Resources

As I previously stated, leveling is the process you use to resolve resource overallocations. There are many ways to level overallocated resources in Microsoft Project, including each of the following:

- Substitute an available resource for the overallocated resource.
- Increase the availability of overallocated resources.
- Schedule overtime for the overallocated resource.
- Manually delay tasks with overallocated resources.
- Delay the start of a resource assignment on a task.
- Adjust the project schedule using task constraints to eliminate resource assignment conflicts.
- Use a task split to interrupt the work on a task to make resources available for other assignments.
- Adjust dependencies and add *Lag* time.
- Add resources to an *Effort Driven* task to shorten the duration of the task.
- Look for potential overlapping work opportunities, such as Finish-to-Start dependencies that do not have a true "finish to start" relationship.
- Negotiate with your project sponsor or customer to delay the *Finish* date of the project.
- Negotiate with your project sponsor or customer to reduce the feature set (scope) of the project.
- Use the built-in leveling tool in Microsoft Project.

Notice that using the built-in leveling tool in Microsoft Project appears last on the preceding list! Each of the preceding leveling methods is powerful and useful for leveling overallocated resources; however, most users assume the only way to level is to use the built-in leveling tool found in Microsoft Project. Given the complexity of using the software's leveling capabilities, the average user of Microsoft Project is far better off using any of the other manual leveling methods. The key to using any method for resource leveling is to remember that you must take complete control of all leveling decisions.

Using a Leveling Methodology

Many Microsoft Project users attempt to level all of their overallocated resources simultaneously in the *Gantt Chart* view using the built-in leveling tool. Although this approach can work in some situations, most often it leads to frustration. This approach does not give insight into how the software actually leveled the overallocations, and may lead to failure because you did not take control over the leveling process. A much better approach is to level overallocated resources using the following methodology:

1. Level each overallocated resource individually in the *Resource Usage* view.
2. Study the results of the leveling process in the *Leveling Gantt* view.
3. Clear unacceptable leveling results and then level the overallocated resource using any other method.
4. Repeat steps #1-3 for each overallocated resource.

Setting Leveling Options

Before you begin the process of leveling overallocated resources in the *Resource Usage* view, you should specify your leveling options by clicking the **Leveling Options** button in the *Level* section of the *Resource* ribbon. Microsoft Project displays the *Resource Leveling* dialog shown in Figure 7 - 30.

In the *Leveling calculations* section of the *Resource Leveling* dialog, you find the *Look for overallocations on a _____ basis* option. This option controls the granularity used by Microsoft Project to determine whether resources are overallocated. The default setting is *Day by Day*, which means that the software only shows you overallocations where resources are overallocated in full day time periods. Click the **Look for overallocations on a _____ basis** pick list and select any other option, as needed. The finest level of granularity is the *Minute by Minute* setting and the coarsest level of granularity is the *Month by Month* setting. When Microsoft Project locates a resource overallocation at the granularity level you specify, the software formats the name of the resource in red, and displays a special indicator in the *Indicators* column to the left of the resource name. If you float your mouse pointer over the indicator, the software simply displays a *This resource is overallocated* message in the tooltip.

 Warning: Be very cautious in using either the *Minute by Minute* or *Hour by Hour* settings in the *Look for overallocations on a _____ basis* pick list. Using either setting forces the software to show many overallocations that are not worth the time and effort to level, including overallocations in 15-minute time periods.

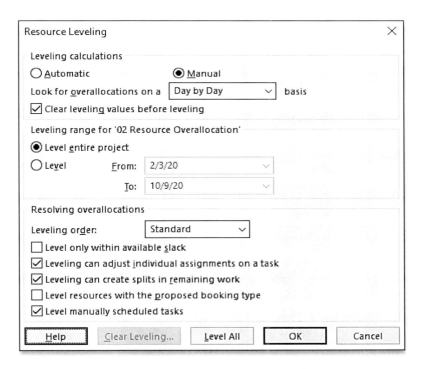

Figure 7 - 30: Resource Leveling dialog

In the *Resource overallocations* section of the *Resource Leveling* dialog, there are several options that you may want to change from the default setting, including:

- Set the **Leveling order** option to the **Priority, Standard** value. By selecting this option, you force the software to consider first the *Priority* number of each task in the software's algorithm of five leveling factors. The additional leveling factors include whether the task is a *Predecessor* to other tasks, the *Start* date of each task, the *Total Slack* value for each task, and whether the task has an inflexible constraint.

- Select the **Level only within available slack** option to guarantee that the leveling operation does not change the *Finish* date of your project. Using this option, Microsoft Project levels overallocations until it reaches the point where it must delay the *Finish* date of your project. At this point, the system discontinues the leveling process and displays a warning dialog. From this point forward, you must select an alternate method for leveling remaining overallocations.

- The **Leveling can adjust individual assignments on a task** option control how Microsoft Project handles leveling on tasks that have multiple resource assignments. If one of the resources is overallocated and another resource is not, selecting this option allows the software to set an *Assignment Delay* on the overallocated resource's work on the task, while leaving untouched the assignment for the non-overallocated resource. Deselecting this option allows Microsoft Project to set a *Leveling Delay* on the entire task, delaying work for both resources. Select this option if the multiple resources assigned to a task can work independently of one another. Deselect this option if the resources must work together at the same time.

- The **Leveling can create splits in remaining work** option allows Microsoft Project to "break up" or split the work for an overallocated resource to resolve the resource overallocation. In essence, a task split causes work to start, stop completely, and then start again. Select this option if work on a task can start and stop without consequences. Deselect this option if the work on a task must run without interruption from the *Start* date to the *Finish* date of the task.

- If your project contains *Manually Scheduled* tasks with overallocated resources, and you want to reschedule the tasks manually to resolve these overallocations, then you should *deselect* the **Level manually scheduled tasks** option. If you leave this option selected, Microsoft Project delays or splits any *Manually Scheduled* tasks with overallocated resources assigned to them. Keep in mind, however, that Projility recommends that you *do not* use *Manually Scheduled* tasks in your enterprise projects.

> **Warning**: Projility recommends that you *never* select the *Automatic* option in the *Leveling Calculations* section of the *Resource Leveling* dialog. When applied, the *Automatic* leveling option causes Microsoft Project to level all overallocated resources automatically the moment an overallocation occurs. It does this in all open projects without asking your permission! This means you lose control over the leveling process, which is never a good idea.

After you select your leveling options in the *Resource Leveling* dialog, click the **OK** button. Microsoft Project saves your option selections in this dialog so that you do not need to reselect them every time you want to level resource overallocations. The beauty of the *Resource Leveling* dialog, however, is that you can change the options in this dialog for each project or even each overallocated resource, as needed.

> **Warning**: *Do not* click the *Level All* button in the *Resource Leveling* dialog. If you click this button, you lose control over the leveling process because the software levels all the overallocated resources in your project in a single operation.

Assignment Planning

 Information: If you want to level an overallocated resource across multiple projects, you must open each of these enterprise projects before you begin the leveling process.

 Hands On Exercise

Exercise 7 - 7

Set leveling options in your Microsoft Project application.

1. In the *Level* section of the *Resource* ribbon, click the **Leveling Options** button.

2. In the *Resource Leveling* dialog, specify the following leveling options:

    - *Leveling calculations* – leave the **Manual** option selected
    - *Look for overallocations on a _____ basis* – select the **Day by Day** option
    - *Clear leveling values before leveling* – **deselect** this option
    - *Leveling range* – select the **Level entire project** option
    - *Leveling order* – select the **Priority, Standard** option
    - *Level only within available slack* – leave this option deselected
    - *Leveling can adjust individual assignments on a task* – **deselect** this option
    - *Leveling can create splits in remaining work* – **deselect** this option
    - *Level resources with the proposed booking type* – leave this option deselected
    - *Level manually scheduled tasks* – **deselect** this option

3. Click the **OK** button to close the *Resource Leveling* dialog.

Module 07

Leveling an Overallocated Resource

Best Practice: Prior to leveling overallocated resources, Projility recommends that you exit Microsoft Project and then re-launch the application. In the *Login* dialog, **deselect** the **Load Summary Resource Assignments** checkbox and then click the **OK** button. With this option deselected, the system focuses the leveling process on only those projects currently open and does not include assignments from other projects not currently open.

To start the process of leveling overallocated resources, select the most critical resource in the project. Your most critical resource is the one whose skills and availability are the most limited in your organization. After selecting this resource, in the *Level* section of the *Resource* ribbon, click the **Level Resource** button. Microsoft Project displays the *Level Resources* dialog, such as the one shown in Figure 7 - 31.

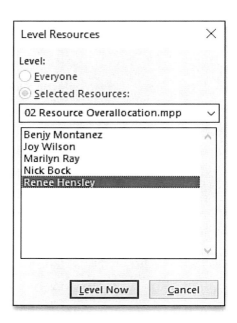

Figure 7 - 31: Level Resources dialog

The *Level Resources* dialog selects the same resource you selected in the *Resource Usage* view. Click the **Level Now** button in the dialog to level the overallocations for the first selected resource using the leveling options you set in the *Resource Leveling* dialog. When you use the built-in leveling tool to level an overallocated resource, Microsoft Project resolves the overallocation using one or both of the following methods:

- The software delays tasks or assignments.
- The software splits tasks or assignments.

Viewing Leveling Results

To see the results of leveling the first overallocated resource, you must apply the *Leveling Gantt* view. The best way to apply the *Leveling Gantt* view is to open a new window containing this view. This process leaves the *Resource Usage* view displayed in one window and the *Leveling Gantt* view displayed in another window. You can use the **Alt + Tab** keyboard shortcut to switch back and forth between the windows as you level an overallocation in the

Assignment Planning

Resource Usage window and view the results in the *Leveling Gantt* window. To display the *Leveling Gantt* view in a new window, complete the following steps:

1. Click the **View** tab to display the *View* ribbon.

2. In the *Window* section of the *View* ribbon, click the **New Window** button. Microsoft Project displays the *New Window* dialog.

3. Click the **View** pick list button and select the **Leveling Gantt** item on the list, shown in Figure 7 - 32.

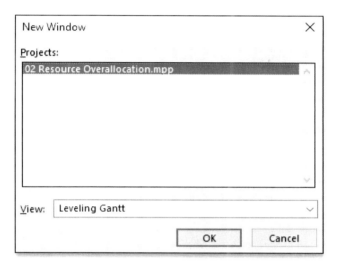

Figure 7 - 32: New Window dialog

4. Click the **OK** button. Microsoft Project displays the *Leveling Gantt* view shown in Figure 7 - 33.

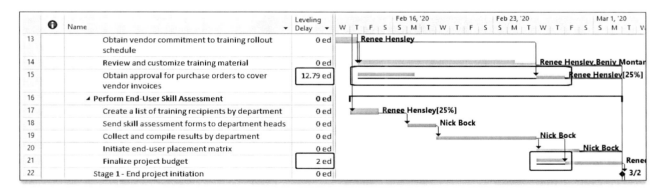

Figure 7 - 33: Leveling Gantt view

The *Leveling Gantt* view includes the *Delay* table on the left side and a special *Gantt Chart* pane on the right. The symbols used in the *Leveling Gantt* view are as follows:

- The **Tan Gantt bars** represent the pre-leveled schedule for each task assigned to the overallocated resource. Figure 7 - 33 shows multiple tasks assigned to Renee Hensley, several of which run in parallel.

209

- The **Light Blue Gantt bars** represent the schedule of the tasks after the software levels the resource overallocation. Figure 7 - 33 shows that Microsoft Project delayed two tasks, which the resource overallocation for Renee Hensley.

- The **Brown underscore** to the left of any Gantt bar represents the amount of delay applied to the task schedule to level the resource overallocation. Figure 7 - 33 shows the delay symbol to the left of the Gantt bar for tasks ID #15 and #21.

- The **Teal underscore** to the right of any Gantt bar represents the amount of time you can delay the task without delaying the *Finish* date of the entire project.

The *Delay* table contains the *Leveling Delay* column to the right of the *Task Name* column. This column shows the amount of delay the software applies to a task to level a resource overallocation. By default, Microsoft Project measures the amount of *Leveling Delay* in **elapsed days** (displayed as **edays** or **ed**). Each elapsed day is a 24-hour calendar day that ignores nonworking time, such as weekends and holidays.

Clearing Leveling Results

As you study the results of leveling an overallocated resource, you may find that Microsoft Project did not level the overallocation as you expected. In these situations, you must clear the unacceptable leveling result and then level using another method. To clear an unacceptable overallocation result, complete the following steps:

1. Click the **Resource** tab to display the *Resource* ribbon.

2. Select any tasks leveled in an unacceptable manner.

3. In the *Level* section of the *Resource* ribbon, click the **Clear Leveling** button. Microsoft Project displays the *Clear Leveling* dialog shown in Figure 7 - 34.

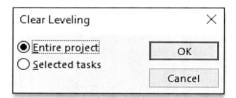

Figure 7 - 34: Clear Leveling dialog

4. In the *Clear Leveling* dialog, select the **Selected tasks** option and then click the **OK** button.

The software sets the *Leveling Delay* value back to the default value to *0d* for each selected task. At this point, you must level the resource overallocation using another method. You have many options available to you, including using one of the manual leveling methods I previously discussed. Another option is to set a *Priority* number on tasks showing the relative importance of each task, and then relevel the overallocations in the *Resource Usage* view.

Setting Task Priority Numbers

When you set task *Priority* numbers for tasks assigned to an overallocated resource, Microsoft Project levels the resource overallocation by factoring in task *Priority* numbers you assign. The software delays tasks with lower *Priority* numbers while maintaining the original schedule of the task with the highest *Priority* number. To set a *Priority* number on tasks with overallocated resources assigned, complete the following steps:

Assignment Planning

1. In the *Leveling Gantt* view (or any task view), double-click a task assigned to an overallocated resource. Microsoft Project displays the *Task Information* dialog shown in Figure 7 - 35.

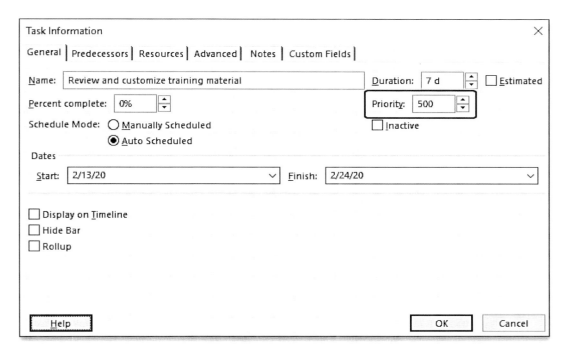

Figure 7 - 35: Task Information dialog,
Set a Priority number

2. Click the *General* tab, if necessary.
3. Set a value between **0** and **1000** in the **Priority** field.

Remember that a *0* value signifies the lowest priority and a *1000* value signifies the highest priority for any task. The default value is *500* which means that the software consider all tasks equally when leveling overallocations.

4. Click the **OK** button.
5. Repeat steps #1-4 for each task to which you assigned the overallocated resource.

When setting *Priority* numbers on multiple tasks, be sure to specify a **unique** *Priority* number for each task. After setting task *Priority* numbers, return to the *Resource Usage* window and level the overallocated resource again.

An alternate method for setting *Priority* numbers to tasks is to insert the *Priority* field temporarily in the *Leveling Gantt* view by completing the following steps:

1. Right-click the *Leveling Delay* column header, select the **Insert Column** item on the shortcut menu.
2. In the pick list of available fields, select the **Priority** field.
3. Enter values in the *Priority* column for each task with an overallocated resource assigned.

 Information: Not only can you set a *Priority* number at the task level; Microsoft Project also allows you to set a *Priority* number at the project level. To level across multiple projects, open each project, display the *Project Information* dialog, specify a value in the **Priority** field for the project, and then click the **OK** button.

211

Hands On Exercise

Exercise 7 - 8

Level resource overallocations, view leveling results, clear leveling, and set task *Priority* numbers.

1. For the *Test 1* task assigned to *Amy McKay*, change the **Assignment Units** value to **50%** to match her *Max. Units* value.

> **Quiz Question:** Notice that this action resolves the overallocation for Amy McKay, but how does this change impact the schedule of the *Test 1* task?

2. For the overallocated resource, *Sarah Baker*, compare her *Peak* value with her *Max. Units* value, and determine the cause of the overallocation.

3. Scroll the timephased grid and look for periods of overallocation for *Sarah Baker*.

4. Select the name, **Sarah Baker**, and then click the **Level Resource** button in the *Level* section of the *Resource* ribbon.

5. In the *Level Resources* dialog, click the **Level Now** button.

Notice that Sarah Baker is no longer overallocated.

6. Click the **View** tab to display the *View* ribbon.

7. In the *Window* section of the *View* ribbon, click the **New Window** button.

8. In the *New Window* dialog, click the **View** pick list and select the **Leveling Gantt** view.

9. In the *New Window* dialog, click the **OK** button.

10. In the *Leveling Gantt* view, notice that Microsoft Project applied a *Leveling Delay* of *5 edays* to the *Test 4* task to resolve Sarah Baker's overallocation.

11. Press the **Alt + Tab** keyboard shortcut and select the window that displays the **Resource Usage** view.

12. For the overallocated resource, *Randy Parker*, compare his *Peak* value with his *Max. Units* value, and determine the cause of the overallocation.

13. Scroll the timephased grid and look for periods of overallocation for *Randy Parker*.

14. Click the **Resource** tab to display the *Resource* ribbon.

15. Select the name, **Randy Parker**, and then click the **Level Resource** button in the *Level* section of the *Resource* ribbon.

16. In the *Level Resources* dialog, click the **Level Now** button.

Assignment Planning

Notice that Randy Parker is no longer overallocated, but that Sarah Baker is overallocated again. This is a situation you may face with leveling resource overallocations in your own projects, a situation that may require several passes of leveling before you finally resolve all overallocations in the project.

17. Press the **Alt + Tab** keyboard shortcut and select the window that displays the **Leveling Gantt** view.

Notice that Microsoft Project applied a *Leveling Delay* of *8 edays* to the *Build 2* and another *Leveling Delay* of *12 edays* to the *Build 3* task to resolve the overallocation for Randy Parker. You have determined that this leveling result is not acceptable.

18. Select the *Build 2* task, press and hold the **Control** key on your computer keyboard, then select the *Build 3* task as well.
19. In the *Level* section of the *Resource* ribbon, click the **Clear Leveling** button.
20. In the *Clear Leveling* dialog, choose the **Selected tasks** option, and then click the **OK** button.
21. Double-click the *Build 2* task.
22. In the *Task Information* dialog, set the **Priority** value to **600**, and then click the **OK** button.
23. Press the **Alt + Tab** keyboard shortcut and select the window that displays the **Resource Usage** view.
24. Select the name, **Randy Parker**, and then click the **Level Resource** button in the *Level* section of the *Resource* ribbon.
25. In the *Level Resources* dialog, click the **Level Now** button.

Notice that Randy Parker is no longer overallocated and that Sarah Baker is not overallocated either.

26. Press the **Alt + Tab** keyboard shortcut and select the window that displays the **Leveling Gantt** view.

Notice that Microsoft Project applied a *Leveling Delay* of *21 edays* to the *Design 3* task to resolve the overallocation for Randy Parker.

27. Close both of the windows in the **Level Resource Overallocations 2019.mpp** sample project and *do not* save the changes.

Creating Resource Engagements for Human Resources

Your organization may have a methodology that requires project managers to secure formal permission from resource managers before using a human resource in an enterprise project. If this describes your organization, you can use Resource Engagements with human resources to create an "electronic paper trail" of the resource request and approval process.

The steps needed for creating Resource Engagements with human resources is nearly identical to the process for creating them with Generic resources, but with a few differences. In the interest of brevity, I omit the screenshots shown previously in Module 04 for creating Resource Engagements with Generic resources, but include several new screenshots. To create Resource Engagements for the human resources in your enterprise project, complete the following steps:

1. Click the **View** tab to display the *View* ribbon.
2. In the *Resource Views* section of the *View* ribbon, click the **Resource Plan** button to display the *Resource Plan* view.
3. Right-click on the **Proposed Start** column header and select the **Insert Column** item on the shortcut menu.
4. In the list of available resource columns, select the **Start** column.
5. Right-click on the **Proposed Finish** column header and select the **Insert Column** item on the shortcut menu.
6. In the list of available resource columns, select the **Finish** column.
7. Right-click on the **Proposed Max Units** column header and select the **Insert Column** item on the shortcut menu.
8. In the list of available resource columns, select the **Peak** column.

Figure 7 - 36 shows the customized *Resource Plan* view with the three additional columns. The *Start* column displays the *Start* date of the earliest-starting task assigned to each resource. The *Finish* column displays the *Finish* date of the latest-finishing task assigned to each resource. The *Peak* column shows the largest *Units* value for any task assigned to each resource. You can use the values in these three columns to define your Resource Engagements for each of your human resources, as needed.

		Name	Start	Proposed Start	Finish	Proposed Finish	Peak	Proposed Max Units	Engagement Status
1		Joy Wilson	12/9/19	NA	2/12/20	NA	100%		
2		Carole Madison	11/13/19	NA	3/9/20	NA	100%		
3		Myrta Hansen	12/9/19	NA	2/17/20	NA	100%		
4		Ruth Andrews	12/9/19	NA	1/21/20	NA	100%		
5		Sharon Brignole	10/7/19	NA	3/9/20	NA	100%		
6		Richard Sanders	11/12/19	NA	11/12/19	NA	100%		
7		Bob Siclari	11/14/19	NA	2/18/20	NA	100%		
8		Gary Manche	10/23/19	NA	3/5/20	NA	100%		

Figure 7 - 36: Customized Resource Plan view

9. Click the **Engagements** tab to display the *Engagements* ribbon.
10. Select the name of a human resource that requires resource manager approval for use in your enterprise project.
11. In the *Engagements* section of the *Engagements* ribbon, click the **Add Engagement** button.
12. In the **Description** field, enter an accurate description of the work to be performed by the resource.

Assignment Planning

Best Practice: If your organization requires resource managers to formally approve the use of the human resource using a Resource Engagement, Projility strongly recommends that you provide an accurate description of the resource's work in the *Description* field. The resource manager can use this description information to determine whether to grant the formal approval.

13. In the **Start** field, enter the date the resource begins work in the project, and in the **Finish** field, enter the date the resource finishes work in the project.

14. In the **Units** field, select or manually enter a value that represents the percentage of the resource's working time to the project work.

Information: In the *Allocate Resources by* section of the *Engagement Information* dialog, you can select the *Work* option and then enter a *Work* estimate measured in hours, rather than as a percentage value using the *Units* field. If you select the *Work* value, you must enter the total number of hours of work performed by the resource over the time span between the *Start* and *Finish* dates you enter.

15. In the **Comments** field, enter an optional comment for the resource manager.

Figure 7 - 37 shows the completed *Engagement Information* dialog for Joy Wilson. Notice how the *Start* and *Finish* dates I entered loosely correspond to the dates in her *Start* and *Finish* columns, as shown previously in Figure 7 - 36.

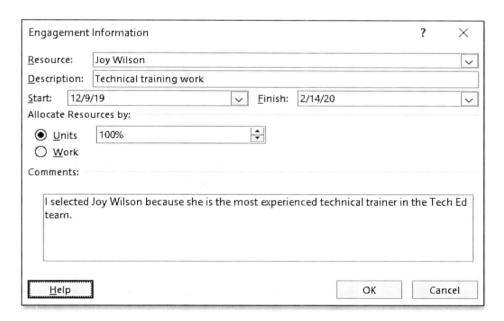

Figure 7 - 37: Engagement Information dialog

16. Click the **OK** button to create the Resource Engagement for the human resource.

215

Module 07

Information: To edit an existing Resource Engagement, double-click the engagement or click the *Information* button in the *Properties* section of the *Engagements* ribbon for the selected engagement.

17. Continue creating Resource Engagements for each of the human resources that require resource manager approval.
18. In the *Engagements* section of the *Engagements* ribbon, click the **Submit** pick list button and select the **Submit All Engagements** item.
19. Save the changes to your enterprise project.

Hands On Exercise

Exercise 7 - 9

Create a Resource Engagement for one of the human resources in your enterprise project.

1. Click the **Open** button in your Quick Access Toolbar.
2. In the *Open* dialog, select the enterprise project you created for this class, and then click the **Open** button.
3. In the *READ-ONLY* band at the top of the project schedule, click the **Check Out** button to check out the project for editing.
4. Click the **View** tab to display the *View* ribbon.
5. In the *Resource Views* section of the *View* ribbon, click the **Resource Plan** button to display the *Resource Plan* view.
6. Drag the split bar to the right side of the *Add New Column* virtual column.
7. Right-click on the **Proposed Start** column header and select the **Insert Column** item on the shortcut menu.
8. In the list of available resource columns, select the **Start** column.
9. Right-click on the **Proposed Finish** column header and select the **Insert Column** item on the shortcut menu.
10. In the list of available resource columns, select the **Finish** column.
11. Right-click on the **Proposed Max Units** column header and select the **Insert Column** item on the shortcut menu.
12. In the list of available resource columns, select the **Peak** column.
13. Drag the split bar to the right side of the *Proposed Max Units* column.

14. Click the **Engagements** tab to display the *Engagements* ribbon.

15. Select the name of one of your fellow human resources on the project team.

16. In the *Engagements* section of the *Engagements* ribbon, click the **Add Engagement** button.

17. In the **Description** field, enter a description of the work to be performed by the resource.

18. Enter dates in the **Start** and **Finish** fields that roughly correspond with the dates shown in the *Start* and *Finish* columns for the selected resource.

19. In the **Units** field, enter the value shown in the *Peak* column for the selected resource.

20. In the *Comments* field, enter an optional comment for the resource manager.

21. Click the **OK** button to create the Resource Engagement for the human resource.

22. In the *Engagements* section of the *Engagements* ribbon, click the **Submit** pick list button and select the **Submit All Engagements** item.

23. Save the changes to your enterprise project.

Module 08

Project Execution

Learning Objectives

After completing this module, you will be able to:

- Analyze the Critical Path in a project
- Save a baseline and back up a baseline for historical purposes
- Understand the publishing process
- Work with project Deliverables

Inside Module 08

Analyzing the Critical Path **221**
    Viewing the "Nearly Critical" Path *223*
Working with Project Baselines **224**
    Saving a Project Baseline *225*
    Backing Up a Baseline *226*
    Clearing the Project Baseline *227*
Understanding Publishing **228**
    Publishing an Enterprise Project *229*
    Setting Custom Permissions for a Project *230*
Working with Project Deliverables **235**
    Adding a New Deliverable in Microsoft Project *235*
    Editing and Deleting Deliverables in Microsoft Project *239*
    Viewing Deliverables in Project Web App *242*
    Adding a New Deliverable Dependency *244*
    Updating the Baseline and Republishing the Project *247*

Analyzing the Critical Path

Microsoft Project defines the **Critical Path** as "The series of tasks that must be completed on schedule for a project to finish on schedule." Every task on the Critical Path is a **Critical task**. By default, all tasks on the Critical Path have a *Total Slack* value of *0 days*, which means they cannot slip without delaying the project *Finish* date. If the *Finish* date of any Critical task slips by even 1 day, the project *Finish* date slips as well.

A **non-Critical task** is any task that is 100% complete or that has a *Total Slack* value greater than *0 days*. A non-Critical task can slip by its amount of *Total Slack* before it affects the *Finish* date of the project. For example, if a task has *5 days* of *Total Slack*, the task must finish more than 5 days late before the resulting slippage changes the project *Finish* date. To manage your project well, you should focus your energy on managing the tasks on the Critical Path, but you should also be aware of the non-Critical tasks in your project as well.

Information: Microsoft Project automatically calculates the *Total Slack* value for each task to determine the Critical Path of the project. To view the *Total Slack* for any task, click the **View** tab to display the *View* ribbon. In the *Data* section of the *View* ribbon, click the **Tables** pick list button, and then select the **Schedule** table. The *Total Slack* column is the last column on the right side of the *Schedule* table.

In Microsoft Project, the Critical Path may run from the *Start* date to the *Finish* date of the project, or it may begin anywhere in the project and run to the *Finish* date of the project. This behavior is a key difference from the traditional Critical Path Method (CPM) definition of the Critical Path.

Information: If you make changes to your project, either by entering actual progress or by making plan revisions, keep in mind that the Critical Path may change as a consequence.

Although there are a number of ways to determine the Critical Path in any project in Microsoft Project, the simplest method is to format the *Gantt Chart* view to display red Gantt bars for Critical tasks. To format the *Gantt Chart* view to show the Critical Path, complete the following steps:

1. Apply the **Gantt Chart** view.

2. Click the **Format** tab to display the *Format* ribbon.

3. In the *Bar Styles* section of the *Format* ribbon, select the **Critical Tasks** checkbox.

In the formatted *Gantt Chart* view shown in Figure 8 - 1, notice that Microsoft Project displays the following:

- Red bars represent Critical tasks on the Critical Path. These tasks have a *Total Slack* value of *0 days*.

- Blue bars represent non-Critical tasks. These tasks have a *Total Slack* value greater than *0 days*.

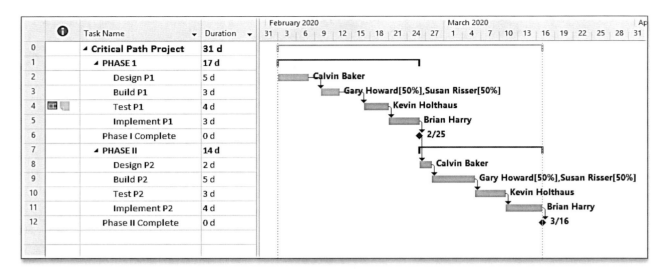

Figure 8 - 1: Gantt Chart view shows Critical Path and Total Slack

In the *Gantt Chart* view shown previously in Figure 8 - 1, the Test P1 task has a Start No Earlier Than constraint applied to it, which creates slack between it and the Build P1 predecessor task. Because of this, Microsoft Project calculates a *Total Slack* value greater than *0 days* for the Design P1 and Build P1 tasks, and then formats their Gantt bars in blue. You can confirm the *Total Slack* values for these two non-Critical tasks in the *Schedule* table shown in Figure 8 - 2. Notice that each of these two tasks has *2 days* of *Total Slack*, which means that they are non-Critical tasks.

	Task Name	Start	Finish	Late Start	Late Finish	Free Slack	Total Slack	Add New Column
0	⊿ Critical Path Project	2/3/20	3/16/20	2/5/20	3/16/20	0 d	0 d	
1	⊿ PHASE 1	2/3/20	2/25/20	2/5/20	2/25/20	0 d	0 d	
2	Design P1	2/3/20	2/7/20	2/5/20	2/11/20	0 d	2 d	
3	Build P1	2/10/20	2/12/20	2/12/20	2/14/20	2 d	2 d	
4	Test P1	2/17/20	2/20/20	2/17/20	2/20/20	0 d	0 d	
5	Implement P1	2/21/20	2/25/20	2/21/20	2/25/20	0 d	0 d	
6	Phase I Complete	2/25/20	2/25/20	2/26/20	2/26/20	0 d	0 d	
7	⊿ PHASE II	2/26/20	3/16/20	2/26/20	3/16/20	0 d	0 d	
8	Design P2	2/26/20	2/27/20	2/26/20	2/27/20	0 d	0 d	
9	Build P2	2/28/20	3/5/20	2/28/20	3/5/20	0 d	0 d	
10	Test P2	3/6/20	3/10/20	3/6/20	3/10/20	0 d	0 d	
11	Implement P2	3/11/20	3/16/20	3/11/20	3/16/20	0 d	0 d	
12	Phase II Complete	3/16/20	3/16/20	3/16/20	3/16/20	0 d	0 d	

Figure 8 - 2: Schedule table reveals non-Critical tasks

 Information: You can also view the Critical Path in any project by applying the *Tracking Gantt* view. Be aware, however, that the *Tracking Gantt* view displays Gantt bars for both the Critical Path and for the baseline schedule of the project. Remember that red Gantt bars show Critical tasks, blue Gantt bars show non-Critical tasks, and gray Gantt bars show the original baseline schedule of each task.

Project Execution

Viewing the "Nearly Critical" Path

In Microsoft Project, you can change the software's definition of a Critical task by clicking the **File** tab and then clicking the **Options** tab in the *Backstage*. In the *Project Options* dialog, click the **Advanced** tab and then scroll down to the *Calculation options for this project* section at the bottom of the dialog. To change the software's definition of a Critical task, change the **Tasks are critical if slack is less than or equal to** option to a value *greater than 0 days*, and then click the **OK** button.

Using this technique is a helpful way to see the "nearly Critical tasks" in your project. "Nearly Critical tasks" are those tasks that are not on the true Critical Path, but are close enough to impact the *Finish* date of the project if they slip by an amount greater than their *Total Slack* value. For example, I have a task with only *1 day* of *Total Slack*, so this task is not a true Critical task since it has a *Total Slack* value greater than *0 days*. However, if this task slips by only 2 days, the *Finish* date of the project must slip by 1 day as a consequence. Therefore, it is not a bad idea to identify the "nearly Critical tasks" in any enterprise project.

Hands On Exercise

Exercise 8 - 1

Format the *Gantt Chart* view to display the Critical Path in your enterprise project.

1. Click the **Task** tab to display the *Task* ribbon.
2. In the *View* section of the *Task* ribbon, click the **Gantt Chart** pick list button and select the **Gantt Chart** view.
3. Click the **Format** tab to display the *Format* ribbon.
4. In the *Bar Styles* section of the *Format* ribbon, select the **Critical Tasks** checkbox.
5. Study the Critical Path in your enterprise project.

Notice that the Critical Path does not begin with the first task in the project schedule as you might expect. Can you explain why task IDs #1, 2, 7, 8, and 9 are non-Critical tasks? And why are all of the other tasks Critical tasks?

6. Click the **View** tab to display the *View* ribbon.
7. In the *Data* section of the *View* ribbon, click the **Tables** pick list button and select the **Schedule** table.
8. Widen the *Task Name* column to fit the column contents and then drag the split bar to the right edge of the *Total Slack* column.

Notice that task IDs #1, 2, 7, 8, and 9 all have a *Total Slack* value greater than *0 days*, which means that they are non-Critical tasks. Notice that all of the other tasks have a *Total Slack* value of *0 days*, which means that they are all Critical tasks.

9. In the *Data* section of the *View* ribbon, click the **Tables** pick list button and select the **Entry** table.
10. Drag the split bar to the right edge of the *Duration* column.
11. Save the latest changes to your enterprise project.

Working with Project Baselines

Prior to executing a project, you should save a **baseline** for your project. A baseline represents a snapshot of the work, cost, and schedule estimates in your initial project plan. Your baseline should represent the schedule your stakeholders approved before you begin tracking progress. All of the variance that Microsoft Project calculates is dependent on the existence of a baseline. Saving a project baseline provides you with a way to analyze project variance by comparing the current state of the project against the original planned state of the project (the baseline).

When you save a baseline in Microsoft Project, the software captures the current values for five important task fields and two important resource fields, and then saves these values in a corresponding set of baseline fields. Table 8 - 1 shows the original fields and their corresponding baseline fields.

Data Type	Field	Baseline Field
Task	Duration	Baseline Duration
Task	Start	Baseline Start
Task	Finish	Baseline Finish
Task	Work	Baseline Work
Task	Cost	Baseline Cost
Resource	Work	Baseline Work
Resource	Cost	Baseline Cost

Table 8 - 1: Baseline information

In addition to the five important task fields captured in the baseline, Microsoft Project also captures some additional task information that includes:

- Task cost information in the *Fixed Cost* and *Fixed Cost Accrual* fields, saving this information in the *Baseline Fixed Cost* and *Baseline Fixed Cost Accrual* fields, respectively
- Task and resource budget information in the *Budget Cost* and *Budget Work* fields, saving this information in the *Baseline Budget Cost* and *Baseline Budget Work* fields, respectively
- Schedule information for Deliverables in the *Deliverable Start* and *Deliverable Finish* fields, saving this information in the *Baseline Deliverable Start* and *Baseline Deliverable Finish* fields

Project Execution

Information: Microsoft Project also saves the timephased *Work* and *Cost* information for both tasks and resources in the timephased *Baseline Work* and *Baseline Cost* fields. You can view these timephased values in the timephased grid portion of either the *Task Usage* or *Resource Usage* views.

Saving a Project Baseline

To save a baseline for the entire project in Microsoft Project, complete the following steps:

1. Click the **Project** tab to display the *Project* ribbon.

2. In the *Schedule* section of the *Project* ribbon, click the **Set Baseline** pick list button and select the **Set Baseline** item. The software displays the *Set Baseline* dialog shown in Figure 8 - 3.

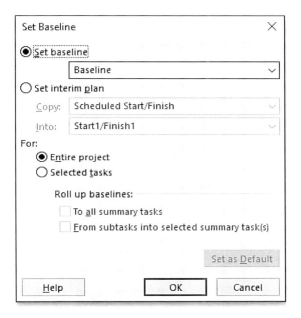

Figure 8 - 3: Set Baseline dialog

3. In the **Set baseline** pick list, leave the **Baseline** item selected.
4. In the *For: section* of the dialog, leave the **Entire project** option selected.
5. Click the **OK** button.

Information: To view the task baseline information in any project, right-click on the **Select All** button and select the **More Tables** item on the shortcut menu. In the *More Tables* dialog, select the **Baseline** table and then click the **Apply** button.

225

Module 08

Best Practice: Projility recommends that you save an original baseline for the entire project only once during the life of a project. After a change control procedure that adds new tasks to your project, you should update the baseline with the information from only the new tasks. This maintains the integrity of your original project baseline.

Backing Up a Baseline

When you save a baseline in the *Baseline* set of fields in Microsoft Project, this becomes the operating baseline for the entire life of the project. It is this baseline that Microsoft uses to calculate all variance, and the software uses this baseline to create the gray Gantt bars in the *Tracking Gantt* view. The *Baseline* set of fields is the baseline you should *always use* for capturing the operating or most 'up to date' baseline for your projects.

If you click the *Set baseline* pick list in the *Set Baseline* dialog, however, you may notice that the pick list includes ten additional sets of baselines, named *Baseline 1* through *Baseline 10*. You can leverage these additional baselines to "back up" your current operating baseline at any point in the project life cycle. In fact, Projility recommends that you back up your initial baseline in the *Baseline 1* set of fields, and that you back up your updated baseline into the *Baseline 2* through *Baseline 10* sets of fields after each change control process that adds new tasks to your project schedule.

Information: In **Module 11**, I discuss the process for updating your baseline after a change control process adds new tasks to the project. For the moment, I discuss only how to capture the original baseline for the project and to back it up into the *Baseline 1* set of fields.

The process for backing up your operating baseline is very similar to the process for capturing your original baseline, but with one major difference. To back up your operating baseline, complete the following steps:

1. Click the **Project** tab to display the *Project* ribbon.

2. In the *Schedule* section of the *Project* ribbon, click the **Set Baseline** pick list button and select the **Set Baseline** item.

3. Click the **Set baseline** pick list and select the **Baseline 1** item, as shown in Figure 8 - 4.

Information: In the *Set baseline* pick list shown in Figure 8 - 4, notice the *(last saved on _____)* text appended to the right of the *Baseline* field. This information reveals whether a user saved a baseline in a particular *Baseline* field and when the user saved the baseline information. If a baseline does not include the *(last saved on _____)* text to the right of its name, this indicates that the baseline is currently unused.

4. In the *For:* section of the dialog, leave the **Entire project** option selected.

5. Click the **OK** button.

6. Save the latest changes to your enterprise project.

Project Execution

Figure 8 - 4: Back up the baseline
to the Baseline 1 set of fields

Clearing the Project Baseline

At some point you may need to clear the baseline information in a project, such as when management decides to delay the *Start* date of your project indefinitely. When your executives finally determine a new project *Start* date, your current baseline information is no longer valid and should be cleared. To clear the baseline values for your project complete the following steps:

1. Click the **Project** tab to display the *Project* ribbon.

2. In the *Schedule* section of the *Project* ribbon, click the **Set Baseline** pick list button and select the **Clear Baseline** item. Microsoft Project displays the *Clear Baseline* dialog shown in Figure 8 - 5.

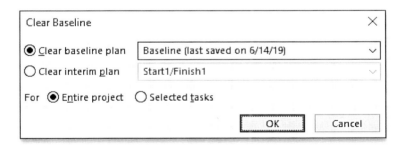

Figure 8 - 5: Clear Baseline dialog

3. Select the **Clear baseline plan** option.

4. On the *Clear baseline plan* pick list, leave the **Baseline** item selected.

5. Leave the **Entire project** option selected.

6. Click the **OK** button.

227

7. Save the new baseline in the *Baseline* set of fields using the steps documented previously.

Hands On Exercise

Exercise 8 - 2

Save an original baseline for your enterprise project and then back up your baseline.

1. Click the **Project** tab to display the *Project* ribbon.
2. In the *Schedule* section of the *Project* ribbon, click the **Set Baseline** pick list button and select the **Set Baseline** item.
3. In the **Set baseline** pick list, leave the **Baseline** item selected.
4. In the *For:* section of the dialog, leave the **Entire project** option selected, and then click the **OK** button.
5. In the *Schedule* section of the *Project* ribbon, click the **Set Baseline** pick list button and select the **Set Baseline** item again.

In the *Set baseline* field, notice the "last saved on" date for the *Baseline* set of fields.

6. Click the **Set baseline** pick list and select the **Baseline 1** item.
7. In the *For:* section of the dialog, leave the **Entire project** option selected, and then click the **OK** button.
8. Save the latest changes to your enterprise project.

Understanding Publishing

When you are ready to "go live" with an enterprise project, so that your team members can begin work on their tasks and submit task progress, you must publish the project. When you publish a project for the very first time, the Project Online system performs a number of major operations. In the simplest sense, the system performs each of the following:

- Saves the project data in the Project Web App database to make it visible in views and reports
- Sends an e-mail to all team members in the project, notifying them of tasks in a new project
- Creates a Project Site for the project in SharePoint, which allows all members of the project team to collaborate on risks, issues, documents, and deliverables associated with the project

Project Execution

Information: When you create a new enterprise project from the *Project Center* page in PWA using an Enterprise Project Type (EPT), Project Online publishes your project initially during the process of creating the project. The main purpose of this initial publishing operation is to create the Project Site for the project in SharePoint.

The result of the publishing operation is that all parties can now see information about the project in Project Web App. For example, executives see the project on the *Project Center* page, resource managers see resource assignment information on the *Resource Assignments* and *Capacity Planning* pages, and team members see their task assignments on both the *Tasks* and *Timesheet* pages.

Publishing a project for the first time also determines who receives task updates from team members working on the project. The system designates the user who publishes the project initially as the **Status Manager** for every task in the project. The person designated as the *Status Manager* is the one who receives and approves task updates from team members for their task assignments. Refer back to **Module 04** for documentation on how to change the *Status Manager* for tasks in an enterprise project.

Publishing an Enterprise Project

To publish an enterprise project, click the **File** tab and then click the **Info** tab in the *Backstage*. On the *Info* page, click the **Publish** button. Figure 8 - 6 shows the *Publish* button for an unpublished project. Notice that the *Publish Project Progress* section of the page indicates the consequences of this project not being published to Project Web App.

Figure 8 - 6: Publish button for an unpublished project

Information: If you customized your ribbon and Quick Access Toolbar using the directions in Exercise 2 – 1, you can click the **Publish** button in your Quick Access Toolbar.

Microsoft Project displays the *Publish Project* dialog for the unpublished project, such as shown in Figure 8 - 7.

229

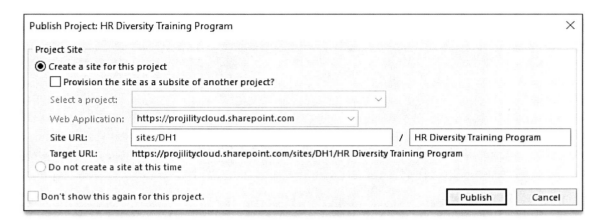

Figure 8 - 7: Publish Project dialog for a project

In the *Publish Project* dialog shown previously in Figure 8 - 7, notice that Microsoft Project disabled the *Do not create a site at this time* option. This is because our application administrator configured a setting in Project Web App that requires the creation of a Project Site during the first publishing operation. Depending on how your application administrator configured your Project Online system, the *Do not create a site at this time* option may either be enabled or disabled. If this option is enabled and you select it, Microsoft Project publishes the project, but does not create a Project Site for the project. Until you actually create the Project Site for a project, Microsoft Project continues to display the *Publish Project* dialog every time you publish your project.

 Information: If you work in an organization where creating a Project Site is optional, and you do not wish to see the *Publish Project* dialog every time you publish your project, select the **Don't show this again for this project** checkbox.

In the *Publish Project* dialog, click the **Publish** button to publish your new project. As the system publishes your project and creates the Project Site, Microsoft Project displays progress information in the *Status* bar at the bottom of the application window. When the publishing operation completes, the system displays a *Publish completed successfully* message in the *Status* bar.

Setting Custom Permissions for a Project

After publishing a new enterprise project for the first time, Project Online allows you to designate custom permissions for the project. For example, the default permissions in the system allow project managers to see and edit only their own projects, and do not allow project managers to see and edit projects they do not own or manage.

Suppose that I need a fellow project manager named Grace Adeyimi to be able to open, edit, save, and publish a new enterprise project using Microsoft Project, but I do not need her to be able to see and edit the project in Project Web App. To set special permissions in Microsoft Project, open and check out the enterprise project and complete the following steps:

1. Click the **File** tab and then click the **Info** tab in the *Backstage*.
2. In the *Permissions* section of the *Info* page, click the **Manage Permissions** button as shown in Figure 8 - 8.

Project Execution

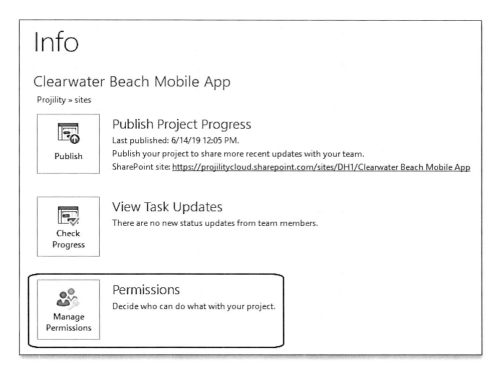

Figure 8 - 8: Set special permissions using the
Manage Permissions button in the Backstage

The system displays the *Project Permissions* page in Project Web App for the enterprise project. Notice in the *Project Permissions* page in in Figure 8 - 8 that there are currently no special permissions created for this project.

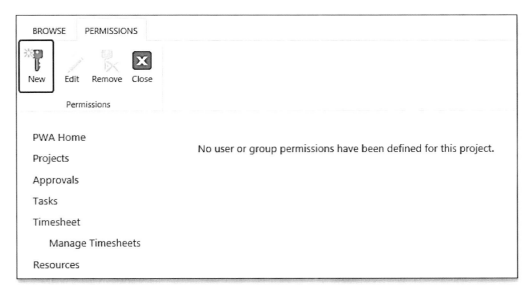

Figure 8 - 9: Project Permissions page

231

Module 08

Information: You can also access the *Project Permissions* page by navigating to the *Project Center* page in Project Web App. On the *Project Center* page, click the row header at the left end of the project name to select the project. In the *Navigate* section of the *Projects* ribbon, click the **Project Permissions** button.

3. On the *Project Permissions* page, click the **New** button in *Permissions* section of the *Permissions* ribbon. The system displays the *Edit Project Permissions* page for the enterprise project, such as shown in Figure 8 - 10.

4. Select one or more users or groups in the *Available Users and Groups* list and then click the **Add** button to add the selected users and groups to the *Users and Groups with Permissions* list.

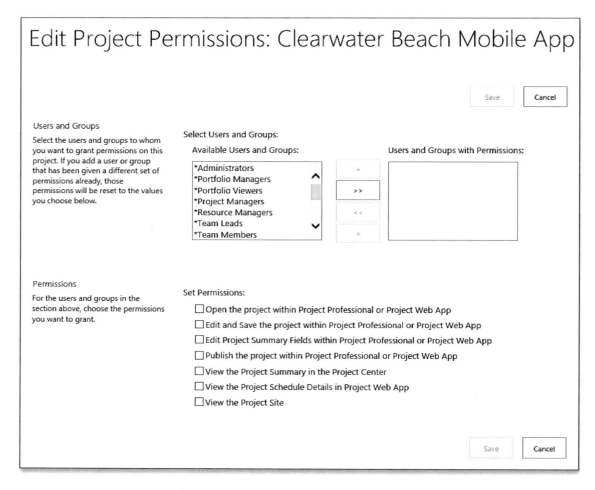

Figure 8 - 10: Edit Project Permissions page

Information: If you want to grant special permissions to individual users, add their names to the *Users and Groups with Permissions* list. If you want to grant special permissions to a group of people with the same role, such as project managers for example, then add one of the group names (prefixed with the asterisk character) to the *Users and Groups with Permissions* list to grant special permissions to everyone in the selected group.

232

Project Execution

 Information: A user must be a member of the *Project Managers* security group to be able to open, edit, save, and publish an enterprise project. If you grant special permissions to a user who is a member of the *Team Members* group, for example, the system does not allow the user to open, edit, save, and publish the enterprise project because the *Team Members* security Group lacks the proper permissions to allow team members to perform these actions in Project Online.

5. In the *Permissions* section of the page, select the checkboxes for the specific permissions you want to grant to the users for the selected project.

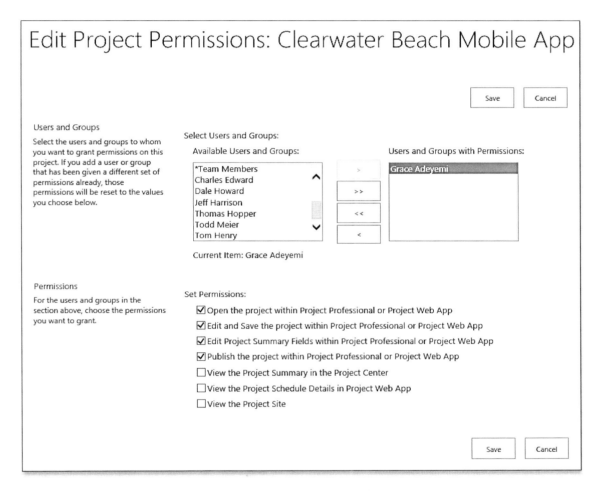

Figure 8 - 11: Edit Project Permissions page
with permissions set for Grace Adeyimi

The available permissions allow a user to open, edit and save, publish, and edit project summary fields for an enterprise project in either Microsoft Project or Project Web App, to view the project in either the *Project Center* page or *Project Details* page in Project Web App, or to view the Project Site. In the *Edit Project Permissions* page shown previously in Figure 8 - 11, notice that I selected only the following permissions to meet the special permissions needs for Grace Adeyimi:

- Open the project within Project Professional or Project Web App
- Edit and Save the project within Project Professional or Project Web App

Module 08

- Edit Project Summary Fields within Project Professional or Project Web App
- Publish the project within Project Professional or Project Web App

By selecting the preceding options, I allow Grace Adeyimi to open, edit, save, and publish the project in Microsoft Project. Grace Adeyimi cannot edit the project using Project Web App, nor can she access the Project Site for the project because she cannot see the project in the *Project Center* page in Project Web App.

6. When you finish selecting special permissions for the selected users and groups, click the **Save** button. The system refreshes the *Project Permissions* page as shown in Figure 8 - 12. Notice that the *Project Permissions* page shows the special permissions for Grace Adeyimi.

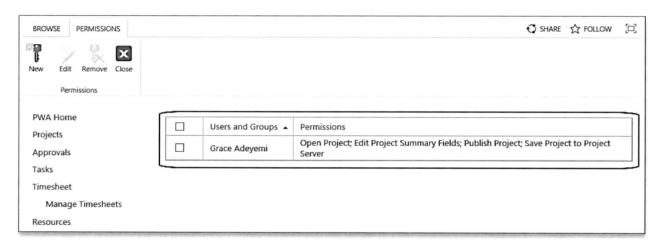

Figure 8 - 12: Project Permissions page with
special permissions for Grace Adeyimi

After creating at least one set of special permissions on the *Project Permissions* page, you can edit or remove special permissions, or close the page using the buttons in *Permissions* ribbon. To edit the special permissions for a project, select the checkbox for the special permissions and then click the **Edit** button in the *Permissions* ribbon. To remove the special permissions for a project, select the checkbox for the special permissions and then click the **Remove** button in the *Permissions* ribbon. To navigate away from the *Project Permissions* page, click the **Close** button in the Permissions ribbon, and the system displays the *Project Center* page in PWA.

 Hands On Exercise

Exercise 8 - 3

Save and publish your enterprise project.

1. Save the latest changes to your enterprise project if you have not done so already.
2. Click the **File** tab and then click the **Info** tab in the *Backstage*.

Project Execution

3. In the *Publish Project Progress* section of the *Info* page, click the **Publish** button.

4. If Microsoft Project displays the *Publish Project* dialog, select the **Create a site for this project** option, and then click the **Publish** button.

5. Watch the *Status* bar at the bottom of your application window to monitor the progress of the Publish operation.

Working with Project Deliverables

In project management terminology, a **deliverable** is the product of a completed project or an intermediate result within a project, such as at the end of each deliverable section of the project. In the world of projects, deliverables can take the form of hardware, software, services, processes, documents, or even ideas.

Project Online offers you a formal feature known as **Deliverables** that allows you to designate project deliverables, along their negotiated delivery dates and any external dependencies that may influence the delivery dates. The purpose of creating Deliverables is so that your executives and team members can see the agreed upon deliverables for the project in the Project Site. Other Project managers can even link tasks within their project to a deliverable within your project schedule.

Before you can use this Deliverables feature with any enterprise project, the project must be published, and a Project Site must exist for the project. This is because the *Deliverables* feature is SharePoint functionality that is a part of the Project Site.

Adding a New Deliverable in Microsoft Project

After publishing your enterprise project, you can create Deliverables in the project by completing the following steps:

1. Click the **Task** tab to display the *Task* ribbon.

2. Select any task to which you want to link a Deliverable in SharePoint. As a general rule, milestone tasks are a good target for linking to a Deliverable since they usually designate the completion of a project deliverable section.

Warning: You cannot create Deliverables in your enterprise project unless a Project Site exists for your project. This means your project must be a published project.

Warning: Microsoft Project does not allow you to create a Deliverable for the Project Summary Task (Row 0 or Task 0). If you attempt to do so, the software disables the *Create Deliverables* item on the *Deliverable* pick list.

3. In the *Insert* section of the *Task* ribbon, click the **Deliverable** pick list and select the **Create Deliverables** item, as shown in Figure 8 - 13.

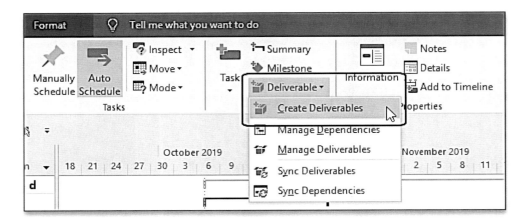

Figure 8 - 13: Create Deliverables option

Microsoft Project creates a new Deliverable in the Project Site, links it to the selected task, and then displays a special indicator (it looks like an open cardboard box) in the *Indicators* column for the task, such as shown in Figure 8 - 14. The system only allows you to create one Deliverable per task in your project schedule, by the way.

 Information: To quickly create Deliverables in your Microsoft Project schedule, use the **Control** key on your computer keyboard to select multiple tasks for which you want to create Deliverables. For example, you might select all of the milestones that represent the completion of deliverable sections in your project. In the *Insert* section of the *Task* ribbon, click the **Deliverables** pick list and select the **Create Deliverables** item. For each task selected, the software creates a new Deliverable in the Project Site and then links it to the selected task

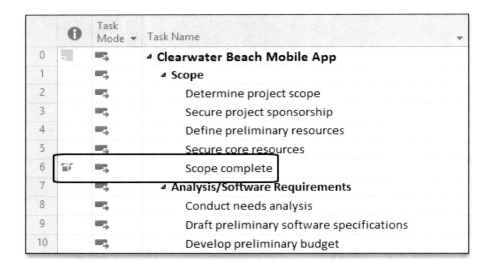

Figure 8 - 14: New Deliverable created

 Information: If you accidentally create a new Deliverable on the wrong task, leave the task selected. In the *Insert* section of the *Task* ribbon, click the **Deliverable** pick list and select the **Create Deliverables** item again. Microsoft Project deletes the erroneous Deliverable from the Project Site and then removes the indicator in the *Indicators* column for the task.

Project Execution

To work with the Deliverables you created in your Microsoft Project schedule, in the *Insert* section of the *Task* ribbon, click the **Deliverable** pick list and select the **Manage Deliverables** item, as shown in Figure 8 - 15.

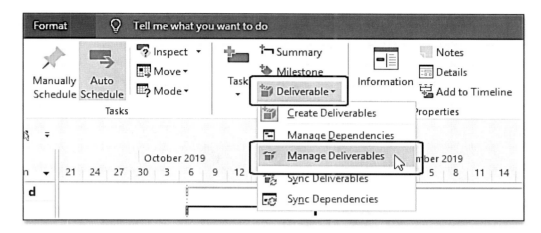

Figure 8 - 15: Manage Deliverables

Microsoft Project displays the *Deliverables* sidepane on the left side of the application window, such as the one shown in Figure 8 - 16. For the purposes of clarity, I manually widened the sidepane, and would recommend you do this as well in your own enterprise projects.

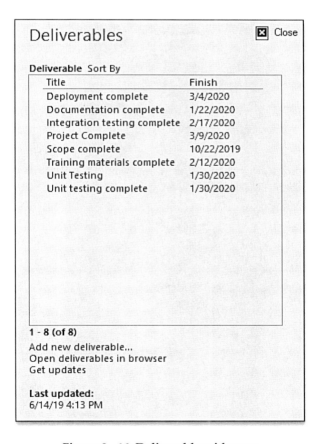

Figure 8 - 16: Deliverables sidepane

237

Module 08

When you link a Deliverable to a task in Microsoft Project, the software automatically sets the *Title* of the Deliverable to the name of the task, sets the *Deliverable Start* date to the *Start* date of the task, and sets the *Deliverable Finish* date to the *Finish* date of the task. By default, the *Deliverables* sidepane sorts the Deliverables alphabetically in ascending order by the *Title* of each Deliverable. To sort them in a different manner, click the **Sort By** pick list at the top of the list of Deliverables and then select the manner of sorting, such as by *Start Date* as shown in Figure 8 - 17.

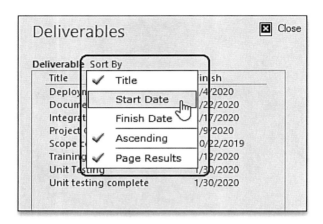

Figure 8 - 17: Sort Deliverables by Start Date

You can use the *Deliverables* sidepane to create new Deliverables by completing the following steps:

1. Select any task in your Microsoft Project schedule.

2. In the *Deliverables* sidepane click the **Add new deliverable** link. The software displays the *Add Deliverable* sidepane on the left side of the application window.

3. In the *Add Deliverable* sidepane, select the **Link to selected task** checkbox. Microsoft Project automatically fills in the Deliverable information from the selected task, such as shown in Figure 8 - 18.

4. Edit the information in the *Title*, *Start*, and *Finish* fields of the *Add Deliverable* sidepane, as needed.

5. Click the **Done** link at the bottom of the *Add Deliverable* sidepane. The software redisplays the *Deliverables* sidepane with the new Deliverable added to the list.

Project Execution

Figure 8 - 18: New Deliverable created in
the Add Deliverable sidepane

Editing and Deleting Deliverables in Microsoft Project

To display all information about any Deliverable, float your mouse pointer over the name of a Deliverable. Microsoft Project displays a tooltip to the right of the Deliverable name, which shows the *Title, Start, Finish, Task ID,* and *Task Name* fields.

To edit a Deliverable in the *Deliverables* sidepane, complete the following steps:

1. Click the name of the Deliverable you want to edit.

The software displays the pick list shown in Figure 8 - 19. The items on the pick list allow you to do the following:

- Edit the Deliverable in Microsoft Project

- Edit the Deliverable in the Project Site using your web browser

- Delete the Deliverable

- Accept changes made to the Deliverable on the Project Site

Module 08

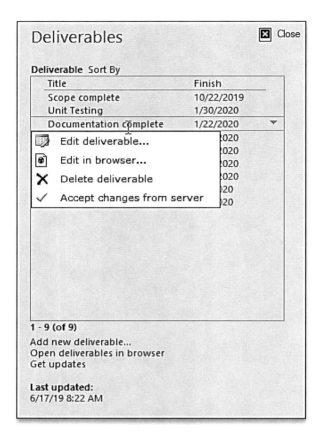

Figure 8 - 19: Prepare to edit a Deliverable

2. Select the **Edit Deliverable** item in the pick list. Microsoft Project displays the *Edit Deliverable* sidepane shown in Figure 8 - 20.

3. Edit the information in the *Title*, *Start*, and/or *Finish* fields as needed.

 Information: Even though the Deliverable automatically inherits its *Title* from the name of the task to which it is linked, Microsoft Project allows you to change the *Title* to meet your project management and reporting requirements. For example, you might want to remove the word "Complete" from the end of each *Title*. In addition, you can change the *Start* and *Finish* dates of the Deliverable to dates that are independent of the task to which the Deliverable is linked.

4. Click the **Done** link at the bottom of the *Edit Deliverable* sidepane. The software redisplays the *Deliverables* sidepane with the information shown for the Deliverable.

Project Execution

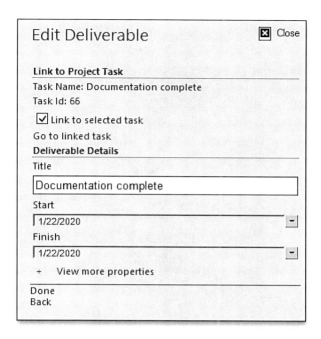

Figure 8 - 20: Edit Deliverable sidepane

 Warning: When the *Finish* date of a task slips, and a Deliverable is linked to the task, the *Finish* date of the Deliverable ***does not slip*** automatically. Microsoft made this programming decision many years ago on the assumption that if your project is going to miss a *Finish* date for a Deliverable, you must negotiate with your client to select a new *Finish* date for the Deliverable. In a situation such as this, you must open the *Deliverables* sidepane for the enterprise project and edit the *Start* and *Finish* dates of the slipping Deliverable using the steps documented in this topical section.

To delete a Deliverable, click anywhere on the Deliverable name and then select the **Delete deliverable** item on the pick list. The software displays the confirmation dialog shown in Figure 8 - 21. Click the **OK** button to complete the deletion process.

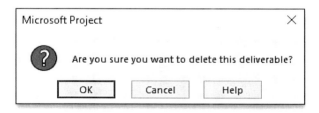

Figure 8 - 21: Delete a Deliverable confirmation dialog

 Information: When you delete a Deliverable in Microsoft Project, the SharePoint system moves the deleted Deliverable to the Recycle Bin of the Project Site for your enterprise project. The deleted Deliverable remains in the Recycle Bin for 30 days, after which the system deletes the Deliverable permanently. At any time during the 30 days, you may navigate to the Recycle Bin in the Project Site and restore the Deliverable.

Module 08

 Hands On Exercise

Exercise 8 - 4

Create a Deliverable for each milestone task in your project and then edit a Deliverable.

1. Click the **View** tab to display the *View* ribbon.
2. In the *Data* section of the *View* ribbon, click the **Filter** pick list and select the **Milestones** filter.
3. Select all five of the displayed milestone tasks.
4. Click the **Task** tab to display the *Task* ribbon.
5. In the *Insert* section of the *Task* ribbon, click the **Deliverable** pick list button, and select the **Create Deliverables** item.

Notice that Microsoft Project simultaneously created Deliverables for each of the milestone tasks. Notice also the special indicator in the *Indicators* column for each of the milestone tasks.

6. Press the **F3** function key to clear the *Milestones* filter and redisplay all tasks.
7. In the *Insert* section of the *Task* ribbon, click the **Deliverable** pick list button, and select the **Manage Deliverables** item.
8. Widen the *Deliverables* sidepane until the name of each Deliverable fits on a single line.
9. Click the **Course Manual Complete** deliverable name and then select the **Edit deliverable** item on the pick list.
10. Remove the word, **Complete**, at the right end of the Deliverable name in the *Title* field.
11. Click the **Done** link at the bottom of the *Edit Deliverable* sidepane.
12. Save the latest changes to your enterprise project.

Viewing Deliverables in Project Web App

In addition to using the Microsoft Project client, you can also view deliverables in Project Web App using the Project Site of the selected project. To view project Deliverables in Project Web App, complete the following steps:

1. In the *Deliverables* sidepane in Microsoft Project, click the **Open deliverables in browser** link at the bottom of the sidepane. The system launches a new web browser application window and displays the *Deliverables* page on the Project Site, such as the one shown in Figure 8 - 22..

Project Execution

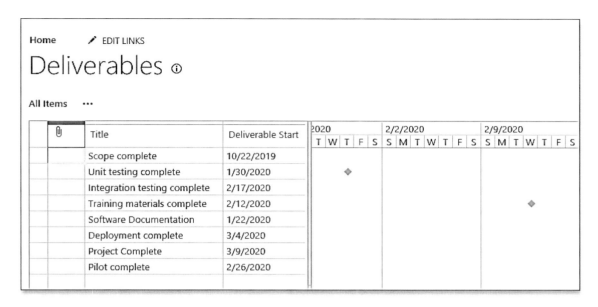

Figure 8 - 22: Deliverables page in the Project Site

Notice that the *Deliverables* page shown previously in Figure 8 - 22 shows the eight Deliverables created so far for the selected project. Project Web App represents each Deliverable using either a milestone symbol or a Gantt bar in the *Gantt Chart* portion of the *Deliverables* page. The system displays a Gantt bar when the *Deliverable Finish* date is later than the *Deliverable Start* date, such as when a deliverable has an "acceptable" date range for delivery of the product, or when you create a new Deliverable in the *Deliverables* sidepane in Microsoft Project.

2. Widen the *Title, Deliverable Start,* or *Deliverable Finish* columns, as needed.

3. Drag the split bar to the desired location on the page.

4. Click the **List** tab to display the *List* ribbon.

5. In the *Gantt View* section of the *List* ribbon, click the **Zoom Out** button shown in Figure 8 - 24 until you see all of the Deliverables for the project in the *Gantt Chart* pane.

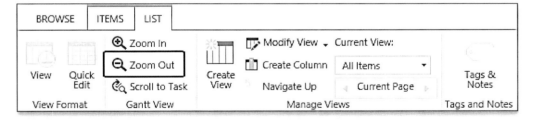

Figure 8 - 23: Click the Zoom Out button

 Warning: Although Project Web App allows you to create new Deliverables using the *New Item* button in the *Items* ribbon, Projility recommends you **do not** create Deliverables from the Project Site. This recommendation is because Deliverables created in the Project Site are not linked to any tasks in the Microsoft Project schedule, nor can they be linked using the Project Site to create them. Always create Deliverables using the Microsoft Project application so that you have full control over the creation process.

243

Module 08

6. When you finish viewing your project Deliverables on the *Deliverables* page in the Project Site, close your web browser.

7. Click the **Close** button in the *Deliverables* sidepane to close the sidepane in your Microsoft Project application window.

8. Save the latest changes to your enterprise project.

Hands On Exercise

Exercise 8 - 5

View your project Deliverables in the Project Site.

1. Click the **Open deliverables in browser** link at the bottom of the *Deliverables* sidepane.
2. Adjust the column widths for the *Title*, *Deliverable Start*, and *Deliverable Finish* fields.
3. Drag the split bar to the right edge of the *Deliverable Finish* field.
4. Click the **List** tab to expand the *List* ribbon.
5. In the *Gantt view* section of the *List* ribbon, click the **Zoom Out** button until you can see the symbols for each Deliverable in the *Gantt Chart* pane.

Notice how Project Web App displays the Deliverables in the Project Site.

6. Close your web browser to return to your Microsoft Project application window.
7. Click the **Close** button in the upper right corner of the Deliverables sidepane to close the sidepane.
8. Save the latest changes to your enterprise project.

Adding a New Deliverable Dependency

In the world of enterprise project management, it is entirely possible for the task schedule in one project to be dependent on a Deliverable date in another project. To address this situation, you can specify a task dependency on an external Deliverable in another project. To create a Deliverable dependency, complete the following steps in Microsoft Project:

1. Click the **Task** tab to display the *Task* ribbon.
2. In the *Insert* section of the *Task* ribbon, click the **Deliverable** pick list and select the **Manage Dependencies** item. The software displays the *Dependency* sidepane on the left side of the application window, such as shown in Figure 8 - 24.

Project Execution

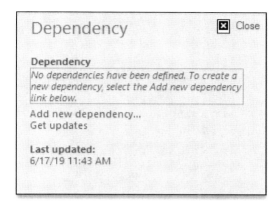

Figure 8 - 24: Dependency sidepane
for an enterprise project

Figure 8 - 24 shows the *Dependency* sidepane for Clearwater Beach Mobile App project. The *Conduct needs analysis* task in this project is dependent on the *Analysis complete* deliverable in the AI for Railcar Allocation System project.

3. Select the task in the active project that is dependent on a Deliverable date.

4. In the *Dependency* sidepane, click the **Add new dependency** link. The system displays the *Add Dependency* sidepane.

5. In the *Add Dependency* sidepane, click the **Select project** pick list and select a project on which the selected task is dependent, as shown in Figure 8 - 25.

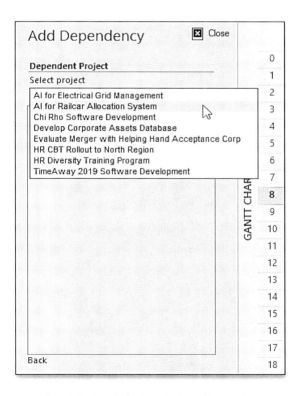

Figure 8 - 25: Select an enterprise project
in the Select project pick list

245

Module 08

 Information: The *Select project* pick list shows all of the enterprise projects which you have permission to access, and which contain at least one Deliverable.

Figure 8 - 26 shows the *Add Dependency* sidepane after I selected the *AI for Railcar Allocation System* project. Notice that the sidepane displays all of the Deliverables in that enterprise project, sorted alphabetically in ascending order by the *Title* of each Deliverable.

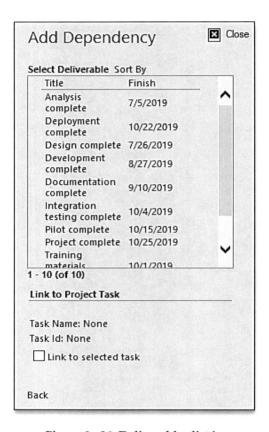

Figure 8 - 26: Deliverables list in
the Add Dependency dialog

6. In the *Select Deliverable* list in the *Add Dependency* sidepane, select the name of an external Deliverable. In my example, I selected the *Analysis complete* item.

7. Select the **Link to selected task** checkbox.

8. Click the **Done** link when finished. The system displays the external Deliverable dependency in the *Dependency* sidepane as shown in Figure 8 - 27.

246

Project Execution

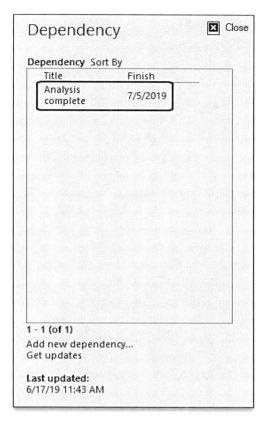

Figure 8 - 27: Dependency on an external Deliverable

9. Save the latest changes to your enterprise project.

10. Click the **Close** button to close the *Dependency* sidepane.

11. Save the latest changes to your enterprise project

Updating the Baseline and Republishing the Project

After creating your Deliverables in the Microsoft Project schedule, keep in mind that your project baseline *does not* include the Deliverable information. This means that if you create at least one Deliverable in your enterprise project, you must save the baseline again and back up the backline again using the steps documented previously in this module, and then you must also republish the project.

247

Module 08

 Hands On Exercise

Exercise 8 - 6

Update the project baselines to capture information about Deliverables, and then republish the project.

1. Click the **Project** tab to display the *Project* ribbon.
2. In the *Schedule* section of the *Project* ribbon, click the **Set Baseline** pick list button and select the **Set Baseline** item.
3. In the **Set baseline** pick list, leave the **Baseline** item selected.
4. In the *For:* section of the dialog, leave the **Entire project** option selected, and then click the **OK** button.
5. In the confirmation dialog, click the **Yes** button to update the baseline information in the *Baseline* set of fields.
6. In the *Schedule* section of the *Project* ribbon, click the **Set Baseline** pick list button and select the **Set Baseline** item again.
7. Click the **Set baseline** pick list and select the **Baseline 1** item.
8. In the *For:* section of the dialog, leave the **Entire project** option selected, and then click the **OK** button.
9. In the confirmation dialog, click the **Yes** button to update the baseline information in the *Baseline 1* set of fields.
10. Save the latest changes to your enterprise project.
11. Click the **File** tab and then click the **Info** tab in the *Backstage*.
12. In the *Publish Project Progress* section of the *Info* page, click the **Publish** button.
13. When the Publish job completes, *close and check in* your enterprise project.
14. Exit Microsoft Project completely.

Module 09

Tracking Time and Task Progress

Learning Objectives

After completing this module, you will be able to:

- Understanding time and task tracking methods available in Project Web App
- Understand best practices for using the Timesheet page in PWA
- Enter time and task progress in the Timesheet page of PWA
- Enter only task progress in the Tasks page of PWA
- Manually enter task progress in the Microsoft Project schedule

Inside Module 09

Tracking Time and Task Progress in Project Web App 251
Accessing the Timesheet Page 251
    Understanding the Timesheet Page 253
    Best Practices for Setting Up the Timesheet Page 255
Using the Timesheet Page 259
    Entering Time in the Timesheet 260
    Adding a New Row to a Timesheet 261
    Reassigning a Task to a Fellow Team Member 268
    Removing a Timesheet Line 269
    Submitting a Timesheet for Approval 270
    Recalling a Submitted Timesheet 273
    Deleting a Timesheet 274
    Responding to a Rejected Timesheet 275
    Submitting Future Planned PTO 275
Using the Tasks Page 278
    Understanding the Tasks Page 278
    Understanding the Assignment Details Page 280
    Reporting Progress from the Tasks Page 283
    Adding Tasks to the Tasks Page 289

Removing a Task 292
Reassigning a Task to a Fellow Team Member 293
Manually Entering Task Progress in the Microsoft Project Schedule **294**
Entering Progress at the Task Level 295
Entering Progress at the Resource Assignment Level 296

Tracking Time and Task Progress in Project Web App

The Project Online system offers three approaches for tracking time and task progress:

- **Timesheet** page in Project Web App
- **Tasks** page in Project Web App
- **Manual entry of task progress** by the project manager in the Microsoft Project schedule

The *Timesheet* page in PWA offers a robust timekeeping system that allows an organization to track all types of time spent on project work, non-project work, and nonworking time such as vacation and sick leave. Using the *Timesheet* page requires team members to enter time and task progress on a daily basis, and to submit their progress on a weekly basis. It is the most demanding method of tracking time and task progress if the organization has never used any type of timesheet system previously, but it offers the most accurate method of tracking task progress.

The *Tasks* page in PWA offers an organization three simpler methods for tracking task progress only. These methods include tracking task progress by percent complete, by actual hours of work performed, and by actual work entered on a daily basis. Each of these three methods offers team members an easier way to get started with tracking task progress, but none of the three methods offers the accuracy found with using the *Timesheet* page.

Organizations that do not intend to use either the *Timesheet* page or the *Tasks* page in PWA can use the manual entry of task progress by the project manager in the Microsoft Project schedule. Using this approach requires a project manager to gather task progress from team members on a weekly basis, such as in a weekly project status meeting, and then to enter the progress directly in the Microsoft Project schedule.

In this module, I individually discuss each method of tracking time and task progress.

Accessing the Timesheet Page

If your organization uses the *Timesheet* page for tracking time and task progress, Project Web App offers two ways to access the *Timesheet* page, as shown in Figure 9 - 1:

- In the *Quick Launch* menu, click the **Timesheet** link.
- In the *Track your work* carousel, click the **Timesheets** tile.

When you click the *Timesheet* link in the *Quick Launch* menu as shown, the system navigates you directly to the *Timesheet* page and creates a new timesheet for the current week. The *Timesheets* tile in the *Track your work* carousel shows you how many unsubmitted timesheets you currently have. Ideally, the *Timesheets* tile should show 0 unsubmitted timesheets. When you click the *Timesheets* tile, the system navigates you to the *Manage Timesheets* page shown in Figure 9 - 2.

 Best Practice: Projlity recommends as a best practice that all users access the *Timesheet* page by first navigating to the *Manage Timesheets* page and then clicking the *Click to create* link for the correct timesheet period. Following this process guarantees that users always create a timesheet for the correct time period, avoiding costly mistakes caused by entering time and task progress in the wrong timesheet period. Notice in the *Home* page of Project Web App shown in Figure 9 - 1 that our application administrator added a custom *Manage Timesheets* link in the *Quick Launch* menu to make sure our users can easily access the *Manage Timesheets* page.

Module 09

Figure 9 - 1: Home page of Project Web App

Manage Timesheets ⓘ

Timesheet Name	Period ▼	Total Hours	Status	Next Approver	Transaction Comment
Click to Create	2019-Wk 25 (6/17/2019 - 6/23/2019)		Not Yet Created		
My Timesheet	2019-Wk 24 (6/10/2019 - 6/16/2019)	8h	Approved		System generated automatic approval
Period Closed	2019-Wk 23 (6/3/2019 - 6/9/2019)		Period Closed		
Period Closed	2019-Wk 22 (5/27/2019 - 6/2/2019)		Period Closed		
Period Closed	2019-Wk 21 (5/20/2019 - 5/26/2019)		Period Closed		
Period Closed	2019-Wk 20 (5/13/2019 - 5/19/2019)		Period Closed		

Figure 9 - 2: Manage Timesheets page

By default, the *Manage Timesheets* page displays the current timesheet period at the top of the page, along with timesheet periods for the last three months of. Notice in the *Manage Timesheets* page shown in Figure 9 - 2 that several time periods contain a *Period Closed* link in the *Timesheet Name* column. According to our organization's methodology for using the *Timesheet* page, our application administrator must close timesheet periods that are more than 2 weeks old to prevent users from editing old timesheet information. Notice also that *Timesheet Name* column contains a *Click to create* link for the current timesheet period. Always click the *Click to create* link to create a timesheet for either the current week, or for last week if you failed to fill out at timesheet for last week.

Understanding the Timesheet Page

The *Timesheet* page enables you to capture your actual time spent on all planned and administrative activities for the current time reporting period. Figure 9 - 3 shows the default layout of the *Timesheet* page.

Figure 9 - 3: Default Timesheet page in PWA

By default, the *Timesheet* page displays only **current tasks**. A "current task" is any task with work scheduled during the current timesheet period, or an incomplete task scheduled in a previous timesheet period (such as a task that you did not complete last week). The *Timesheet* page also displays administrative time categories which an organization uses to capture non-project work and nonworking time. The three default administrative time categories include *Administrative*, *Sick time*, and *Vacation*. Your application administrator can add other administrative time categories based on the needs of your organization. Notice in Figure 9 - 3: Default Timesheet page in PWA shown previously that our application administrator added two additional administrative time categories: *Meetings* and *Training*.

By default, the *Timesheet* page includes the following fields (columns) in the *My Work* view:

- The **Task Name/Description** field provides a descriptive name of the timesheet line.

- The **Project Name** field provides the name of the project. For administrative time categories, this field contains the term *Administrative*.

- The **Comment** field allows the user to add a comment related to this timesheet line entry.

- The **Billing Category** field is an optional field that allows the user to designate a billing category for a particular timesheet line. For example, if your application administrator set up a billing category named *Merger*, you can set this value to designate a timesheet line as a Merger effort.

- The **Process Status** field that provides the current state of the timesheet line in regard to the approval cycle.

- The **Start** field displays the scheduled *Start* date for the task assignment.

- The **Finish** field displays the scheduled *Finish* date for the task assignment.

- The **Remaining Work** field that displays the calculated *Remaining Work* value for the task assignment. If allowed in your organization, team members can also adjust the *Remaining Work* estimate on any task assignment, as appropriate.

- The **% Work Complete** field displays the calculated *% Work Complete* value for the task assignment.

- The **Work** field displays the scheduled *Work* value for the task assignment.

- The **Actual Work** field displays the calculated value for *Actual Work* entered by the user on the task assignment.

- The **Duration** field displays the current *Duration* value for the task.

On the right side of the *Timesheet* page, you also see a timesheet grid containing one week of daily cells for reporting time and task progress.

> **Information**: The default settings in Project Web App configure the *Timesheet* page for the **daily** entry of time and task progress. If required by your organization, the application administrator can change the configuration to allow for the **weekly** entry of time and task progress, where users enter an entire week of work for a seven-day period in a single cell. For best use of the *Timesheet* functionality, Projility strongly recommends that your organization use the default daily method of entering of time and task progress.

When you click anywhere in the *Timesheet* page, Project Web App expands the *Timesheet* ribbon shown in Figure 9 - 4. This ribbon offers you several important tools for working with and submitting timesheets. You can also click the *Options* tab to expand the *Options* ribbon shown in Figure 9 - 5. This ribbon offers additional options for working with timesheets, such as applying grouping to the timesheet lines.

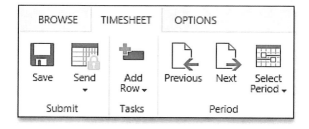

Figure 9 - 4: Timesheet ribbon

Tracking Time and Task Progress

Figure 9 - 5: Options ribbon

By default, each task in the timesheet includes two types of time: **Actual** and **Planned**, with each displayed on its own line. You use the *Actual* line to enter actual work hours for a task or administrative time category. You use the *Planned* line to view the planned or scheduled work hours for each task assignment.

The gray **Information** bar, displayed immediately below the ribbon, shows you information about the following:

- Status of the timesheet

- Total number of hours entered in the timesheet

- Dates for the timesheet period

By default, the timesheet opens in the current time reporting period unless you selected a different time period on the *Manage Timesheets* page. If you need to switch to another time reporting period, use the **Previous** and **Next** buttons in the *Period* section of the *Timesheet* ribbon to navigate to a specific time period. You can also navigate to another timesheet period by clicking the *Select Period* pick list button and selecting an option on the pick list.

In the *Data* section of the *Options* ribbon, you can use the **Group By** pick list to group your timesheet lines by the column name you select. For example, select the *Project Name* grouping to group your tasks by the name of the associated project.

Best Practices for Setting Up the Timesheet Page

To make it the easiest for team members to enter time and task progress on the *Timesheet* page in PWA, Projility recommends that your team members set up the page by completing the following steps:

1. Double-click the right edge of the **Task Name/Description** column header to "best fit" the column width to accommodate longer task names.

2. Float your mouse pointer over the **Project Name** column header, click the pick list button, and then select the **Hide Column** item.

3. Float your mouse pointer over the **Comment** column header, click the pick list button, and then select the **Configure Columns** item. Project Web App displays the *Configure Columns* dialog shown in Figure 9 - 6.

255

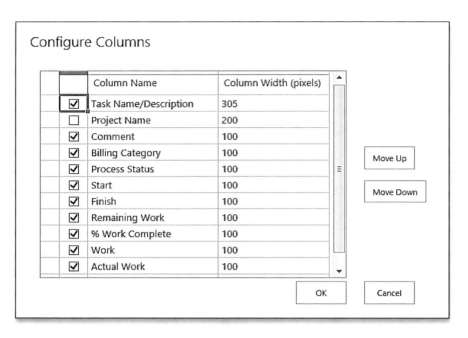

Figure 9 - 6: Configure Columns dialog

4. In the *Configure Columns* dialog, click the *Remaining Work* row to select it, but ***do not deselect*** the checkbox at the left end of the row.

5. With the *Remaining Work* row selected, repeatedly click the **Move Up** button until the *Remaining Work* row appears immediately ***above*** the *Comment* row, such as shown in Figure 9 - 7.

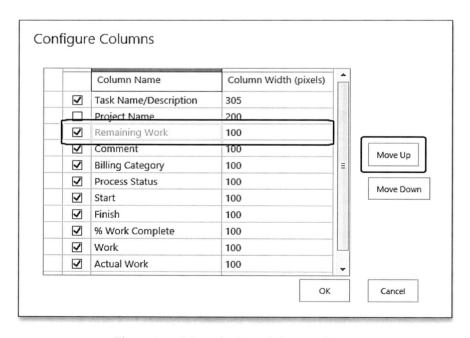

Figure 9 - 7: Move the Remaining Work row

6. In the *Configure Columns* dialog, click the **OK** button.

Tracking Time and Task Progress

7. Double-click the right edge of the **Remaining Work** column header to "best fit" the column width.
8. Drag the split bar to the right edge of the *Remaining Work* column.
9. In the *Data* section of the *Options* ribbon, click the **Group By** pick list button and select the **Project Name** item.

 Information: In your own Project Web App, it is entirely possible that you may not need to complete any of the previous steps because your application administrator configured the layout of the *Timesheet* page to follow the preceding best practice recommendations.

Figure 9 - 8 shows the *Timesheet* page configured according to the steps in this topical section. Users only need to complete the preceding set of steps once, as Project Web App saves each user's custom layout of the *Timesheet* page individually.

	Task Name/Description ↑	Remaining Work	Time Type	Mon 6/17	Tue 6/18	Wed 6/19	Thu 6/20	Fri 6/21
	▲ **Project Name: Administrative**							
☐	Administrative		Actual					
			Planned					
☐	Meetings		Actual					
			Planned					
☐	Sick time		Actual					
			Planned					
☐	Training		Actual					
			Planned					
☐	Vacation		Actual					
			Planned					
	▲ **Project Name: Chi Rho Software Development**	44h						
☐	Develop preliminary budget	16h	Actual					
			Planned				8h	8h
☐	Draft preliminary software specifications	16h	Actual					
			Planned	8h	8h			
☐	Incorporate feedback on software specifications	8h	Actual					
			Planned					4h
☐	Review software specifications/budget with team	4h	Actual					
			Planned					4h
	▲ **Project Name: Develop Corporate Assets Database**	24h						
☐	Incorporate feedback into functional specifications	8h	Actual					
			Planned					4h
☐	Review functional specifications	16h	Actual					
			Planned				4h	8h	4h
			Actual					
			Planned		8h	8h	12h	16h	16h

Figure 9 - 8: Timesheet page layout configured with Projility best practices

The final step in the best practice setup of the *Timesheet* page is a step that users must repeat each time they access the *Timesheet* page. In the upper right corner of every page, PWA displays the **Focus on Content** button shown in Figure 9 - 9. This little button is not very obvious (I like to describe it as "hidden in plain sight"), but it is incredibly useful on wide PWA pages such as the *Timesheet* page. Click the *Focus on Content* button to hide the header information above the ribbon and the *Quick Launch* menu on the left side of the page. Doing this allows you to display the *Timesheet* page with the maximum height and width possible. Because the *Focus on Content* button is a "toggle" button, you can click the button a second time to display the *Quick Launch* menu again.

Figure 9 - 9: Focus on Content button

Hands On Exercise

Exercise 9 - 1

Navigate to the *Timesheet* page in PWA and set it up using Projility recommended best practices.

1. Launch your preferred web browser and navigate to the *Home* page of your organization's Project Web App.

Notice in the *Track your work* carousel that the *Timesheets* tile shows how many unsubmitted timesheets you have. Ideally, you should see no unsubmitted timesheets.

2. In the *Track your work* carousel, click the **Timesheets** tile.

3. In the *Manage Timesheets* page, click the **Click to Create** link for the current week's timesheet period.

 Information: If you are working this exercise using self-study, click the **Click to Create** link for the timesheet period that represents the first week of your enterprise project used for this class.

4. Double-click the right edge of the **Task Name/Description** column header to "best fit" the column width to accommodate longer task names.

5. Float your mouse pointer over the **Project Name** column header, click the pick list button, and then select the **Hide Column** item.

6. Float your mouse pointer over the **Comment** column header, click the pick list button, and then select the **Configure Columns** item.

Tracking Time and Task Progress

7. In the *Configure Columns* dialog, click the **Remaining Work** row to select it.
8. With the *Remaining Work* row selected, repeatedly click the **Move Up** button until the *Remaining Work* row appears immediately *above* the *Comment* row.
9. In the *Configure Columns* dialog, click the **OK** button.
10. Double-click the right edge of the **Remaining Work** column header to "best fit" the column width.
11. Drag the split bar to the right edge of the *Remaining Work* column.
12. Click the **Options** tab to display the *Options* ribbon.
13. In the *Data* section of the *Options* ribbon, click the **Group By** pick list button and select the **Project Name** item.
14. Leave the *Timesheet* page open in Project Web App for the next exercise.

Using the Timesheet Page

When you enter actual work hours for a task, Project Online assumes all hours are billable regular work (non-overtime work) and the system calculates the task cost using the actual work multiplied by your *Standard Rate* value from the Enterprise Resource Pool. Some organizations, however, need to track overtime work in addition to regular work, and to differentiate between billable and non-billable work. When you enter overtime work, the system calculates the task cost using the actual overtime work multiplied by your *Overtime Rate* value from the Enterprise Resource Pool. The system calculates the task cost for both billable and non-billable work using the actual work multiplied by your *Standard Rate* value, by the way.

To track overtime work for each project task, select the **Overtime** checkbox in the *Show/Hide* section of the *Options* ribbon. To track non-billable time, select the *Non-Billable* checkbox. Figure 9 - 10 shows the *Options* ribbon with the *Overtime* and *Non-Billable* checkboxes selected. Figure 9 - 11 shows how the *Overtime, Actual Non-Billable,* and *Overtime Non-Billable* time types appear in the timesheet for several tasks.

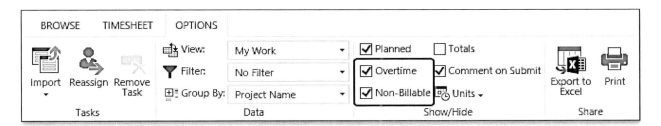

Figure 9 - 10: Overtime and Non-Billable
options in the Options ribbon

 Information: In the *Show/Hide* section of the *Options* ribbon, you can select the **Totals** checkbox to add the *Totals* row to the timesheet. This row displays the total amount of time entered each day for each task individually.

259

Develop preliminary budget	16h	Actual		
		Planned		
		Overtime		
		Actual Non-Billable		
		Overtime Non-Billable		
		Total		
Draft preliminary software specifications	16h	Actual		
		Planned	8h	8h
		Overtime		
		Actual Non-Billable		
		Overtime Non-Billable		
		Total		

Figure 9 - 11: Additional time types added to Timesheet page

By default, Project Web App selects two options in the *Show/Hide* section of the *Options* ribbon: the *Planned* option and the *Comment on Submit* option. When selected, the *Planned* option causes the system to display the *Planned* row for every task. You can use the *Planned* row to see the schedule of planned work for the week for any task. When selected, the *Comment on Submit* option causes the system to display a *Comment* dialog when you submit your timesheet for approval, which allows you to add a comment about the timesheet period as a whole.

Best Practice: For easiest use with entering actual work in the timesheet, at the beginning of each week, select the **Planned** checkbox, and then study the planned work for each task. When you understand the schedule of work for the current week, ***deselect*** the **Planned** checkbox to remove the *Planned* row in the timesheet. Doing so reduces the likelihood of accidentally entering actual work for the wrong task.

Entering Time in the Timesheet

On the *Timesheet* page, you can enter actual time spent on project work, non-project work, and nonworking time. The process for entering actual work in the daily timesheet grid is fairly simple. At the end of each day, enter the amount of time you spent that day on any task or administrative time category listed in the timesheet. Add a comment in the *Comment* field for any task requiring additional information. After you enter your time and optional comments each day, click the **Save** button in the *Submit* section of the *Timesheet* ribbon to save the latest changes to your timesheet. On the last day of the timesheet period (usually Friday), adjust the **Remaining Work** value for any task according to your organization's methodology for tracking task progress. Drag the split bar to the right edge of the *Comment* column to add a comment about any task that needs addition documentation about your update, such as for tasks on which you increased the *Remaining Work* value, for example.

Figure 9 - 12 shows my timesheet for the week of June 17-23. Notice that I entered a combination of task work and administrative time for the week. The *Information* bar shows that I entered a total of *32 hours* through Thursday of the current week.

Tracking Time and Task Progress

	Task Name/Description ↑	Remaining Work	Time Type	Mon 6/17	Tue 6/18	Wed 6/19	Thu 6/20	Fri 6/21
	▲ **Project Name: Administrative**							
☐	Administrative		Actual					
☐	Meetings		Actual					
☐	Sick time		Actual				8h	
☐	Training		Actual					
☐	Vacation		Actual					
	▲ **Project Name: Chi Rho Software Development**	24h						
☐	Develop preliminary budget	12h	Actual				4h	
☐	Draft preliminary software specifications	0h	Actual	8h	8h			
☐	Incorporate feedback on software specifications	8h	Actual					
☐	Review software specifications/budget with team	4h	Actual					
	▲ **Project Name: Develop Corporate Assets Database**	20h						
☐	Incorporate feedback into functional specifications	8h	Actual					
☐	Review functional specifications	12h	Actual				4h	
			Actual	8h	8h	8h	8h	

Figure 9 - 12: Actual time entered in the Timesheet page

Information: Notice in the *Timesheet* page shown previously in Figure 9 - 12 that the page shows the results of completing Projility recommended best practices for setting up the page. Notice that I hid the *Planned* row, which makes it a little easier to enter actual time for each task and administrative time category.

Adding a New Row to a Timesheet

You may occasionally work on an assignment that is either not on your timesheet or is of a category of work that needs special tracking. To add an extra row to your timesheet, click the **Timesheet** tab to display the *Timesheet* ribbon. In the *Tasks* section of the *Timesheet* ribbon, click the **Add Row** pick list button and select one of the items on the pick list, such as the *Add a New Task* item shown in Figure 9 - 13. Select an item on the *Add Row* pick list button according to the following descriptions:

- Select the **Select from Existing Assignments** item to add an existing task not currently shown on your timesheet. You can use this option to display a completed task in the past that requires additional work this week, or to display a future task on which you need to enter early progress this week.

- Select the **Add New Task** item to propose a task not currently included in a Microsoft Project schedule. The new task might represent work that the project manager missed when planning the project schedule, or it might represent unexpected work that arose this week and for which you need to enter progress.

- Select the **Add Yourself to a Task** item when you need to work on a task to which you are not currently assigned. For example, you might want to assist a fellow project team member with a task to which the project manager did not assign you originally.

- Select the **Add Team Tasks** item when you need to take ownership of a task assigned to a team of resources rather than to an individual resource.

- Select the **Add Non-Project Line** when you need to add a line for an administrative time category not currently included in Project Web App, or to add a duplicate line for an existing administrative time category.

- Select the **Add Personal Task** item when you need to enter a task not included in any Microsoft Project schedule, but which constitutes working time rather than nonworking time.

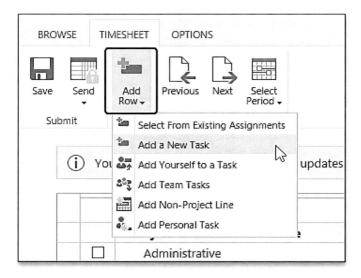

Figure 9 - 13: Click the Add Row pick list button

Adding an Existing Assignment

Click the **Add Row** pick list button and select the **Select from Existing Assignments** item to display a task not currently shown on your timesheet. Project Web App displays the *Add an Existing Task* dialog shown in Figure 9 - 14. You can use this feature when you need to display a completed task in the past that requires additional work this week or to display a future task on which you need to enter early progress this week.

The *Add an Existing Task* dialog displays a collapsed list of projects in which you have at least one task assignment. Expand the project containing the task you want to add to the timesheet, then expand successive summary sections until you locate the task in question. Select the checkbox for the task you want to add to the timesheet and then add an optional comment in the *Comment* field to explain why you added the task. Click the **OK** button to add the selected task to your timesheet. Once added, you can begin entering progress on the task.

Tracking Time and Task Progress

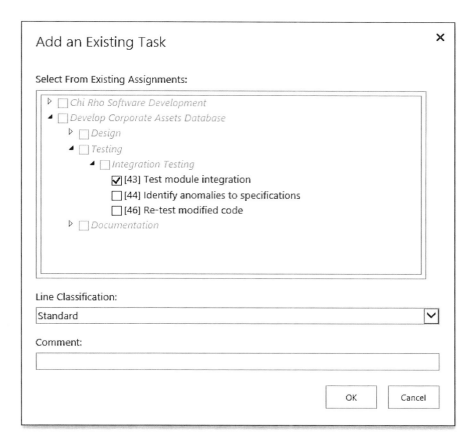

Figure 9 - 14: Add an Existing Task dialog

Adding a New Task

Click the **Add Row** pick list button and select the **Add New Task** item to propose a task not currently included in a Microsoft Project schedule. The new task might represent work that the project manager missed when planning the project schedule, or it might represent unexpected work that arose this week and for which you need to enter progress. Project Web App displays the *New Task* page shown in Figure 9 - 15. For clarity, I captured the image after selecting or entering all of the information in the page so that you can easily see the desired result.

Begin the process of filling out this page by clicking the **Project** pick list and selecting the project in which you want to propose the new task. The system refreshes the page with information relevant to the selected project. Click the **Summary task** pick list and select the summary section in which you want to propose the new task. The pick list includes every summary task in the selected project, by the way. Once again, the system refreshes the page with information relevant to the selected summary section.

In the **Task name** field, enter the name of the proposed task. In the **Start** field, enter your estimated start date for the proposed task. To determine the finish date of the project, the system needs you to enter either an estimated finish date in the **Finish** field or the number of hours of estimated work in the **Total work** field. If you enter a date in the *Finish* field, the system calculates the *Total work* value; if you enter a value in the *Total work* field, the system calculates the *Finish* date of the proposed task. In the **Comments** field, enter a comment to explain why you need to add the proposed task to the project. Click the **Send** button to complete the process of proposing a new task.

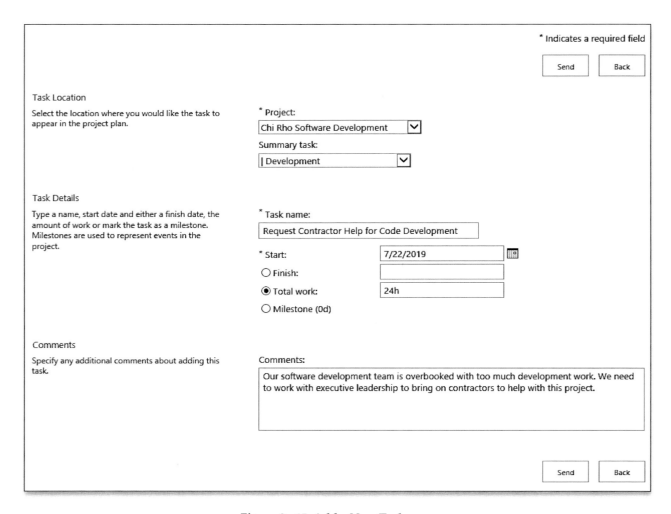

Figure 9 - 15: Add a New Task page

Adding Yourself to a Task

Click the **Add Row** pick list button and select the **Add Yourself to a Task** item when you need to assist a fellow project team member with work on a task to which the team member is already assigned but which you are not currently assigned. Project Web App displays the *New Task* page shown in Figure 9 - 16. For clarity, I captured the image after selecting or entering all of the information in the page so that you can easily see the desired result. Notice that this particular *New Task* page is very similar to the *New Task* page shown previously in Figure 9 - 15.

Begin the process of filling out this page by clicking the **Project** pick list and selecting the project containing the task to which you want to add yourself. The system refreshes the page with information relevant to the selected project. Click the **Summary task** pick list and select the summary section containing the task to which you want to add yourself. Once again, the system refreshes the page with information relevant to the selected summary section.

Click the **Assign myself to the following task** pick list and select the task to which you want to add yourself. The system refreshes the *Start* and *Finish* fields with the current schedule of the selected task. If you want to help with the task during the entire time period, leave the dates shown in the *Start* and *Finish* fields. Change the dates in the **Start** and/or **Finish** fields if you want to help with only a certain time period for the task. Enter a value in the **Total work** field if you only want to contribute a limited number of working hours to help with the task.

Tracking Time and Task Progress

In the **Comments** field, enter a comment to explain why you need to add yourself to help on the task. Click the **Send** button to complete the process of adding yourself to the existing task.

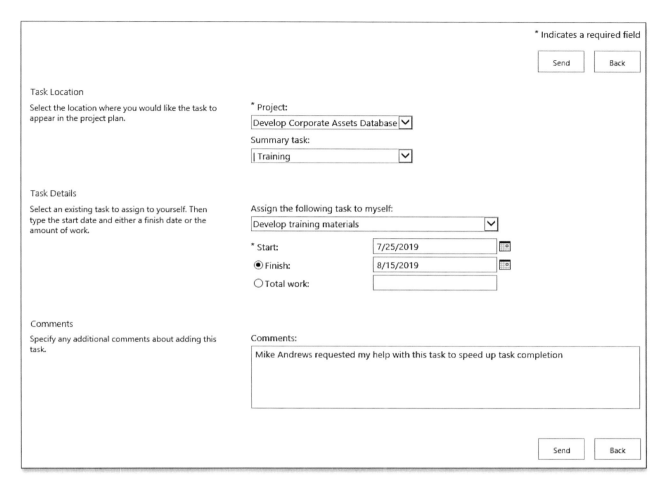

Figure 9 - 16: New Task page - add yourself to a task

Adding Team Tasks

A **Team task** is a task assigned to a **Team resource**. A *Team* resource is a special type of Generic resource that represents a specific group of resources. Project managers can assign a *Team* resource to a task when they want the members of that team to decide among themselves which of them will do the work on the task.

Click the **Add Row** pick list button and select the **Add Team Tasks** item when you need to take ownership of a task assigned your team. The system displays the *Team Tasks* page shown in Figure 9 - 17. Select the checkbox for each task for which you want to take ownership and then click the **Assign to me** button in the *Tasks* section of the *Tasks* ribbon. Click the **Close** button to return to the *Timesheet* page. By the way, Project Web App does not display the new task on your *Timesheet* page unless the task occurs during the current timesheet period.

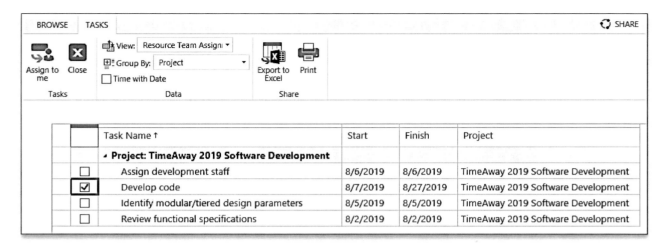

Figure 9 - 17: Team Tasks page

Adding a Non-Project Line

Suppose that the judicial system called me for one day of jury duty this week. My organization considers jury duty as *Administrative* time and as non-project work; therefore, I need to add a *Jury Duty* line to my timesheet so that I can report time against it.

Click the **Add Row** pick list button and select the **Add Non-Project Line** item when you need to add additional administrative time rows to your timesheet. The system displays the *Administrative Time* dialog shown in Figure 9 - 18. For clarity, I captured the image after selecting or entering all of the information in the dialog so that you can easily see the desired result.

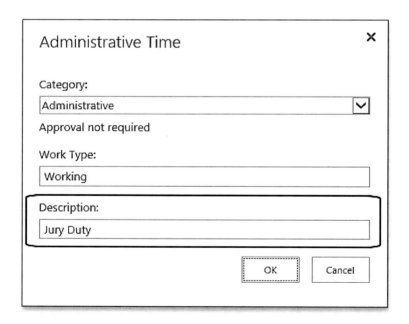

Figure 9 - 18: Administrative Time dialog

Click the **Category** pick list and select the category of administrative time. In my case, I selected the **Administrative** item in the pick list. Leave the **Work Type** value selected and then enter a description in the **Description** field. In my case, I entered **Jury Duty** in the *Description* field. Click the **OK** button to add the administrative line to your timesheet.

Adding a Personal Task

Our Technical Education team asked me to provide a "Train the Trainer" class on how to use our company's newly updated timekeeping system known as TimeAway 2019. Our organization considers a task like this as non-project task work, and does not consider it as administrative time. Therefore, so that I can log the hours of work on this special task, I need to add it to my timesheet as a personal task.

Click the **Add Row** pick list button and select the **Add Personal Task** item when you need to add a non-project line that is not administrative time to your timesheet. The system displays the *Add New Personal Task* dialog shown in Figure 9 - 18. For clarity, I captured the image after selecting or entering all of the information in the dialog so that you can easily see the desired result.

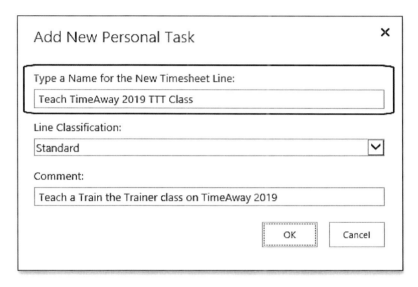

Figure 9 - 19: Add New Personal Task dialog

In the *Add New Personal Task* dialog, enter the name of the personal task in the **Type a Name for the New Timesheet Line** field. Click the **Line Classification** pick list and select a value, if required by your organization. Enter a comment in the **Comment** field to explain why you added the personal task. Click the **OK** button to add the new personal task to your timesheet and sets the *Project Name* value for the task to *Personal Tasks*.

 Information: Because of standardized reporting requirements, some organizations do not allow timesheet users to add personal tasks to their timesheets. If you click the Add Row pick list button and the Add Personal Task item is disabled, this means your organization does not allow you to add personal tasks to your timesheet.

Module 09

Reassigning a Task to a Fellow Team Member

After a project manager assigns a team member to a task, Project Web App offers the team member the option to reassign the task to someone else. This feature is useful in organizations where the project manager assigns a team leader to tasks in an enterprise project, and then the team leader is responsible for reassigning the tasks to members of their team.

 Information: Many organizations disable the *Reassign Task* feature in Project Web App because it should be the responsibility of the project manager to assign resources to tasks and not the responsibility of team members.

To reassign a task to someone else, click the **Options** tab to display the *Options* ribbon. Select the checkbox *for a single task* and then click the **Reassign** button in the *Tasks* section of the *Options* ribbon. Project Web App displays the *Task Reassignment* page shown in Figure 9 - 20. For clarity, I captured the image after selecting or entering all of the information in the page so that you can easily see the desired result.

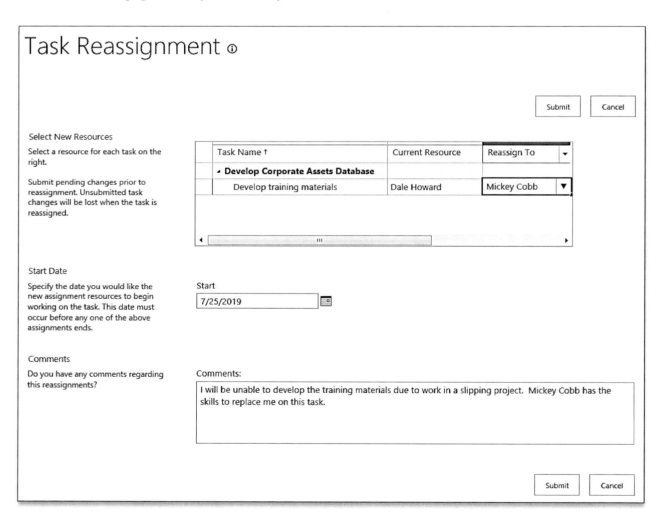

Figure 9 - 20: Task Reassignment page

Warning: If you select the checkboxes for multiple tasks, the *Task Reassignment* dialog displays a list of **every task assigned to you in every enterprise project**. To keep the reassignment process as simple as possible, Projility strongly recommends that you select only one task at a time and then reassign it.

The *Select New Resources* section of the *Task Reassignment* dialog shows you the task assignment you selected. The *Reassign To* column in the data grid allows you to select the resource to whom you want to reassign the task. When you click the pick list button in the **Reassign To** field, the list displays all of the current members of the project team.

Information: The *Reassign To* pick list contains **only** the members of your project team. If you need to reassign a task to someone outside your project team, you must first ask your project manager to add the new resource to the project team.

In in the *Reassign To* field, select the name of the team member to whom you want to reassign the selected task. In the *Start Date* section of the *Task Reassignment* page, click the **Start** pick list and select the date on which the fellow team member begins work on the task. The date in the *Start* field defaults *Start* date of the task in the Microsoft Project schedule. If you want to reassign the entire task to the fellow team member, ***do not*** change the date in the *Start* field. If you want to reassign only a part of the task to the fellow team member, choose a **Start** date that is later than the task *Start* date and earlier than the task *Finish* date.

Enter any relevant comment in the **Comments** field. When you complete your entries, click the **Submit** button. When you click the *Submit* button, Project Web App immediately removes you from the task and displays a task reassignment request on the *Approval Center* page of the project manager. Before the system finalizes the task reassignment request, the project manager must approve it.

Removing a Timesheet Line

At some point you may want to remove a line on your timesheet, such as when you do not need to report on a particular administrative time item such as *Vacation*. To remove a timesheet line, click the **Options** tab to display the *Options* ribbon. Select the checkbox to the left of the *Task Name/Description* column for the line you want to remove, and then click the **Remove Task** button in the *Tasks* section of the *Options* ribbon. Project Web App immediately removes the line from the timesheet.

Information: The system does not prompt you prior to removing a timesheet line, so use this option carefully.

Warning: ***Do not*** use this method to remove project tasks from your timesheet as a way of "hiding" tasks on which you have no progress to report this week. When you delete a task, this action sends a *Delete Assignment* request to the project manager who is the *Status Manager* of the task. If the project manager erroneously approves this request on the *Approval Center* page in PWA, the system completely removes you from the task in the Microsoft Project schedule. Once removed, you no longer see this task on future timesheets.

Submitting a Timesheet for Approval

At the end of each reporting period (usually Friday afternoon), you must submit your timesheet for approval. Click the **Timesheet** tab to display the *Timesheet* ribbon. In the *Submit* section of the *Timesheet* ribbon, click the **Send** pick list button, and then select the **Turn in Final Timesheet** item, such as shown in Figure 9 - 21.

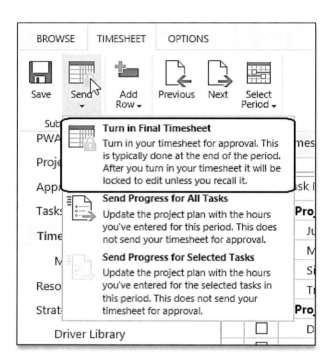

Figure 9 - 21: Select the Turn in Final Timesheet item

The system displays the *Send Timesheet* dialog shown in Figure 9 - 22. Enter an optional comment in the **Comment** field and then click the **OK** button.

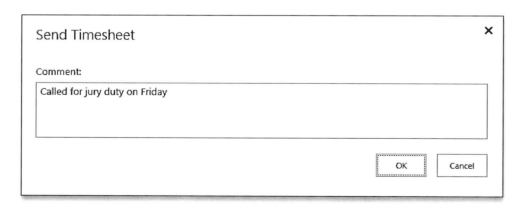

Figure 9 - 22: Send Timesheet dialog

After submission, Project Web App sends the timesheet data for two types of approval:

- The system routes your timesheet to your **Timesheet Manager**, who can approve or reject your timesheet as a whole. If your application administrator designated you as your own *Timesheet Manager*, the system automatically approves your submitted timesheets.

- The system routes each task update to the **Status Manager** of the task, who can approve or reject each task update individually.

The *Send* pick list shown previously in Figure 9 - 21 contains two other options: the *Send Progress for All Tasks* and *Send Progress for Selected Tasks* options. You may need to use one of these options when your company needs you to submit "interim progress" on tasks in your timesheet, such as when a reporting period ends in the middle of a timesheet period. In a situation such as this, you would need to select the checkboxes for every task on which you applied progress and then select the *Send Progress for Selected Tasks* item in the *Send* pick list. Alternatively, you could simply click the *Send* pick list and select the *Send Progress for All Tasks* item instead. At the end of the week, you would still need to submit your timesheet by selecting the *Turn in Final Timesheet* item in the *Send* pick list.

Recommended Best Practices for Using the Timesheet

To summarize the content presented in the preceding subtopical sections, Projility recommends that you use the following best practice approach for entering and submitting time on the *Timesheet* page in Project Web App:

1. At the end of each day, enter the amount of time you spent that day on any task or administrative time category listed in the timesheet.

2. Add a comment in the *Comment* field for any task requiring additional information.

3. After you enter your time and optional comments each day, click the **Save** button in the *Submit* section of the *Timesheet* ribbon to save the latest changes to your timesheet.

4. On the last day of the timesheet period (usually Friday), adjust the **Remaining Work** value for any task according to your organization's methodology for submitting time.

5. Submit your timesheet for approval.

Hands On Exercise

Exercise 9 - 2

Enter time and task progress in the *Timesheet* page of Project Web App and then submit your timesheet for approval.

1. In the upper right corner of the *Timesheet* page, click the **Focus on Content** button.

2. Study the information in the *Planned* row for the *Write Chapter 1* task.

3. Click the **Options** tab to display the *Options* ribbon, if necessary.

4. In the *Show/Hide* section of the *Options* ribbon, ***deselect*** the **Planned** checkbox.

5. In your *Timesheet* page, enter the following time on tasks and administrative time categories:

Task Name	M	T	W	Th	F
Write Chapter 1	8h	6h	4h	0h	8h
Administrative		2h	4h		
Sick time				8h	

6. Click the **Timesheet** tab to display the *Timesheet* ribbon.

7. In the *Submit* section of the *Timesheet* ribbon, click the **Save** button.

8. Change the **Remaining Work** value for the **Write Chapter 1** task to **62h**.

9. Drag the split bar to the right edge of the *Comment* field.

10. In the **Comment** field, enter the following text to describe the reason for increasing the *Remaining Work* value:

    Increased Rem Work due of new features needed in Chapter 1

11. In the *Submit* section of the *Timesheet* ribbon, click the **Save** button.

12. In the *Tasks* section of the *Timesheet* ribbon, click the **Add Row** pick list button and select the **Select from Existing Assignments** item.

13. In the *Add an Existing Task* dialog, first expand your enterprise project, and then expand the **Course Manual** summary section.

14. Select the checkbox for the **Write Chapter 2** task and then click the **OK** button.

Notice that the *Timesheet* page now displays the *Write Chapter 2* future task on your timesheet for the current reporting period.

15. Select the checkbox for the **Vacation** administrative time item.

16. Click the **Options** tab to display the *Options* ribbon.

17. In the *Tasks* section of the *Options* ribbon, click the **Remove Task** button.

18. Click the **Timesheet** tab to display the *Timesheet* ribbon.

19. In the *Submit* section of the *Timesheet* ribbon, click the **Send** pick list button, and select the **Turn in Final Timesheet** item.

20. In the **Comment** field of the Send Timesheet dialog, enter the text for the following comment:

    Sick time on Thursday – no work completed

21. In the *Send Timesheet* dialog, click the **OK** button.

22. In the upper right corner of the *Timesheet* page, click the **Focus on Content** button again to display the entire page.

Tracking Time and Task Progress

Recalling a Submitted Timesheet

There may come a time when you need to edit information on a timesheet that you previously submitted for approval, such as to correct a mistake or make changes. By default, Project Web App does not allow you to edit a submitted timesheet because it locks all of the cells normally open for data entry. Before you can edit a submitted timesheet, you must recall it. To recall a timesheet, do the following:

1. On the *Home* page of Project Web App, click the **Timesheets** tile in the *Track your work* carousel to display the *Manage Timesheets* shown previously in Figure 9 - 2.

2. On the *Manage Timesheets* page, click anywhere in the row of the timesheet period you want to recall, such as clicking the name of the timesheet period shown in the *Period* column. ***Do not*** click the *My Timesheet* link, however, as this navigates you away from the *Manage Timesheets* page.

3. In the *Timesheet* section of the *Timesheets* ribbon, click the **Recall** button shown in Figure 9 - 23.

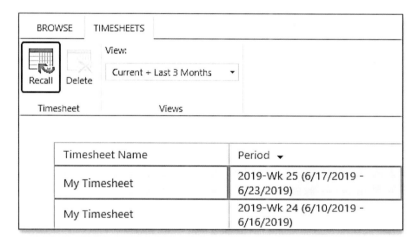

Figure 9 - 23: Click the Recall button

4. In the confirmation dialog shown in Figure 9 - 24, click the **OK** button to recall the timesheet. Project Web App recalls the timesheet, unlocks all data cells in the timesheet, and resets the status of the timesheet to *In Progress*.

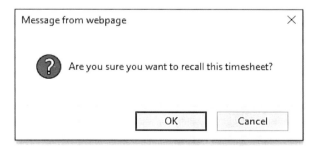

Figure 9 - 24: Confirmation dialog to recall a timesheet

5. Click the **My Timesheet** link for the selected time period to return to the *Timesheet* page for that period, where you can edit and resubmit your timesheet.

273

Module 09

 Warning: Project Web App allows you to recall a timesheet already approved by your *Timesheet Manager*. It even allows you to recall a timesheet containing task updates already approved by their *Status Manager*. If you need to recall a timesheet, Projility strongly recommends that you reach out to your *Timesheet Manager* and to the impacted project managers to warn them about the changes you are about to make to your timesheet after recalling it.

Deleting a Timesheet

There may come a time when you need to delete an entire timesheet and recreate it. This can result from a variety of situations, such as when you totally "mess up" a timesheet and want to start over. Complete the following to delete a timesheet:

1. On the *Home* page of Project Web App, click the **Timesheets** tile in the *Track your work* carousel to display the *Manage Timesheets* shown previously in Figure 9 - 2.

2. On the *Manage Timesheets* page, click anywhere in the row of the timesheet period you want to delete, such as clicking the name of the timesheet period shown in the *Period* column. ***Do not*** click the *My Timesheet* link, however, as this navigates you away from the *Manage Timesheets* page.

3. In the *Timesheet* section of the *Timesheets* ribbon, click the **Delete** button shown in Figure 9 - 25.

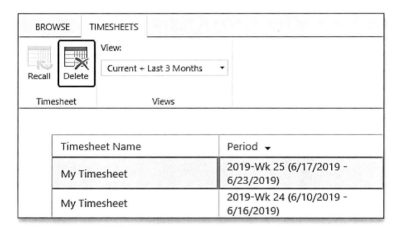

Figure 9 - 25: Click the Delete button

4. In the confirmation dialog shown in Figure 9 - 26, click the **OK** button to delete the timesheet. Project Web App deletes the timesheet and changes the name shown in the *Timesheet Name* column to *Click to Create*.

Tracking Time and Task Progress

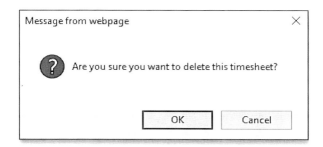

Figure 9 - 26: Confirmation dialog to delete a timesheet

5. To recreate the timesheet, click the **Click to Create** link for the selected timesheet period.

Responding to a Rejected Timesheet

In the normal flow of timesheet approvals, your *Timesheet Manager* approves your timesheets. However, there is always a possibility that your *Timesheet Manager* could reject a timesheet for a variety of reasons, such as:

- You submit a timesheet containing a factual error, such as when you enter the number 23 but you intended to type the number 3.

- You report work that totals greater than 24 hours in a single day.

- You create a personal task to document time that does not fit into any of the tasks on your timesheet, and your *Timesheet Manager* needs you to add the time to an administrative task instead.

- You fail to report time on a day when you actually worked.

In situations like this, your *Timesheet Manager* may reject your timesheet and ask you to correct the troublesome information. When a *Timesheet Manager* rejects a timesheet, Project Web App automatically sends an e-mail message to notify the timesheet user of the rejection.

To respond to a rejected timesheet, use the *Manage Timesheets* page to navigate to the rejected timesheet. The system opens the *Timesheet* page for the rejected timesheet in editing mode. Make corrective action on the timesheet line that shows a **Rejected** value in the *Process Status* column and then resubmit the timesheet for approval.

Submitting Future Planned PTO

Organizations can use the *Timesheet* page for users to submit planned PTO in the future, such as planned vacation time. If your organization uses the *Timesheet* page for this purpose, you can submit planned PTO by completing the following steps:

1. On the *Home* page of Project Web App, click the **Timesheets** tile in the *Track your work* carousel to display the *Manage Timesheets* shown previously in Figure 9 - 2.

2. In the *Timesheet* ribbon at the top of the *Manage Timesheets* page, click the **View** pick list and select a view that displays future timesheet periods, such as the **Next 6 Months + Last 3 Months** view shown in Figure 9 - 27.

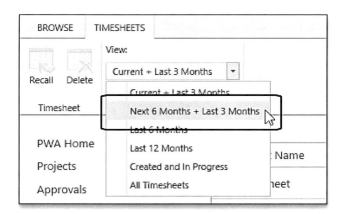

Figure 9 - 27: Select a view that
displays future time periods

3. Click the **Click to Create** link for the timesheet period in which you want to plan future PTO.

4. Click the **Options** tab to display the *Options* ribbon.

5. In the *Show/Hide* section of the *Options* ribbon, select the **Planned** checkbox to display the *Planned* row for each line in your timesheet.

6. In the upper right corner of the page, click the **Focus on Content** button shown previously in Figure 9 - 9 to hide the header information above the ribbon and the *Quick Launch* menu on the left side of the page.

7. Enter the number of hours of planned PTO in the **Planned** row of the **Vacation** line in your timesheet. For example, notice in the timesheet shown in Figure 9 - 28 that I intend to take 40 hours of planned vacation during the week of August 12.

	Task Name/Description ↑	Remaining Work	Time Type	Mon 8/12	Tue 8/13	Wed 8/14	Thu 8/15	Fri 8/16
	▲ Project Name: Administrative							
☐	Administrative		Actual					
			Planned					
☐	Meetings		Actual					
			Planned					
☐	Sick time		Actual					
			Planned					
☐	Training		Actual					
			Planned					
☐	Vacation		Actual					
			Planned	8h	8h	8h	8h	8h
	▲ Project Name: Chi Rho Software	120h						
☐	Develop code	120h	Actual					
			Planned	8h	8h			

Figure 9 - 28: Planned PTO entered for the Vacation line

8. Click the **Timesheet** tab to display the *Timesheet* ribbon.

9. In the *Submit* section of the *Timesheet* ribbon, click the **Save** button.

10. In the upper right corner of the page, click the **Focus on Content** button display the header information above the ribbon and the *Quick Launch* menu on the left side of the page

Information: When you save your timesheet with the planned PTO entered as Vacation (or any other nonworking administrative time category), Project Online immediately adds the PTO as nonworking time on your calendar in the Enterprise Resource Pool. The system automatically reschedules any project work assigned to you during the PTO time period.

When you return from taking the planned PTO, reopen the timesheet containing the planned PTO. Enter your actual PTO time in the **Actual** row of the **Vacation** line and then submit your timesheet for approval.

Hands On Exercise

Exercise 9 - 3

Create planned PTO in a future timesheet period.

Warning: You should only work this exercise if your organization wants users to enter planned PTO in future timesheet periods *and* if you have actual vacation time planned sometime in the next six months.

1. Click the **Browse** tab and then click the **Project Online** icon at the top of the *Timesheet* page to return to the *Home* page of Project Web App.

2. In the *Track your work* carousel, click the **Timesheets** tile to navigate to the *Manage Timesheets* page.

3. In the *Views* section of the *Timesheets* ribbon, click the **View** pick list and select the **Next 6 Months + Last 3 Months** view.

4. Click the **Click to Create** link for a future time period during which you want to plan future PTO.

5. In the upper right corner of the *Timesheet* page, click the **Focus on Content** button.

6. Drag the split bar to the right edge of the *Remaining Work* column, if necessary.

7. Click the **Options** tab to display the *Options* ribbon.

8. In the *Show/Hide* section of the *Options* ribbon, select the **Planned** checkbox.

9. In the **Planned** row for the **Vacation** line, enter **8h** on each working day of the time period.

10. Click the **Timesheet** tab to display the *Timesheet* ribbon.

11. In the *Submit* section of the *Timesheet* ribbon, click the **Save** button.

12. In the upper right corner of the *Timesheet* page, click the **Focus on Content** button again.

Using the Tasks Page

If your organization uses the *Tasks* page to track task progress only, be aware that Project Web App offers three primary methods of tracking task progress. Each organization can select one of the following methods:

- The **Percent of Work Complete** method allows users to enter a cumulative *% Work Complete* value and an optional *Remaining Work* estimate for each task assignment.

- The **Actual Work Done and Work Remaining** method allows users to enter a cumulative *Actual Work* value and to adjust the *Remaining Work* estimate for each task assignment.

- The **Hours of Work Done per Period** method allows users to enter their hours of *Actual Work* in a daily timesheet grid and to adjust the *Remaining Work* estimate for each task assignment.

> **Information**: Project Web App does offer a fourth method of tracking task progress known as **Free Form**. As the name implies, this method of tracking allows a user to enter task progress using any of the three methods. Projility ***does not*** recommend using the *Free Form* method because it requires users to master three different methods of tracking task progress. Therefore, my assumption in this book is that your organization selected one of the three primary methods of tracking progress shown in the bulleted list above.

Your organization may choose to use the default layout of the *Tasks* page, or use a modified layout recommended by Projility. Because of this flexibility, I discuss each method of tracking progress, using both the default layout and the custom Projility layout of the *Tasks* page.

Understanding the Tasks Page

To navigate to the *Tasks* page in Project Web App, use either of the following methods:

- In the *Quick Launch* menu, click the **Tasks** link.
- In the *Track your work* carousel, click the **Tasks** tile.

Project Web App displays the *Tasks* page, such as the one shown in Figure 9 - 29. By default, the *Tasks* page displays every task assigned to you, organized into groups with two levels of grouping. The first level of grouping is by **Planning Window**, while the second level of grouping is by **Project Name**. The *Planning Windows* grouping organizes tasks into up to four groups, which are as follows:

- The **In Progress for Current Period** group includes every incomplete task with work scheduled during the current week, plus incomplete tasks from previous weeks.

- The **Near Future – Next 2 Periods** group includes incomplete tasks scheduled to start in the next two weeks, unless specified otherwise by your application administrator.

- The **Distant Future** group includes incomplete tasks scheduled to start three or more weeks in the future.

- The **Completed** group includes only completed tasks.

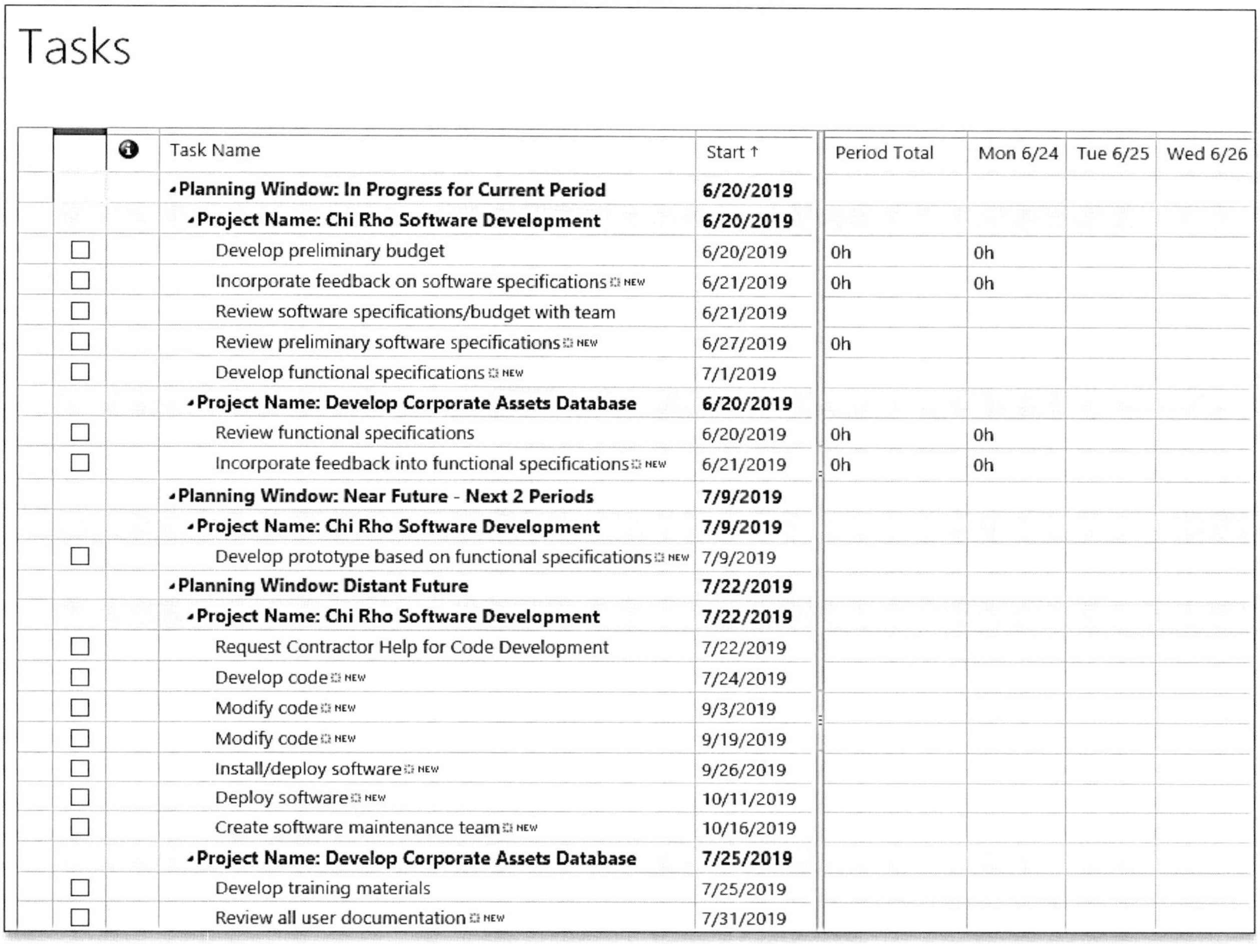

Tasks

	❶	Task Name	Start ↑	Period Total	Mon 6/24	Tue 6/25	Wed 6/26
		▲Planning Window: In Progress for Current Period	6/20/2019				
		▲Project Name: Chi Rho Software Development	6/20/2019				
☐		Develop preliminary budget	6/20/2019	0h	0h		
☐		Incorporate feedback on software specifications ✦ NEW	6/21/2019	0h	0h		
☐		Review software specifications/budget with team	6/21/2019				
☐		Review preliminary software specifications ✦ NEW	6/27/2019	0h			
☐		Develop functional specifications ✦ NEW	7/1/2019				
		▲Project Name: Develop Corporate Assets Database	6/20/2019				
☐		Review functional specifications	6/20/2019	0h	0h		
☐		Incorporate feedback into functional specifications ✦ NEW	6/21/2019	0h	0h		
		▲Planning Window: Near Future - Next 2 Periods	7/9/2019				
		▲Project Name: Chi Rho Software Development	7/9/2019				
☐		Develop prototype based on functional specifications ✦ NEW	7/9/2019				
		▲Planning Window: Distant Future	7/22/2019				
		▲Project Name: Chi Rho Software Development	7/22/2019				
☐		Request Contractor Help for Code Development	7/22/2019				
☐		Develop code ✦ NEW	7/24/2019				
☐		Modify code ✦ NEW	9/3/2019				
☐		Modify code ✦ NEW	9/19/2019				
☐		Install/deploy software ✦ NEW	9/26/2019				
☐		Deploy software ✦ NEW	10/11/2019				
☐		Create software maintenance team ✦ NEW	10/16/2019				
		▲Project Name: Develop Corporate Assets Database	7/25/2019				
☐		Develop training materials	7/25/2019				
☐		Review all user documentation ✦ NEW	7/31/2019				

Figure 9 - 29: Tasks page

By default, the *Tasks* page the following columns of information in the sheet on the left side of the page:

- Attachments
- Task Name
- Start
- Finish
- Remaining Work
- % Work Complete
- Work
- Actual Work
- Process Status

Module 09

The *Attachments* column displays a yellow sticky note icon for tasks that include task notes, and it displays a paperclip icon to denote tasks linked to issues, risks, and documents. You can click either of these icons to view the item attached to a task.

By default, the *Tasks* page displays a task sheet on the left side of the page and a timesheet grid on the right side of the page. Project Web App allows you to change the layout of the page to match the method of tracking task progress selected by your organization. Click the **Layout** pick list button in the *Display* section of the *Tasks* ribbon, as shown in Figure 9 - 30. In the pick list, select the **Gantt Chart** item to display a *Gantt Chart* on the right side, or select the **Sheet** item to display only the columns in the task sheet on the left side.

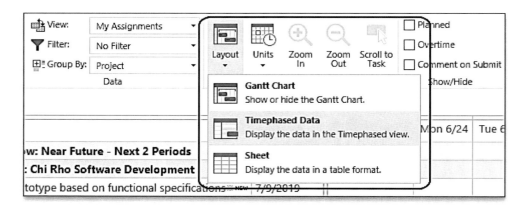

Figure 9 - 30: Change the layout of the Tasks page

If your organization uses either the *Percent of Work Complete* method or the *Actual Work Done and Work Remaining* method, Projility recommends that you initially select the **Gantt Chart** item in the **Layout** pick list so that you can study the schedule of tasks assigned to you, and then you should select the **Sheet** item on the **Layout** pick list to display only the task sheet. If your organization uses the *Hours of Work Done per Period* method, Projility recommends you leave the **Timephased Data** item selected on the **Layout** pick list.

Understanding the Assignment Details Page

In the *Tasks* page, Project Web App displays the name of each task as a hyperlink. When you click the name of any task in the *Tasks* page, Project Web App navigates you to the *Assignment Details* page for the selected task. Because of the extreme length of this page, I elected to display a screenshot of each section of the page for readability purposes. The *Assignment Details* page displays complete information for the selected task assignment, and includes the following sections:

- In the **General Details** section shown in Figure 9 - 31, you can enter task progress information, such as *Start* date, *Remaining Work*, or *Finish* date.

Tracking Time and Task Progress

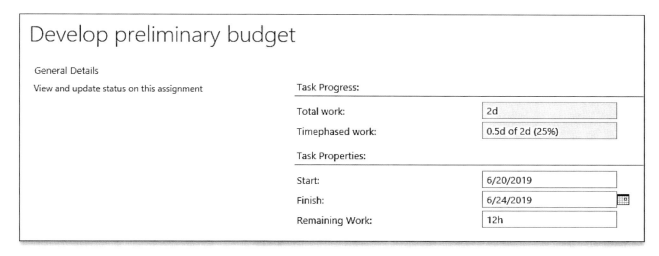

Figure 9 - 31: Assignment Details page,
General Details section

- In the **Recent Task Changes** section shown in Figure 9 - 32, you can view the history of task updates submitted by you, task approvals made by the project manager, and comments added by you or the project manager.

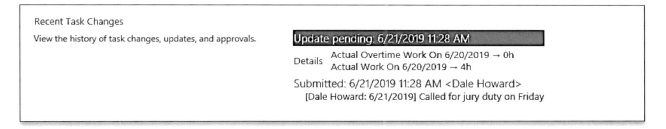

Figure 9 - 32: Assignment Details page,
Recent Task Changes section

- In the **Attachments** section shown in Figure 9 - 33, you can view risks, issues, or documents associated with the task.

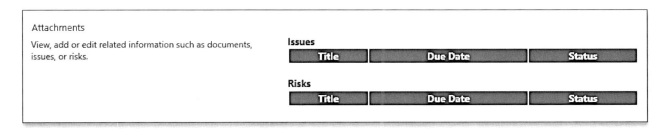

Figure 9 - 33: Assignment Details page,
Attachments section

281

- In the **Contacts** section shown in Figure 9 - 34, you can see the names of your project manager (the *Owner* of the project), the approval manager (the *Status Manager* for the task), the names of people assigned to the task, and the names of your fellow project team members.

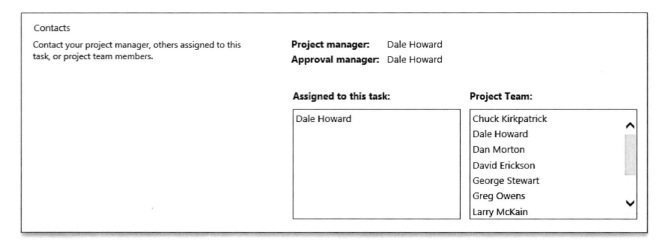

Figure 9 - 34: Assignment Details page,
Contacts section

- In the **Related Assignments** section shown in Figure 9 - 35, you can see predecessor and successor tasks related to the selected task. Notice that the *Tasks scheduled to finish before this task can start* subsection shows the name of one task that is a direct predecessor for the selected task, along with the status of the predecessor as a % complete value. Notice also that the *Tasks dependent on this task's finish date* subsection shows the name of two tasks that are direct successors to the selected task, along with the names of resources assigned to the successor tasks.

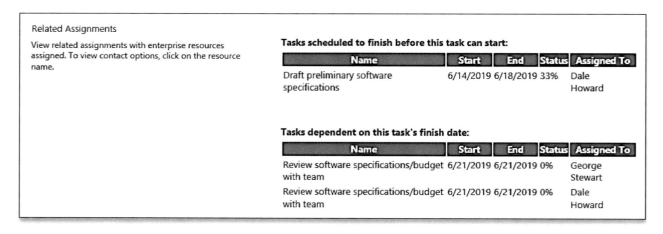

Figure 9 - 35: Assignment Details page,
Related Assignments section

Tracking Time and Task Progress

Best Practice: If you finish early on a task, Projility recommends that you use the information in the *Tasks dependent on this task's finish date* subsection to find out the names of the team members assigned to any direct successor tasks. Once known, notify them of your early finish so that they have a chance to start work early on their successor tasks, if possible.

- In the **Notes** section shown in Figure 9 - 36, you can read any notes added to the task in the Microsoft Project schedule by the project manager, and you can add your own notes to the task as well. The top pane in the *Notes* section displays notes added by the project manager, while the bottom pane allows you to add your own notes to the task. When you add a note to a task, by the way, and the project manager approves your task updates, the system automatically adds your note to the task in the Microsoft Project schedule. The system "name stamps" the notes text with your name to indicate you are the team member who added the note.

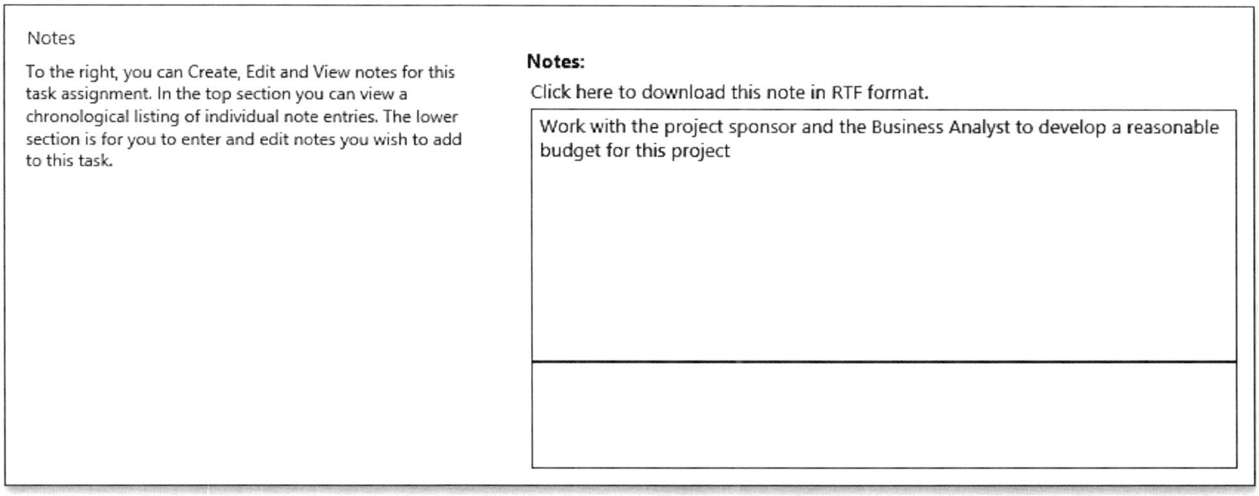

Figure 9 - 36: Assignment Details page,
Notes section

Information: Although it is possible to enter task progress in the *Assignment Details* page, Projility **does not** recommend doing so as most organizations do not configure this page to match the configuration of the *Tasks* page. Instead, always enter task progress in the *Tasks* page, but use the *Assignment Details* page to obtain additional information about any task, or to add a note to the task.

Reporting Progress from the Tasks Page

As stated earlier, Project Web App offers your organization three primary methods for tracking progress. The information you enter on the *Tasks* page varies with the method of tracking your organization selected. Your organization may choose to use the default layout of the *Tasks* page, or use a modified layout recommended by Projility. Because of this flexibility, I discuss each method of tracking progress, using both the default layout and the custom Projility layout of the *Tasks* page.

Using Percent of Work Complete

Although you may report progress from the *Tasks* page using the **Percent of Work Complete** method of tracking progress, Project Web App accepts only a limited amount of information with the default layout of the *Tasks* page. To enter task progress using this method of tracking progress on the *Tasks* page, use the following methodology to report progress on a task:

1. Click in the **% Work Complete** field and enter your estimate of the cumulative percentage of work completed to date on the task.

Information: When you enter *100%* in the *% Work Complete* field for a task, Project Web App assumes that you started and finished the task as originally scheduled. The system has no way of knowing whether you started or finished the task early or late compared to the original schedule.

2. To adjust the remaining work (also known as the ETC or Estimate To Complete) for the task, click in the **Remaining Work** field and enter your remaining work estimate in hours.

3. To add a note to the task, click the task name and scroll to the bottom of the *Assignment Details* page. Enter your note in the bottom pane of the *Notes* section and then click the **Save** button.

Best Practice: Projility recommends that you train team members to "date stamp" their notes text by adopting a standard convention for entering notes. Project Web App "name stamps" each note to show who submitted the note, but does not "date stamp" the note to show when the team member submitted the note.

4. In the *Submit* section of the *Tasks* ribbon, click the **Save** button to save your task progress changes if you are not ready to send them to your project manager for approval

5. When you are ready to submit the task changes to your project manager, click the **Send Status** pick list button in the *Submit* section of the *Tasks* ribbon, and then select the **All Tasks** option, as shown in Figure 9 - 37.

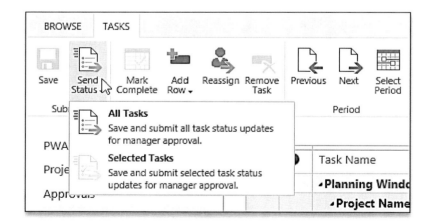

Figure 9 - 37: Click the Send Status pick list button

Best Practice: Using Percent of Work Complete

Projility recommends that your application administrator create a customized *My Assignments* view for the *Tasks* page that optimizes the experience of users entering task progress using the **Percent of Work Complete** method of tracking progress. This custom view should include the following columns displayed in the following order:

- Task Name
- Actual Start
- % Work Complete (with a *% Wk Comp* label applied)
- Remaining Work (with a *Rem Work* label applied)
- Actual Finish
- Start (set to Read Only to prevent data entry)
- Finish (set to Read Only to prevent data entry)

You application administrator should use the *Sheet* layout for this customized *My Assignments* view with the default *Project* grouping applied. Figure 9 - 38 shows the *Tasks* page customized with the Projility recommended best practice layout.

		Task Name	Actual Start	% Wk Comp	Rem Work	Actual Finish	Start ↑	Finish
		▲Planning Window: In Progress for Current Period			122h		6/20/2019	7/11/2019
		▲Project Name: Chi Rho Software Development			40h		6/20/2019	7/1/2019
☐		Develop preliminary budget	6/20/2019	25%	12h		6/20/2019	6/24/2019
☐		Incorporate feedback on software specifications NEW		0%	8h		6/24/2019	6/25/2019
☐		Review software specifications/budget with team		0%	4h		6/24/2019	6/24/2019
☐		Review preliminary software specifications NEW		0%	16h		6/28/2019	7/1/2019
		▲Project Name: Develop Corporate Assets Database			20h		6/20/2019	6/24/2019
☐		Review functional specifications	6/20/2019	25%	12h		6/20/2019	6/24/2019
☐		Incorporate feedback into functional specifications NEW		0%	8h		6/21/2019	6/24/2019
		▲Planning Window: Near Future - Next 2 Periods		0%	152h		7/2/2019	7/22/2019
		▲Project Name: Chi Rho Software Development		0%	72h		7/2/2019	7/15/2019
☐		Develop functional specifications NEW		0%	40h		7/2/2019	7/9/2019
☐		Develop prototype based on functional specifications NEW		0%	32h		7/10/2019	7/15/2019
		▲Planning Window: Distant Future		0%	624h		7/22/2019	10/25/2019
		▲Project Name: Chi Rho Software Development		0%	216h		7/22/2019	10/25/2019
☐		Request Contractor Help for Code Development NEW		0%	24h		7/22/2019	7/24/2019
☐		Develop code NEW		0%	120h		7/25/2019	8/21/2019
☐		Modify code NEW		0%	24h		9/11/2019	9/16/2019

Figure 9 - 38: Customized Tasks page for Percent Work Complete entry

To enter progress using the **Percent of Work Complete** method of tracking using the customized layout of the *Tasks* page, use the following methodology:

1. When you start work on a new task, click in the **Actual Start** field for the selected task and select the date you began work on the task.
2. Click in the **% Work Complete** field and enter your estimate of the cumulative percentage of work completed to date on the task.
3. Enter your estimated amount of remaining work in the **Remaining Work** field.

4. When you complete a task, click in the **Actual Finish** field for the selected task and select the date you finished work on the task.

5. To add a note to the task, click the task name and scroll to the bottom of the *Assignment Details* page. Enter your note in the bottom pane of the *Notes* section and then click the **Save** button.

6. In the *Submit* section of the *Tasks* ribbon, click the **Save** button to save your task progress changes if you are not ready to send them to your project manager for approval.

7. When you are ready to submit the task changes to your project manager, click the **Send Status** pick list button in the *Submit* section of the *Tasks* ribbon, and then select the **All Tasks** option, such as shown previously in Figure 9 - 37.

Information: Although this method of tracking requires a little more work on the part of team members, entering an *Actual Start* date and an *Actual Finish* date provides the project manager much more accurate scheduling information to better forecast schedule slippage.

Using Actual Work Done and Work Remaining

In the **Actual Work Done and Work Remaining** method of tracking task progress, you must enter the cumulative amount of actual work spent on the task and the amount of work remaining. This method of time accrual enables you and your management to gauge progress toward the task outcome and the quality of the initial work estimate. This method also minimizes the somewhat subjective nature of *% Work Complete* tracking by forcing focus on the level of effort expended versus planned. To enter task progress using this method of tracking progress on the *Tasks* page, use the following methodology to report progress on a task:

1. Click in the **Actual Work** field and enter your cumulative amount of actual work completed to date on the task.

2. To adjust the remaining work (also known as the ETC or Estimate To Complete) for the task, click in the **Remaining Work** field and enter your remaining work estimate.

Information: When you enter the initial amount of actual work in the *Actual Work* field, Project Web App assumes that you started work on the task as originally scheduled. When you set the *Remaining Work* value to *0 hours*, the system assumes you finished the task as originally scheduled. Project Web App has no way of knowing whether you started or finished the task early or late compared to the original schedule.

3. To add a note to the task, click the task name and scroll to the bottom of the *Assignment Details* page. Enter your note in the bottom pane of the *Notes* section and then click the **Save** button.

Best Practice: Projility recommends that you train team members to "date stamp" their notes text by adopting a standard convention for entering notes. Project Web App "name stamps" each note to show who submitted the note, but does not "date stamp" the note to show when the team member submitted the note.

4. In the *Submit* section of the *Tasks* ribbon, click the **Save** button to save your task progress changes if you are not ready to send them to your project manager for approval

Tracking Time and Task Progress

5. When you are ready to submit the task changes to your project manager, click the **Send Status** pick list button in the *Submit* section of the *Tasks* ribbon, and then select the **All Tasks** option, as shown previously in Figure 9 - 37.

Best Practice: Using Actual Work Done and Work Remaining

Projility recommends that your application administrator create a customized *My Assignments* view for the *Tasks* page that optimizes the experience of users entering task progress using the **Actual Work Done and Work Remaining** method of tracking progress. This custom view should include the following columns displayed in the following order:

- Task Name
- Actual Start
- Actual Work
- Remaining Work (with a *Rem Work* label applied)
- Actual Finish
- Start (set to Read Only to prevent data entry)
- Finish (set to Read Only to prevent data entry)

You application administrator should use the *Sheet* layout for this customized *My Assignments* view with the default *Project* grouping applied. Figure 9 - 39 shows the *Tasks* page customized with the Projility recommended best practice layout.

Task Name	Actual Start	Actual Work	Rem Work	Actual Finish	Start ↑	Finish
▲Planning Window: In Progress for Current Period		34h	122h		6/20/2019	7/11/2019
▲Project Name: Chi Rho Software Development		4h	40h		6/20/2019	7/1/2019
Develop preliminary budget	6/20/2019	4h	12h		6/20/2019	6/24/2019
Incorporate feedback on software specifications NEW		0h	8h		6/24/2019	6/25/2019
Review software specifications/budget with team		0h	4h		6/24/2019	6/24/2019
Review preliminary software specifications NEW		0h	16h		6/28/2019	7/1/2019
▲Project Name: Develop Corporate Assets Database		4h	20h		6/20/2019	6/24/2019
Review functional specifications	6/20/2019	4h	12h		6/20/2019	6/24/2019
Incorporate feedback into functional specifications NEW		0h	8h		6/21/2019	6/24/2019
▲Planning Window: Near Future - Next 2 Periods		0h	152h		7/2/2019	7/22/2019
▲Project Name: Chi Rho Software Development		0h	72h		7/2/2019	7/15/2019
Develop functional specifications NEW		0h	40h		7/2/2019	7/9/2019
Develop prototype based on functional specifications NEW		0h	32h		7/10/2019	7/15/2019
▲Planning Window: Distant Future		0h	624h		7/22/2019	10/25/2019
▲Project Name: Chi Rho Software Development		0h	216h		7/22/2019	10/25/2019
Request Contractor Help for Code Development NEW		0h	24h		7/22/2019	7/24/2019
Develop code NEW		0h	120h		7/25/2019	8/21/2019
Modify code NEW		0h	24h		9/11/2019	9/16/2019

Figure 9 - 39: Customized Tasks page for Using Actual Work Done and Work Remaining entry

To enter progress using the **Using Actual Work Done and Work Remaining** method of tracking using the customized layout of the *Tasks* page, use the following methodology:

1. When you start work on a new task, click in the **Actual Start** field for the selected task and select the date you began work on the task.

2. Click in the **Actual Work** field and enter your cumulative amount of actual work completed to date on the task.

3. Enter your estimated amount of remaining work in the **Remaining Work** field.

4. When you complete a task, click in the **Actual Finish** field for the selected task and select the date you finished work on the task.

5. To add a note to the task, click the task name and scroll to the bottom of the *Assignment Details* page. Enter your note in the bottom pane of the *Notes* section and then click the **Save** button.

6. In the *Submit* section of the *Tasks* ribbon, click the **Save** button to save your task progress changes if you are not ready to send them to your project manager for approval.

7. When you are ready to submit the task changes to your project manager, click the **Send Status** pick list button in the *Submit* section of the *Tasks* ribbon, and then select the **All Tasks** option, such as shown previously in Figure 9 - 37.

Information: Although this method of tracking requires a little more work on the part of team members, entering an *Actual Start* date and an *Actual Finish* date provides the project manager much more accurate scheduling information to better forecast schedule slippage.

Using Hours of Work Done per Period

Organizations that use the **Hours of Work Done per Period** method of tracking progress need the finer level of detail, typically to meet billing requirements or the more rigorous requirements of a mature project management process. Reporting task progress on the *Tasks* page typically requires the daily entry of task progress in the timesheet grid. To enter task progress using this method of tracking progress on the default layout of the *Tasks* page, use the following methodology:

1. In the upper right corner of the *Tasks* page, click the **Focus on Content** button.

2. At the end of each day, enter the hours you worked on each task in the timesheet grid.

3. To add a note to the task, click the task name and scroll to the bottom of the *Assignment Details* page. Enter your note in the bottom pane of the *Notes* section and then click the **Save** button.

4. In the *Submit* section of the *Tasks* ribbon, click the **Save** button to save your current progress at the end of each day.

5. On the last day of the reporting period (usually Friday), enter your estimated amount of remaining work in the **Remaining Work** field for any tasks that require an adjustment.

6. When you are ready to submit the task changes to your project manager, click the **Send Status** pick list button in the *Submit* section of the *Tasks* ribbon, and then select the **All Tasks** option, such as shown previously in Figure 9 - 37.

7. In the upper right corner of the *Tasks* page, click the **Focus on Content** button again.

Tracking Time and Task Progress

Best Practice: As a best practice for using the *Hours of Work Done per Period* method of tracking task progress, Projility recommends that you enter actual progress on a daily basis and submit progress to your project manager on a weekly basis.

Adding Tasks to the Tasks Page

Along with reporting progress, other important task-related activities on the *Tasks* page include the following:

- Adding a new task
- Adding yourself to an existing task
- Adding Team tasks

To add a new task to the *Tasks* page, click the **Tasks** tab to expand the *Tasks* ribbon. In the *Tasks* section of the *Tasks* ribbon, click the **Add Row** pick list button and select one of the items on the pick list, such as shown in Figure 9 - 40. I discuss each of these items in detail in the topical sections that follow.

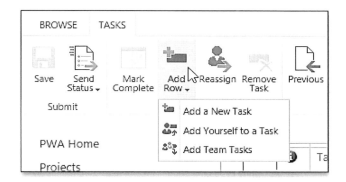

Figure 9 - 40: Add Row pick list button
in the Tasks page

Adding a New Task

Click the **Add Row** pick list button and select the **Add New Task** item to propose a new task not currently included in a Microsoft Project schedule. The new task might represent work that the project manager missed when planning the project schedule, or it might represent unexpected work that arose this week and for which you need to enter progress. Project Web App displays the *New Task* page shown in Figure 9 - 41. For clarity, I captured the image after selecting or entering all of the information in the page so that you can easily see the desired result.

Begin the process of filling out this page by clicking the **Project** pick list and selecting the project in which you want to propose the new task. The system refreshes the page with information relevant to the selected project. Click the **Summary task** pick list and select the summary section in which you want to propose the new task. The pick list includes every summary task in the selected project, by the way. Once again, the system refreshes the page with information relevant to the selected summary section.

In the **Task name** field, enter the name of the proposed task. In the **Start** field, enter your estimated start date for the proposed task. To determine the finish date of the project, the system needs you to enter either an estimated finish date in the **Finish** field or the number of hours of estimated work in the **Total work** field. If you enter a date in the *Finish* field, the system calculates the *Total work* value; if you enter a value in the *Total work* field, the system

289

calculates the *Finish* date of the proposed task. In the **Comments** field, enter a comment to explain why you need to add the proposed task to the project. Click the **Send** button to complete the process of proposing a new task.

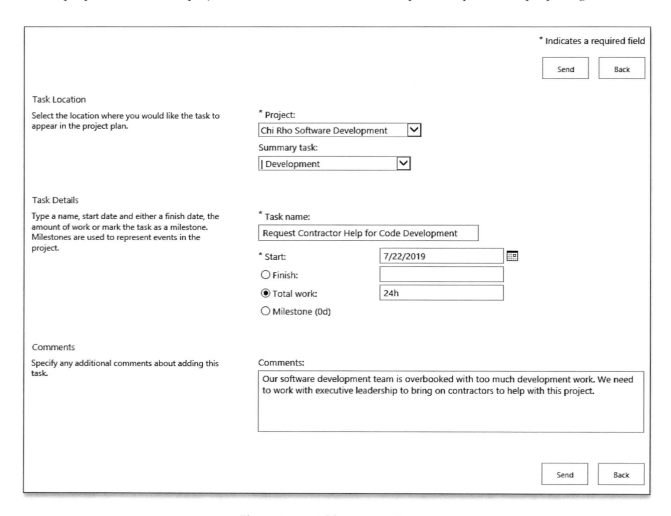

Figure 9 - 41: Add a New Task page

Adding Yourself to an Existing Task

Click the **Add Row** pick list button and select the **Add Yourself to a Task** item when you need to assist a fellow project team member with work on a task to which the team member is already assigned but which you are not currently assigned. Project Web App displays the *New Task* page shown in Figure 9 - 42. For clarity, I captured the image after selecting or entering all of the information in the page so that you can easily see the desired result. Notice that this particular *New Task* page is very similar to the *New Task* page shown previously in Figure 9 - 41.

Begin the process of filling out this page by clicking the **Project** pick list and selecting the project containing the task to which you want to add yourself. The system refreshes the page with information relevant to the selected project. Click the **Summary task** pick list and select the summary section containing the task to which you want to add yourself. Once again, the system refreshes the page with information relevant to the selected summary section.

Click the **Assign myself to the following task** pick list and select the task to which you want to add yourself. The system refreshes the *Start* and *Finish* fields with the current schedule of the selected task. If you want to help with the task during the entire time period, leave the dates shown in the *Start* and *Finish* fields. Change the dates in the

Tracking Time and Task Progress

Start and/or **Finish** fields if you want to help with only a certain time period for the task. Enter a value in the **Total work** field if you only want to contribute a limited number of working hours to help with the task.

In the **Comments** field, enter a comment to explain why you need to add yourself to help on the task. Click the **Send** button to complete the process of adding yourself to the existing task.

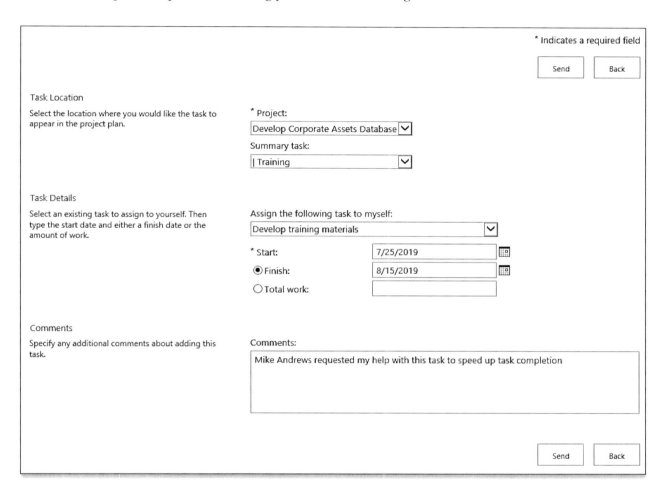

Figure 9 - 42: New Task page - add yourself to a task

Adding Team Tasks

A **Team task** is a task assigned to a **Team resource**. A *Team* resource is a special type of Generic resource that represents a specific group of resources. Project managers can assign a *Team* resource to a task when they want the members of that team to decide among themselves which of them will do the work on the task.

Click the **Add Row** pick list button and select the **Add Team Tasks** item when you need to take ownership of a task assigned your team. The system displays the *Team Tasks* page shown in Figure 9 - 43. Select the checkbox for each task for which you want to take ownership and then click the **Assign to me** button in the *Tasks* section of the *Tasks* ribbon. Click the **Close** button to return to the *Tasks* page.

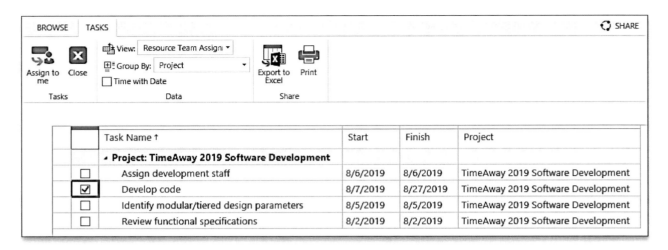

Figure 9 - 43: Team Tasks page

Removing a Task

Project Web App allows you to request the removal of task assignments from the *Tasks* page. The *Remove Task* action sends a removal request that the project manager must review and approve. Good targets for removal are rejected tasks, such as a new proposed task that the project manager rejected.

 Warning: You cannot use the *Remove Task* action to "hide" tasks on the *Tasks* page. When you remove a task, this action sends a *Delete Assignment* request to the project manager who is the *Status Manager* of the task. If the project manager erroneously approves this request on the *Approval Center* page in PWA, the system completely removes you from the task in the Microsoft Project schedule. Once removed, you no longer see this task on the *Tasks* page.

To remove a task from the *Tasks* page, select the checkbox to the left of the task you want to remove, and then click the **Remove Task** button in the *Tasks* section of the *Task* ribbon. Project Web App displays the confirmation dialog shown in Figure 9 - 44.

Figure 9 - 44: Remove Task confirmation dialog

In the confirmation dialog, click the **OK** button to complete the task removal process. The system refreshes the *Tasks* page, formats the selected task using the strikethrough font, and leaves the checkbox selected for the task you want to remove.

With the removed task still selected, in the *Submit* section of the *Tasks* ribbon, click the **Send Status** pick list button, and then select the **Selected Tasks** item. Project Web App sends your project manager a *Delete Assignment* request

Tracking Time and Task Progress

for the deleted task. After your project manager approves the task removal and publishes the changes to the project, the system removes you from the task in the Microsoft Project schedule. However, the system does not remove the task on the *Tasks* page.

Reassigning a Task to a Fellow Team Member

After a project manager assigns a team member to a task, Project Web App offers the team member the option to reassign the task to someone else. This feature is useful in organizations where the project manager assigns a team leader to tasks in an enterprise project, and then the team leader is responsible for reassigning the tasks to members of their team.

To reassign a task to someone else, select the checkbox *for a single task* and then click the **Reassign** button in the *Tasks* section of the *Tasks* ribbon. Project Web App displays the *Task Reassignment* page shown in Figure 9 - 45. For clarity, I captured the image after selecting or entering all of the information in the page so that you can easily see the desired result.

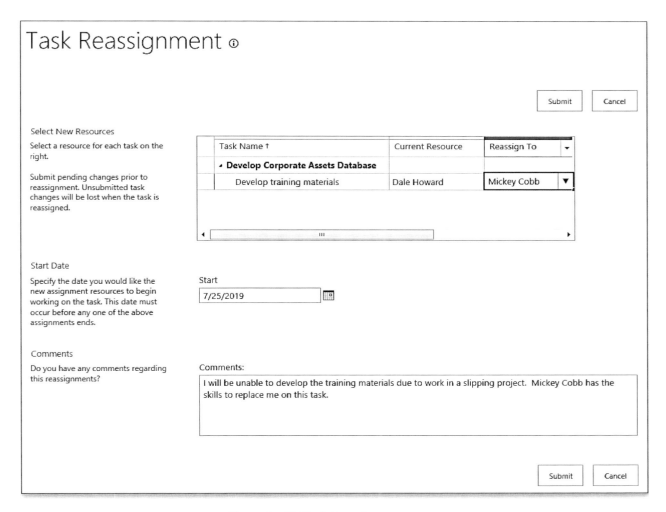

Figure 9 - 45: Task Reassignment page

293

Warning: If you select the checkboxes for multiple tasks, the *Task Reassignment* dialog displays a list of ***every task assigned to you in every enterprise project***. To keep the reassignment process as simple as possible, Projility strongly recommends that you select only one task at a time and then reassign it.

The *Select New Resources* section of the *Task Reassignment* dialog shows you the task assignment you selected. The *Reassign To* column in the data grid allows you to select the resource to whom you want to reassign the task. When you click the pick list button in the **Reassign To** field, the list displays all of the current members of the project team.

 Information: The *Reassign To* pick list contains ***only*** the members of your project team. If you need to reassign a task to someone outside your project team, you must first ask your project manager to add the new resource to the project team.

In in the *Reassign To* field, select the name of the team member to whom you want to reassign the selected task. In the *Start Date* section of the *Task Reassignment* page, click the **Start** pick list and select the date on which the fellow team member begins work on the task. The date in the *Start* field defaults *Start* date of the task in the Microsoft Project schedule. If you want to reassign the entire task to the fellow team member, ***do not*** change the date in the *Start* field. If you want to reassign only a part of the task to the fellow team member, choose a **Start** date that is later than the task *Start* date and earlier than the task *Finish* date.

Enter any relevant comment in the **Comments** field. When you complete your entries, click the **Submit** button. When you click the *Submit* button, Project Web App immediately removes you from the task and displays a task reassignment request on the *Approval Center* page of the project manager. Before the system finalizes the task reassignment request, the project manager must approve it.

 Information: Many organizations disable the *Reassign Task* feature in Project Web App because it should be the responsibility of the project manager to assign resources to tasks and not the responsibility of team members.

Manually Entering Task Progress in the Microsoft Project Schedule

Some organizations choose not to use either the *Timesheet* page or the *Tasks* page in Project Web App to collect task progress from project team members. Examples of reasons for this decision could include the following:

- The project management maturity level of the organization is to low that it cannot support the use of Project Web App to capture time and task progress.
- The organization uses an external timekeeping system for financial purposes, and does not want to require users to enter time in two different timekeeping systems.

When an organization does not use the time and task tracking capabilities of Project Web App, the organization must rely on the project manager to manually enter task progress in the Microsoft Project schedule. If this describes your organization, you should use some type of process for collecting the task updates each week, such as through weekly project status meetings or even through e-mail messages. Once collected, you can manually enter the task updates in the project schedule.

Although there are a number of methods for manually entering task progress in a Microsoft Project schedule, I recommend only two approaches to keep the process as simple as possible. These approaches are:

- Enter progress at the task level.
- Enter progress at the resource assignment level.

I discuss each of these approaches individually.

Entering Progress at the Task Level

The simplest method for entering progress is at the task level using the *% Complete* value for each task. Using this method, you must ask your team members to estimate their cumulative percentage of work completed to date on each task that occurred during the previous reporting period. Keep in mind, however, that there are two primary limitations for using this method of tracking:

- This tracking method is not date sensitive. When you enter a *Percent Complete* value for a task in Microsoft Project, the software assumes that the task started, even if it started late in reality. If you mark a task as 100% complete, the software assumes the task finished as scheduled, even if it finished late in reality.

- This tracking method does not allow team members to provide their Estimate To Complete (ETC) for any task. Without the ETC information, the software has no knowledge of whether the task will take longer than its original planned *Duration* value.

To work around these two limitations, you should gather the following progress information for each task by asking your project team the following questions:

- **Actual Start date** – "If you started work on a new task this week, what day did you actually start?"
- **Percent Complete** – "As a percentage, how much work have you completed on the task to date?"
- **Remaining Duration** – "How many days of work do you think you have left on the task?"
- **Actual Finish date** – "If you completed work on a task this week, what day did you actually finish?"

Based on the answers provided by your team members for the preceding questions you can manually enter actual progress by completing the following steps in your Microsoft Project schedule:

1. Apply the *Gantt Chart* view.
2. Right-click on the **Select All** button and select the **Tracking** table on the shortcut menu.
3. Widen the *Task Name* column, as needed.
4. Drag the split bar to the right of the *% Complete* column.
5. "Drag and drop" the **% Complete** column to the right of the *Actual Start* column.
6. "Drag and drop" the **Remaining Duration** column to the right of the *% Complete* column.
7. Drag the split bar to the right of the *Actual Finish* column.
8. Enter the actual start date of each task in the **Actual Start** column.
9. Enter the estimated percentage of completion for each task in the **% Complete** column.

10. Enter the number of days left for the task (the Estimate To Complete) for each task in the **Remaining Duration** column.

11. When the team members complete work on task, enter the actual completion date for the task in the **Actual Finish** column.

> **Information**: Using the preceding steps, you never need to enter *100%* in the *% Complete* column for any task. Instead, when you enter a date in the *Actual Finish* column, Microsoft Project automatically calculates a *100%* complete value in the *% Complete* column.

> **Warning**: After entering the *% Complete* value, if you increase the *Remaining Duration* value, Microsoft Project automatically **reduces** the *% Complete* value. If you decrease the *Remaining Duration* value, Microsoft Project automatically **increases** the *% Complete* value.

Figure 9 - 46 shows the *Tracking* table set up to track progress at the task level, and with progress entered for the first two tasks. Notice that the Design task is *100%* complete with both an *Actual Start* date and an *Actual Finish* date entered for the task. Notice that the Build task is only *50%* complete, with an *Actual Start* date and a *% Complete* value entered, with no adjustment to the *Remaining Duration* value, and with no *Actual Finish* date entered yet.

	Task Name	Act. Start	% Comp.	Rem. Dur.	Act. Finish
0	▲ **Tracking Project**	2/3/20	43%	8.5 d	NA
1	▲ PHASE 1	2/3/20	43%	8.5 d	NA
2	Design	2/3/20	100%	0 d	2/7/20
3	Build	2/10/20	50%	1.5 d	NA
4	Test	NA	0%	4 d	NA
5	Implement	NA	0%	3 d	NA
6	Phase I Complete	NA	0%	0 d	NA

Figure 9 - 46: Progress entered at the task level

Entering Progress at the Resource Assignment Level

The preceding process works well when you enter progress on tasks with only a single resource assigned. If you want to enter progress on tasks with multiple resources assigned, and want to enter progress for each assigned resource individually, you can use a variation of the preceding set of steps. To enter progress at the resource assignment level, complete the following steps in your Microsoft Project schedule:

1. Apply the *Task Usage* view.
2. Right-click on the **Select All** button and select the **Tracking** table on the shortcut menu.
3. Widen the *Task Name* column, as needed.
4. Drag the split bar to the right of the *Physical % Complete* column.

Tracking Time and Task Progress

5. Right-click on the **Actual Finish** column header and select the **Insert Column** item on the shortcut menu.
6. In the list of available task columns, select the **% Work Complete** column.
7. Right-click on the **Actual Finish** column header and select the **Insert Column** item on the shortcut menu again.
8. In the list of available task columns, select **Remaining Work** column.
9. Drag the split bar to the right of the *Actual Finish* column.
10. Enter the actual start date in the **Actual Start** column for each assigned resource on a task.
11. Enter the estimated percentage of completion in the **% Work Complete** column for each assigned resource on a task.
12. Enter the number of hours of work left in the **Remaining Work** column for each assigned resource on a task.
13. When the team members complete work on task, enter the actual completion date in the **Actual Finish** column for each assigned resource on a task. If all assigned resources finished the task on the same day, you could enter the **Actual Finish** date for the task instead of the assigned resources, by the way.

Information: Using the preceding steps, you never need to enter *100%* in the *% Work Complete* column for any task. Instead, when you enter a date in the *Actual Finish* column, Microsoft Project automatically calculates a *100%* complete value in the *% Work Complete* column.

Warning: After entering the *% Work Complete* value, if you increase the *Remaining Work* value, Microsoft Project automatically **reduces** the *% Work Complete* value. If you decrease the *Remaining Work* value, Microsoft Project automatically **increases** the *% Work Complete* value. This behavior occurs on both *Fixed Units* tasks and *Fixed Work* tasks.

Figure 9 - 47 shows the *Tracking* table set up to track progress at the resource assignment level, and with progress entered for the first two tasks. Notice that the Design task is *100%* complete with both an *Actual Start* date and an *Actual Finish* date entered for each assigned resource on the task. Notice that the Build task is only *38%* complete, with an *Actual Start* date entered, a *% Work Complete* value entered, with no adjustment to the *Remaining Work* value and with no *Actual Finish* date entered yet. Notice that Microsoft Project calculated the *% Work Complete* value of *38%* for the Build task as the average of the values entered for Randy Parker (*50%*) and Ron Appel (*25%*).

Information: In the customized *Tracking* table shown in Figure 9 - 47, I right justified the data in the *% Work Complete* and the *Remaining Work* fields. To right justify the data in any column, right-click in its column header and select the **Field Settings** item on the shortcut menu. In the *Field Settings* dialog, click the **Align data** pick list and select the **Right** item. Optionally change the value in the **Width** field to widen the column, and then click the **OK** button.

297

	Task Name	Act. Start	% Work Complete	Remaining Work	Act. Finish
0	**Tracking Project**	**2/3/20**	**49%**	**142 h**	**NA**
1	**PHASE 1**	**2/3/20**	**49%**	**142 h**	**NA**
2	Design	2/3/20	100%	0 h	2/7/20
	Kate Witkowski	*2/3/20*	*100%*	*0 h*	*2/7/20*
	Susan Tartaglia	*2/3/20*	*100%*	*0 h*	*2/7/20*
	Larry Barnes	*2/3/20*	*100%*	*0 h*	*2/7/20*
3	Build	2/10/20	38%	30 h	NA
	Randy Parker	*2/10/20*	*50%*	*12 h*	*NA*
	Ron Appel	*2/10/20*	*25%*	*18 h*	*NA*
4	Test	NA	0%	64 h	NA
	Kate Witkowski	*NA*	*0%*	*32 h*	*NA*
	Susan Tartaglia	*NA*	*0%*	*32 h*	*NA*
5	Implement	NA	0%	48 h	NA
	Randy Parker	*NA*	*0%*	*24 h*	*NA*
	Larry Barnes	*NA*	*0%*	*24 h*	*NA*
6	Phase I Complete	NA	0%	0 h	NA

Figure 9 - 47: Progress entered at the resource assignment level

Module 10

Approving Time and Task Progress

Learning Objectives

After completing this module, you will be able to:

- Understand the Project Update Cycle
- Use the Approval Center to Approve Pending Task Updates
- Process Applied Updates in the Microsoft Project Schedule

Inside Module 10

Understanding the Project Update Cycle **301**
   Viewing Pending Approvals on the Home Page *301*
Accessing the Approval Center Page **302**
   Understanding the Approval Center Page *302*
   Best Practices for Setting Up the Approval Center Page *304*
Processing Task Updates **306**
   Reviewing Pending Task Updates *309*
   Rejecting Task Updates *311*
   Approving Task Updates *313*
   Creating Rules for Auto Approving Task Updates *314*
   Rescheduling Incomplete Work from the Past *320*
   Updating Expense Cost Resource Information *326*
   Publishing the Latest Schedule Changes *328*
Axioms for Success with Tracking Progress **328**

Understanding the Project Update Cycle

Approvals are a key process if you are responsible for managing enterprise projects using Project Online. In fact, as a project manager, always keep in mind that you are the "gatekeeper" between pending task updates and your Microsoft Project schedule. You are the last line of defense to prevent Project Web App from applying erroneous task updates to your project schedule.

Approving pending task updates is part of a larger, ongoing process for applying task updates to your Microsoft Project schedule. Figure 10 - 1 shows the major steps in the task update process. On a weekly basis, I recommend that you complete each of the following steps for every enterprise project that you manage. In this module, I provide in-depth documentation about each of these steps.

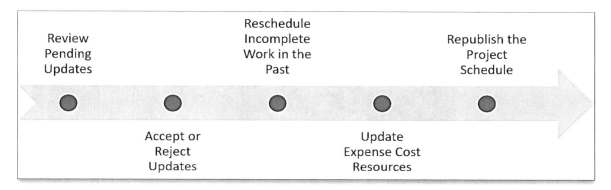

Figure 10 - 1: Task update approval process

Viewing Pending Approvals on the Home Page

On the *Home* page of Project Web App, the *Approvals* tile in the *Track your work* carousel displays the total number of pending task updates waiting for your approval. Notice in Figure 10 - 2 that the *Approvals* tile shows that I have 25 pending task updates ready for me to approve.

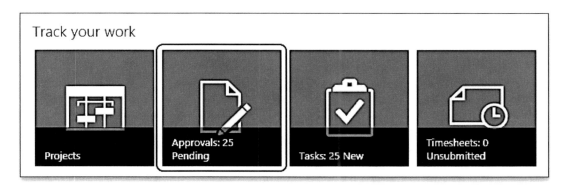

Figure 10 - 2: Pending task updates ready for approval

Module 10

Accessing the Approval Center Page

The *Approval Center* page in Project Web App is where you review and either approve or reject pending task updates. The system provides you with two ways to navigate to the *Approval Center* page, as shown in Figure 10 - 3:

- In the *Quick Launch* menu, click the **Approvals** link.
- In the *Track your work* carousel, click the **Approvals** tile.

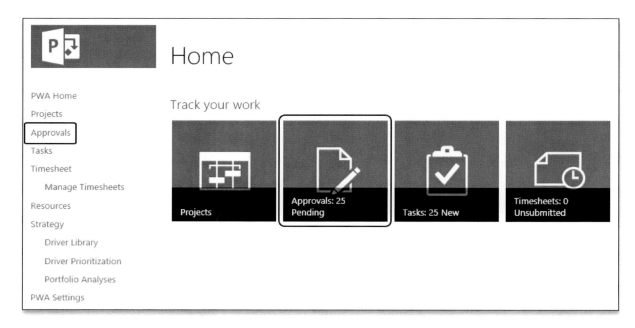

Figure 10 - 3: Navigate to the Approval Center page

Understanding the Approval Center Page

When you click either the *Approvals* link in the *Quick Launch* menu or the *Approvals* tile in the *Track your work* carousel, Project Web App displays the *Approval Center* page shown in Figure 10 - 4. This page includes a table on the left side with columns containing important information about each pending task update, plus a timesheet grid on the right side that displays the *Actual Work* values entered by the team member on each task.

Information: The *Approval Center* page is one of only a few pages in Project Web App that your application administrator cannot customize for the organization. On the other hand, the system does allow you to customize the page manually to meet your own task approval management needs. You can hide unneeded columns, change the display order of the columns in the table, and apply custom grouping and filtering as well. You can apply up to three levels of grouping using the fields included in the table in this page. When you manually customize the layout of the *Approval Center* page, Project Web App retains the customized layout each time you return to the page.

Approving Time and Task Progress

	Approval Type	Name ↑	Project		Total	6/17/2019	6/18/2019
☐	Status Update	Create project org chart	Chi Rho Softwar	Actual	0h		
☐	Status Update	Define all project stakeholders	Chi Rho Softwar	Actual	0h		
☐	Status Update	Define Financial Risks and Opportunities	Evaluate Merger	Actual	0h		
☐	Status Update	Define Financial Scenarios	Evaluate Merger	Actual	0h		
☐	Status Update	Define major milestones	Chi Rho Softwar	Actual	0h		
☐	Status Update	Define standard PM and approval gates	Chi Rho Softwar	Actual	0h		
☐	Status Update	Develop code	TimeAway 2019	Actual	0h		
☐	Status Update	Develop delivery timeline	Chi Rho Softwar	Actual	0h		
☐	Status Update	Develop Financial Synergy Analysis	Evaluate Merger	Actual	0h		
☐	Status Update	Develop integration test plans using product spe	Develop Corpor	Actual	0h		
☐	Status Update	Develop preliminary budget	Chi Rho Softwar	Actual	4h		
☐	Status Update	Develop training materials	Develop Corpor	Actual	0h		
☐	Status Update	Develop unit test plans using product specificatio	Develop Corpor	Actual	0h		
☐	Status Update	Document schedule baseline	Chi Rho Softwar	Actual	0h		
☐	Status Update	Document WBS, milestones, and deliverables	Chi Rho Softwar	Actual	0h		
☐	Status Update	Draft preliminary software specifications	Chi Rho Softwar	Actual	16h	8h	8h
☐	Status Update	Identify Market Synergies	Evaluate Merger	Actual	0h		
☐	Status Update	Identify Product Synergies	Evaluate Merger	Actual	0h		

Figure 10 - 4: Approval Center page in PWA

The *Approval Center* table includes the following columns:

- The **Approval Type** column displays the transaction type of the line. For pending task updates, the system displays a *Status Update* value.

- The **Name** column displays the name of the updated task.

- The **Project** column displays the name of the project containing the updated task.

- The **Update Type** indicates the nature of the transaction. Task update types can include task updates, new task requests, task reassignment requests, new assignment requests, and delete assignment requests.

- The **Resource** column displays the name of the resource to whom the update applies.

- The **Owner** column displays the *Assignment Owner* of the task assignment. Most of the time, the *Resource* column and the *Owner* column display the same name.

- The **Transaction Comment** column displays any comments entered by resource to provide more context to the transaction.

- The **Sent Date** column displays the date of the transaction submission.

- The **Start** column displays current *Start* date of the task assignment.

- The **Finish** column displays the new *Finish* date of the task assignment if you approve it.

- The **Total** column displays the total number of *Actual Work* hours entered by the resource for the transaction.

- The **% Complete** column displays the completion percentage for the task assignment.

- The **Remaining Work** column displays the *Remaining Work* value for the task assignment.

- The **Task Hierarchy** displays a breadcrumb trail that reveals the location of the task in the project WBS.

Module 10

Best Practices for Setting Up the Approval Center Page

As mentioned earlier in this module, Project Web App allows you to manually customize the layout of the *Approval Center* page. Projility recommends you complete the following steps to manually customize this page:

1. Float your mouse pointer over the **Approval Type** column header, click the pick list arrow button in the column header, and select the **Hide Column** item on the pick list.

    Information: As a project manager, the *Approval Type* column contains no useful information for you since it only displays a *Status Update* value for every line. If you play a dual role as a project manager who manages projects and a resource manager who approves timesheets in PWA, however, you should not hide this column as it allows you to differentiate between task updates, timesheet approvals, and nonworking time requests.

2. Float your mouse pointer over the **Project** column header, click the pick list arrow button in the column header, and select the **Hide Column** item on the pick list.

3. Float your mouse pointer over the **Update Type** column header, click the pick list arrow button in the column header, and select the **Configure Columns** item on the pick list. Project Web App displays the *Configure Columns* dialog shown in Figure 10 - 5.

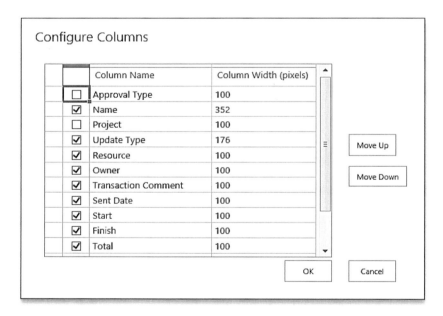

Figure 10 - 5: Configure Columns dialog

4. In the *Configure Columns* dialog, click the *Remaining Work* row to select it, but ***do not deselect*** the checkbox at the left end of the row.

5. With the *Remaining Work* row selected, repeatedly click the **Move Up** button until the *Remaining Work* row appears immediately ***above*** the *Resource* row, and then click the **OK** button.

6. In the *Data* section of the *Approvals* ribbon, click the **Group By** pick list and select the **Project** group, as shown in Figure 10 - 6.

304

Approving Time and Task Progress

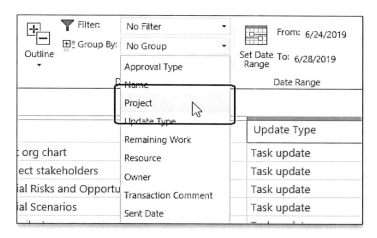

Figure 10 - 6: Group task updates by project

7. Widen the *Name* column, the *Update Type* column, and the *Remaining Work* column to set the width for each column, as needed.

8. Drag the split bar to the right edge of the *Remaining Work* column.

Figure 10 - 7 shows the customized layout of the *Approval Center* page.

Name ↑	Update Type	Remaining Work	Total
▸ Project: Chi Rho Software Development		36h	24h
▸ Project: Chi Rho Software Market Research	Task update	6h	46h
▸ Project: Develop Corporate Assets Database		118h	32h
▸ Project: Evaluate Merger with Helping Hand	Task update	0h	40h
▸ Project: HR CBT Rollout to North Region	Task update	34h	18h
▸ Project: TimeAway 2019 Software Development	Task reassignment	0h	0h

Figure 10 - 7: Customized Approval Center page

Hands On Exercise

Exercise 10 - 1

Customize the *Approval Center* page using Projility recommended best practices.

1. In the *Quick Launch* menu, click the **Approvals** link to navigate to the *Approval Center* page.

2. Float your mouse pointer over the **Approval Type** column header, click the pick list arrow button in the column header, and select the **Hide Column** item on the pick list.

305

3. Float your mouse pointer over the **Project** column header, click the pick list arrow button in the column header, and select the **Hide Column** item on the pick list.

4. Float your mouse pointer over the **Update Type** column header, click the pick list arrow button in the column header, and select the **Configure Columns** item on the pick list.

5. In the *Configure Columns* dialog, click the **Remaining Work** row to select it, but *do not* deselect the checkbox at the left end of the row.

6. With the *Remaining Work* row selected, repeatedly click the **Move Up** button until the *Remaining Work* row appears immediately above the *Resource* row, and then click the **OK** button.

7. In the *Data* section of the *Approvals* ribbon, click the **Group By** pick list and select the **Project** group.

8. Widen the *Name* column, the *Update Type* column, and the *Remaining Work* column to set the width for each column, as needed.

9. Drag the split bar to the right edge of the *Remaining Work* column.

Processing Task Updates

 Warning: Before you begin the process of approving and rejecting pending task updates, close and check in every enterprise project and then close your Microsoft Project application as well. Project Web App cannot apply approved task updates to an enterprise project when it is in a checked out state.

Early in the week, such as on Monday morning, I recommend that you review and approve task updates submitted by your project team members for last week's reporting period. The first step in preparing to review pending task updates is to click the **Focus on Content** button in the upper-right corner of the *Approval Center* page. Clicking this button hides the header information above the ribbon and the *Quick Launch* menu on the left side of the page, which allows you to display the *Approval Center* page with the maximum height and width possible. Because the *Focus on Content* button is a "toggle" button, you can click the button a second time to display the *Quick Launch* menu again.

The second preparatory step is to specify the date range displayed in the timesheet grid on the right side of the *Approval Center* page. By default, the system automatically sets a date range from the date of the earliest timephased *Actual Work* value to the date of the latest timephased *Actual Work* value, but you can change this date range if you wish. As a general rule, you should set the date range to one week if your organization's task update process is weekly, or set the date range to two weeks if your organization's task update process is bi-weekly. In the *Date Range* section of the *Approvals* ribbon, click the **Date Range** button, as shown in Figure 10 - 8.

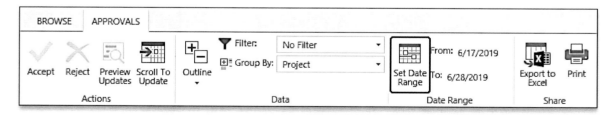

Figure 10 - 8: Click the Date Range button

Project Web App displays the *Select Date Range* dialog shown in Figure 10 - 9. In this dialog, click the **From** button and **To** buttons and select a date in the calendar date pickers to select your date range. Click the **OK** button to apply the custom date range to the timesheet grid on the right side of the *Approval Center* page.

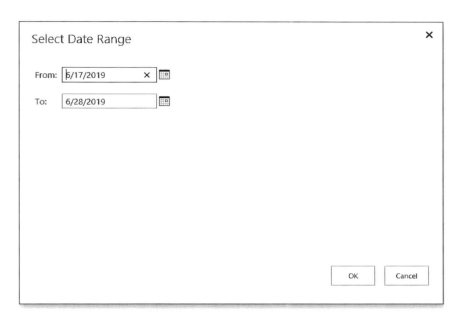

Figure 10 - 9: Select Date Range dialog

The third preparatory step is to expand the first project grouping so that you can see all of the pending task updates for that project. I strongly recommend that you review and approve the updates for only *one project at a time* so that you can maintain full control over the approval process. Figure 10 - 10 shows the pending task updates for the Chi Rho Software Development project.

	Name ↑	Update Type	Remaining Work
	▲ **Project: Chi Rho Software Development**		**36h**
☐	Develop delivery timeline	Task update	0h
☐	Develop preliminary budget	Task update	12h
☐	Draft preliminary software specifications	Task update	0h
☐	Incorporate feedback on software specifications	Task reassignment request	0h
☐	Obtain approvals to proceed	Task update	0h
☐	Request Contractor Help for Code Development	New task request	24h
☐	Review software specifications/budget with team	Task update	0h
☐	Secure required resources	Task update	0h
	▶ **Project: Chi Rho Software Market Research**	Task update	**6h**
	▶ **Project: Develop Corporate Assets Database**		**118h**
	▶ **Project: Evaluate Merger with Helping Hand Acce**	Task update	**0h**
	▶ **Project: HR CBT Rollout to North Region**	Task update	**34h**
	▶ **Project: TimeAway 2019 Software Development**	Task reassignment request	**0h**

Figure 10 - 10: Expanded project grouping

307

Warning: The *Approval Center* page is no place for project managers who are in a hurry to approve task updates because a single mistake here can have adverse consequences in a project. For example, I knew of a project manager who accidentally approved several dozen *Delete assignment requests* because he was in a hurry. When he opened his Microsoft Project schedule, the assigned resource was missing on several dozen tasks because approving the *Delete assignment requests* caused Project Online to remove the assigned resource from those tasks! He asked me for the location of the *Undo* button so that he could undo his mistake. Unfortunately, I had the sad news of telling him that there is no *Undo* button and that he would need to manually correct his mistake.

In the expanded project group, examine the information shown in the *Update Type* column for each pending task update. Table 10 - 1 shows the five possible update types you may see on the *Approval Center* page, along with the triggering action that created the update, and the consequences if you approve the update.

Action by Team Member	Resulting Update Type	Consequence if Approved
Updated task status and/or changed information such as Remaining Work	*Task update* (*Actual Work* values and changed *Remaining Work* values formatted with the red font color)	If approved, the system applies the reported task progress and other changes to the project schedule
Added self to an existing task	*New reassignment request*	If approved, the system adds the resource as a helper on the task
Reassigned a task to another resource	*Task reassignment request*	If approved, the system transfers their remaining work on the task assignment to the new resource
Deleted a task	*Delete assignment request* (name formatted with the red font color)	If approved, the system removes the resource from the task in the project schedule
Proposed a new task for the project schedule	*New task request* (name formatted with the red font color)	If approved, the system adds the new task and assigns the resource to the task

Table 10 - 1: Task update types

Warning: Projility strongly recommends that you **never approve** a *Delete assignment request* because this action automatically removes the assigned resource from the task in the Microsoft Project schedule. A better approach is to reject the *Delete assignment request* and then to communicate directly with the team member about other courses of action.

Approving Time and Task Progress

Reviewing Pending Task Updates

It is your responsibility as project manager to approve or reject all updates submitted by your team members. Approving a task update automatically transfers the updated task information into the appropriate Microsoft Project schedule. Rejecting a task update triggers an automatic e-mail message to the resource and may require the resource to take the appropriate action in response to your rejection. This process guarantees that you, the project manager, always serve as the "gatekeeper" between updates from your project team members and the Microsoft Project schedule.

 Information: Project Online does not allow team members to bypass you and to directly update task progress into the Microsoft Project schedule. Because the system requires your participation in the update process, you must provide at least minimal oversight to all pending task updates, which prevents the system from updating "dirty data" into your project schedule.

Before you process updates, pay special attention to any task update that includes a note in the *Transaction Comment* column. Click the name of the task update to display the *Task Details* dialog, such as the one shown in Figure 10 - 11. A note attached to a task update usually indicates that your team members are trying to give you additional information about their task updates. After reviewing the information shown in the *Task Details* dialog, click the **Cancel** button to close the dialog and return to the *Approval Center* page.

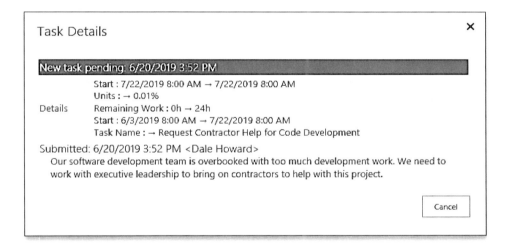

Figure 10 - 11: Task Details dialog with submission note

 Best Practice: Projility recommends that you take the time to read all notes included with task updates. If your team member took the time to write a note, you should take the time to read the note in response.

Before you approve any task updates, you may want to display a preview that shows you the consequences of approving the task updates to your Microsoft Project schedule. Select the checkbox to the left of every task update in the expanded project grouping. In the *Actions* section of the *Approvals* ribbon, click the **Preview Updates** button shown in Figure 10 - 12.

309

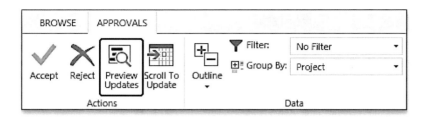

Figure 10 - 12: Click the Preview Updates button

Project Web App opens a new web browser tab and then displays the *Approval Preview* page, such as the one shown in Figure 10 - 13. This page displays every detailed task and milestone in the project, but does not display summary tasks by default. To display summary tasks in the *Approval Preview* page, select the **Summary Tasks** checkbox in the *Show/Hide* section of the *Preview* ribbon, as shown in Figure 10 - 14.

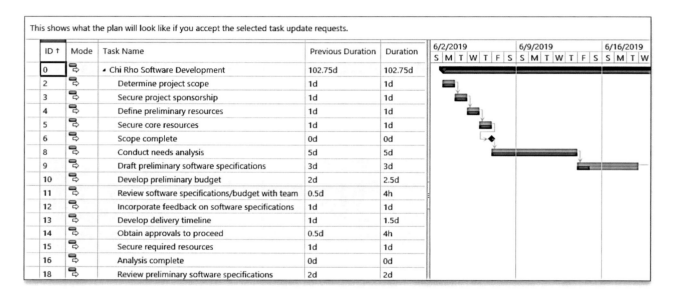

Figure 10 - 13: Approval Preview page

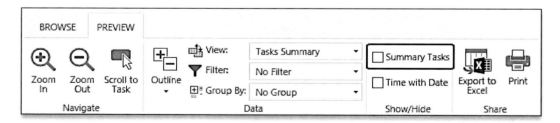

Figure 10 - 14: Select the Summary Tasks checkbox

The system indicates your selected task updates by displaying their names as a hyperlink in the *Task Name* column, and by formatting the tasks using the yellow cell background color. The *Approval Preview* page displays the *Tasks Summary* view for the project, and shows the post-approval state of the project before you actually approve the task updates.

The *Gantt Chart* portion of the *Approval Preview* page allows you to compare current task progress against the previous schedule for each task. The system displays this information as follows:

- The gray Gantt bar represents the previous schedule of each task.
- The blue Gantt bar represents the new schedule of each task after updates.
- The dark blue portion of each Gantt bar represents the *% Complete* value (current progress) for each task after updates.
- The gray diamond represents the previous schedule for each milestone task.
- The black diamond represents the new schedule for each milestone task after task updates.

The table portion of the view shown in the *Approval Preview* page contains columns that you can use to analyze the pending impact of approving the pending task updates. The default columns in the *Approval Preview* page include the following:

- ID
- Mode
- Task Name
- Previous Duration (the pre-approval Duration)
- Duration (the new Duration after approval)
- Previous Start (the pre-approval Start date)
- Start (the new Start date after approval)
- Previous Finish (the pre-approval Finish date)
- Finish (the new Finish date after approval)
- Previous % Complete (the pre-approval % Complete)
- % Complete (the new % Complete after approval)
- Work
- Resource Names

Look for tasks with a *Duration* value greater than the *Previous Duration* value, as this indicates that the task is slipping. Look for tasks with a *Finish* date later than the *Previous Finish* date, it also indicates that the task is slipping. Click the **Close Tab** button (the **X** button) in the upper right corner of the *Approval Preview* browser tab to close the *Approval Preview* page.

Rejecting Task Updates

During the process of reviewing pending task updates, you may need to reject one or more updates. You might reject a task update for reasons such as:

- The team member accidentally types an incorrect *Actual Work* value.
- The team member forgets to update the *Remaining Work* value.

Module 10

- The team member fails to add a note to explain the reason for submitting a *New task request*.

You might also reject a *Task reassignment request* because team members should not make staffing decisions in the project. As indicated previously, Projility recommends that you ***always reject*** a *Delete assignment request*.

To begin the process of rejecting task updates, deselect the checkboxes for all task updates currently selected. The fastest way to do this is to float your mouse pointer over the blank column header at the top of the column containing the checkboxes. Click the pick list arrow button in the blank column header and select the **Clear All** item, as shown in Figure 10 - 15.

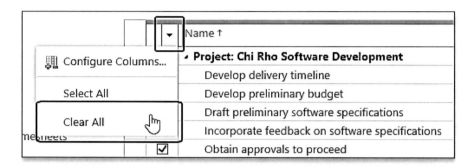

Figure 10 - 15: Click the Clear All item

Next, select the checkboxes for only the task updates you need to reject. In the *Actions* section of the *Approvals* ribbon, click the **Reject** button. The system displays the *Confirm Delete* dialog. In the dialog, enter your reason for rejecting the task update, such as shown Figure 10 - 16, and then click the **OK** button. Project Web App removes the update from the *Approval Center* page.

 Best Practice: When you reject a task update, Project Web App sends an e-mail message to the team member to notify them of the rejection. Do not rely on the e-mail message alone to communicate with your team member. Instead, reach out to your team member by e-mail by phone, or in person. Explain your reason for the rejection and the actions you want them to follow in response. Project Web App can never replace good "old fashioned" human communications!

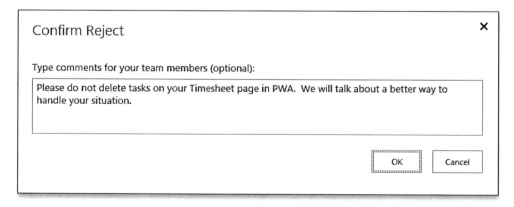

Figure 10 - 16: Confirm Reject dialog

Approving Time and Task Progress

Approving Task Updates

After rejecting any task updates that need to be rejected, you are ready to approve pending task updates. Remember that you should approve task updates for only one project at a time. To begin the process, select the checkbox for every task in the expanded project group. In the *Actions* section of the *Approvals* ribbon, click the **Accept** button. The system displays the *Confirm Approval* dialog shown in Figure 10 - 17. In the dialog, enter an optional approval comment, and then click the **OK** button. Project Web App applies the updates to the Microsoft Project schedule and then removes the approved updates from the *Approval Center* page.

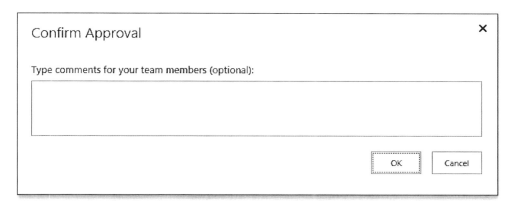

Figure 10 - 17: Confirm Approval dialog

Repeat the preceding process for each project you manage. Expand the project grouping for the next set of pending task updates. Select the checkbox for each task update in the project, preview the impact of the task updates if you approve them, and then approve the selected task updates. When you complete this process for every project in which you have pending task updates, Project Web App completely clears the *Approval Center* page, which indicates you have no pending task updates awaiting your approval.

 Hands On Exercise

Exercise 10 - 2

 Warning: Before you begin this exercise, confirm that you closed and checked in every enterprise project currently open, and that you closed your Microsoft Project application as well.

Review, preview, and then approve pending task updates.

1. Expand the grouping for the enterprise project used with this training.
2. Drag the split bar to the right and then scroll the *Approval Center* table to the right so that you can examine the information shown in each column.

313

Notice that the system formatted the *Remaining Work* value with a red font color, indicated that the team member edited this number.

3. Drag the split bar back to the left and examine the daily *Actual* values in the timesheet grid on the right side of the page.
4. In the *Approval Center* table, click the name of the **Write Chapter 1** task.
5. Examine the update information shown in the *Task Details* dialog, including any *Comments* or *Notes* text.

In the *Task Details* dialog, notice that the *Remaining Work* line shows you the original value, followed by a right-arrow symbol, followed by the new value entered by the team member.

6. Click the **Cancel** button to close the *Task Details* dialog
7. In the *Approval Center* table, select the checkbox for the **Write Chapter 1** task.
8. In the *Actions* section of the *Approvals* ribbon, click the **Preview Updates** button.
9. On the *Approval Preview* page, select the **Summary Tasks** checkbox in the *Show/Hide* section of the *Preview* ribbon.
10. In the table on the left side of the *Approval Preview* page, study the information shown in each pair of columns, such as in the *Previous Duration* and *Duration* columns.
11. Examine the *Gantt Chart* shown on the right side of the *Approval Preview* page and look for signs of impending schedule slippage.

Notice that the *Gantt Chart* pane shows impending task slippage in the *Course Manual* section of the schedule if you approve the pending task update.

12. Click the **Close Tab** button (the **X** button) in the upper right corner of the *Approval Preview* browser tab to close the *Approval Preview* page.
13. In the *Actions* section of the *Approvals* ribbon, click the **Accept** button.
14. In the *Confirm Approval* dialog, click the **OK** button.

Creating Rules for Auto Approving Task Updates

As your organization gains experience with using Project Web App for entering and approving task progress, you may find that some of your team members have a history of providing very reliable task updates. If you assign yourself to tasks in the Microsoft Project schedule and then enter your progress in Project Web App, you may not want the system to require you to approve your own task updates. In situations such as these, you may want Project Web App to automatically approve task updates from your reliable team members and from yourself.

The **Rules** feature in Project Web App allows you to create logical rules for automatically approving task updates into your projects. When you apply a rule, the system approves only those task changes that meet the criteria defined in the rule. You can create rules that run automatically every time a team member submits a task update, or you can create rules that you run manually whenever you want to run them.

To access the *Rules* feature from the *Approval Center* page, click the **Manage Rules** button in the *Navigate* section of the *Approvals* ribbon, as shown in Figure 10 - 18. The system displays the *Rules* page shown in Figure 10 - 19.

Approving Time and Task Progress

Figure 10 - 18: Automatic Approval Rules page

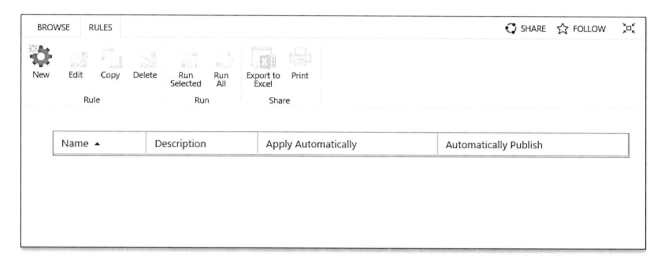

Figure 10 - 19: Rules page in PWA

By default, Project Web App does not contain any pre-defined rules, so the *Rules* page is initially blank. To create a new rule, click the **New** button in the in the *Rule* section of the *Rules* ribbon. The system displays the *Edit/Create Rule* page shown in Figure 10 - 20.

In the *Name* section of the *Edit/Create Rule* page, enter a name for the new rule in the **Name** field. Enter an optional description in the **Description and Comments for Team Members** field. The text you enter in the *Description and Comments for Team Members* field becomes part of the history of approvals displayed on the *My Assignments* page for each task approved by the rule.

In the *Automatic updates* section of the page, select the **Automatically run this rule** checkbox if you want Project Web App to run the rule automatically or when you click the **Run All Rules** button on the *Rules* page. Leave the option deselected if you want to run the rule manually. Select the **Automatically publish the updates** option to automatically publish the updates.

 Warning: Projility recommends that you *do not* select the *Automatically publish the updates* checkbox. You should not publish your enterprise projects until you have completed the other steps in the project update cycle shown previously in Figure 10 - 1.

315

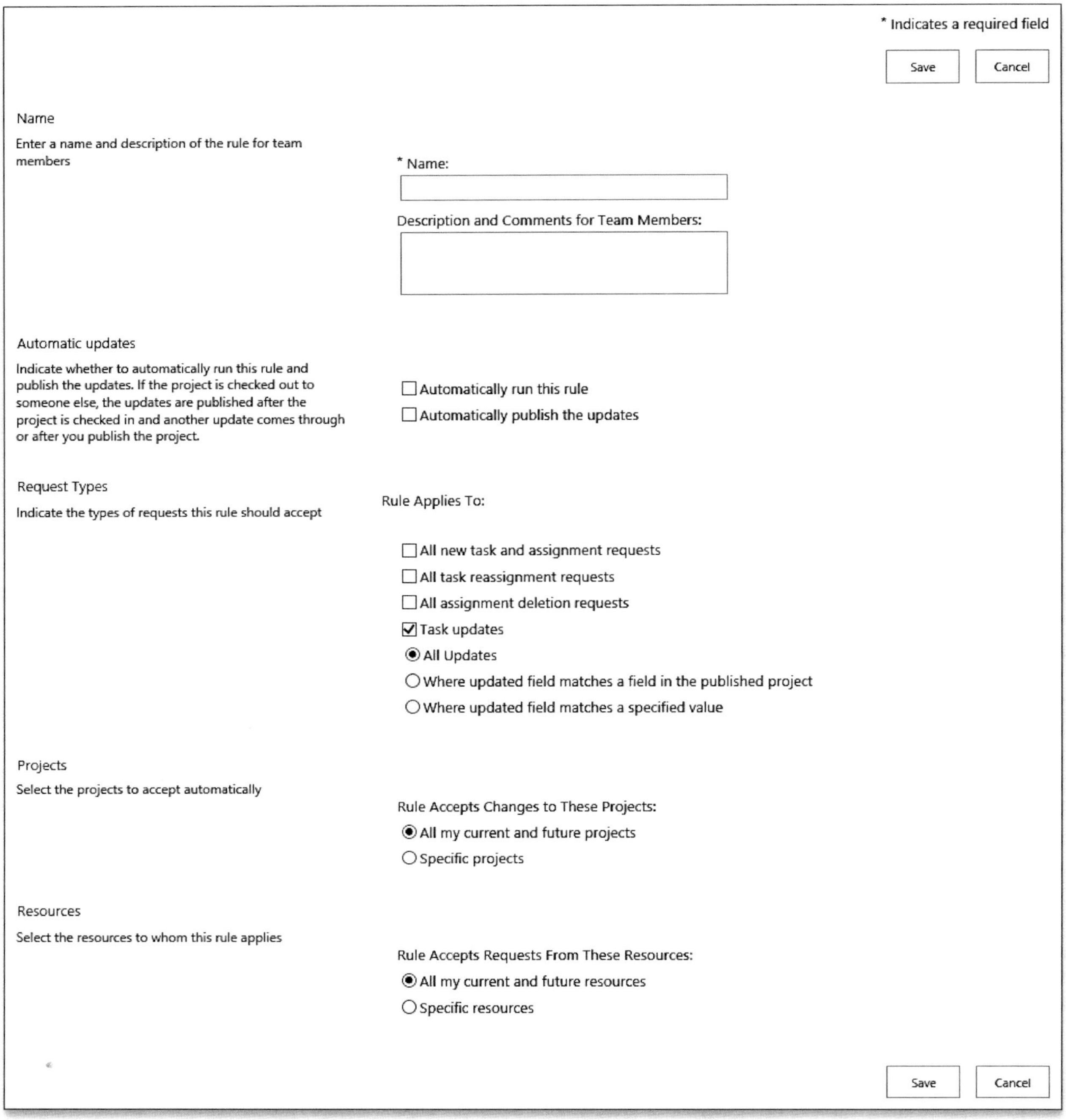

Module 10

Figure 10 - 20: Edit/Create Rule page for a new rule

In the *Request Types* section, select the type of updates you want to process with the rule. Table 10 - 2 shows the types of updates you can select.

316

Setting	Action
All new task and assignment requests	Select this option to automatically approve updates in which a team member proposes a new task or adds him/herself to an existing task.
All task reassignment requests	Select this option to automatically approve updates in which a team member delegates a task to another team member, or assigns him/herself to a *Team* task.
All assignment deletion requests	Select this option to automatically approve updates in which the team member deleted a task.
Task updates	Select this option to automatically approve progress updates from your team members.

Table 10 - 2: List of Request Types

Warning: Projility recommends that you **never** select the *All assignment deletion requests* item, because this action automatically removes the assigned resources from tasks in the Microsoft Project schedule.

If you select the **Task updates** option in the *Request Types* section of the page, the system offers three additional options, which are:

- All Updates
- Where updated field matches a field in the published project
- Where updated field matches a specified value

Select the **All Updates** option to process all progress updates from the resources you select in the *Resources* section of the page. If you select the **Where updated field matches a field in the published project** option, Project Web App expands the *Request Types* section with additional options, as shown in Figure 10 - 21. In the expanded *Request Types* section, create a custom filter by specifying values from the *Updated Field*, *Operator*, and *Published Field* pick lists. For example, you might enable automatic approval of updates where *Assignment Actual Work* is less than or equal to *Assignment Baseline Work* to automatically approve actual work that is less than the planned work.

If you select the **Where updated field matches a specific value** option, Project Web App expands the *Request Types* section with a different set of options, as shown in Figure 10 - 22. In the expanded *Request Types* section, create a custom filter by specifying values from the *Updated Field* and *Operator* pick lists, and entering a value in the *Value* field. For example, you might enable automatic approvals for all updates where the assignment *% Work Complete* value is equal to *100* (completed tasks).

317

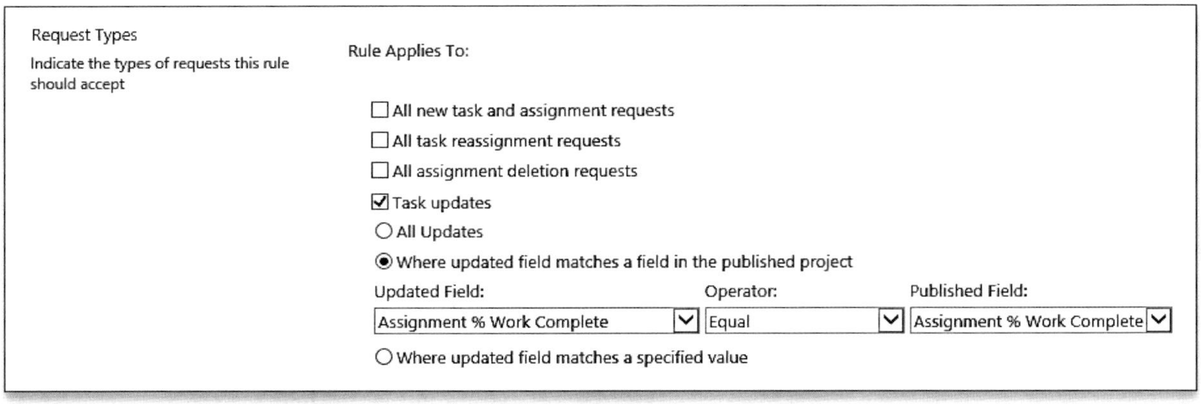

Module 10

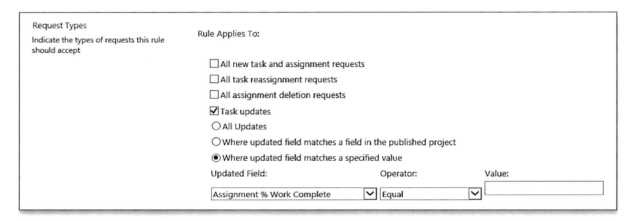

Figure 10 - 21: Request Type custom filter based on field value match

Figure 10 - 22: Request Type custom filter based on specific value match

In the *Projects* section of the page, select an option to determine to which projects the rule must apply. Select the **All my current and future projects** option to apply the rule to all current and future projects. If you select the **Specific projects** option, the system expands the *Projects* section, as shown in Figure 10 - 23.

From the *Available Projects* list, select one or more projects and then click the **Add** button (the **>>** button) to add your selections to the *Selected Projects* list. If you want the rule to process future projects in addition to the selected projects, select the **All projects in the future** checkbox option as well.

Approving Time and Task Progress

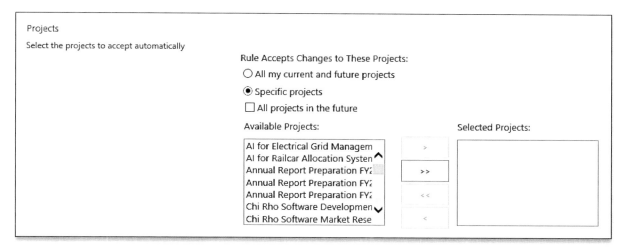

Figure 10 - 23: Expanded Projects section allows selection of specific projects

In the *Resources* section, select an option to determine the resources to which the rule applies. Select the **All my current and future** resources option to apply the rule to updates from all current and future resources. If you select the **Specific resources** option, the system expands the *Resources* section, as shown in Figure 10 - 24.

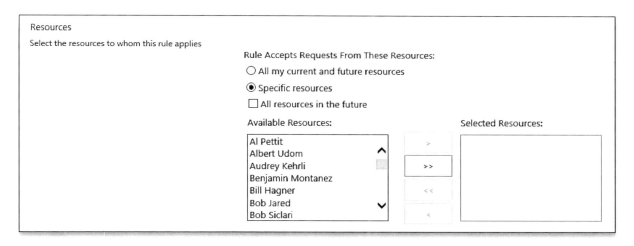

Figure 10 - 24: Expanded Resources section allows selection of specific resources

From the *Available Resources* list, select one or more resources and then click the **Add** button (the >> button) to add them to the *Selected Resources* list. If you want the rule to process updates from future resources in addition to the selected resources, select the **All resources in the future** checkbox as well.

 Information: If you select the *All task reassignment requests* option in the *Request Types* section of the page, the system adds two additional options in the *Resources* section. These options allow you to select the specific resources for task reassignment.

When finished, click the **Save** button to save the new rule. Figure 10 - 25 shows two new rules on the *Rules* page in Project Web App. These rules are as follows:

- **Task Updates from Me** – This rule automatically approves any progress updates submitted by me.

319

- **Task Updates from the PMO Team** – This rule automatically approves any progress updates submitted by members of the PMO staff.

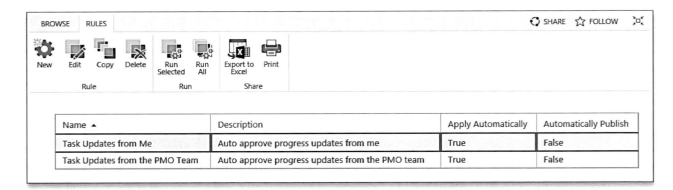

Figure 10 - 25: Two new rules created to approve task updates

In Figure 10 - 25 shown previously, notice the buttons displayed in the *Rules* ribbon. When you select a rule in the data grid, these buttons give you the option to edit, copy, delete, or run the rule individually. Click the **Edit** button to modify the selected rule. Click the **Copy** button to create a new rule based on a copy of the selected rule. After copying a rule, you must then edit the copy, including renaming the copied rule. Click the **Delete** button to delete the selected rule.

Select a rule in the data grid and then click the **Run Selected** button to run only the selected rule. To run all rules in a single operation, click the **Run All** button.

 Best Practice: Projility strongly recommends that you **do not** create a rule that approves all types of updates for all of your team members in all of your projects. Although this is a tempting method for reducing the amount of time spent each week approving task updates, you run the risk of a rule automatically approving of an erroneous update that introduces "dirty data" into your Microsoft Project schedule. Instead, we recommend that you use rules sparingly and in specific situations, such the two examples shown in Figure 10 - 25.

Rescheduling Incomplete Work from the Past

After approving task updates, you should open the updated enterprise project in Microsoft Project, and then locate tasks with incomplete work in the past. There are several causes for incomplete work in the past:

- The assigned team member did not start a task scheduled to begin last week.
- The assigned team member performed no work on a task with work scheduled last week.
- The assigned team member performed work on a task scheduled last week, but the actual work performed was less than the expected progress.

Leaving incomplete work scheduled in the past is a major problem for both project managers and team members. For the project manager, incomplete work in the past leads to an incorrect schedule. For team members, incomplete work in the past makes it difficult to enter progress in the *Timesheet* page of PWA because they cannot see the planned work for the task in their current timesheet.

Figure 10 - 26 shows an example of incomplete work scheduled in the past, where the red dashed line represents the *Status* date, which is Friday of last week (the last day of last week's reporting period). In the figure, the first two Gantt bars represent tasks in which the progress in on schedule, so there is no incomplete work in the past. Notice in these two Gantt bars how the dark blue stripe meets the edge of the *Status* date line, indicating that the progress is on schedule. Notice in the last two Gantt bars how the dark blue strip does not meet the edge of the *Status* date line. In the third Gantt bar, the dark blue stripe extends only halfway to the *Status* date line, indicating that the task is approximately 50% completed when the expectation was 100% completion. In the fourth Gantt bar, there is no dark blue stripe at all, indicating that the team member did not work on the task last week when the expectation was 100% completion.

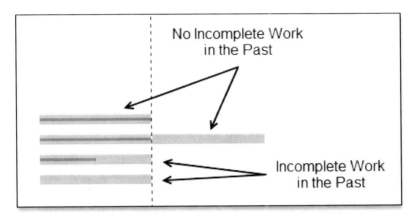

Figure 10 - 26: Incomplete work in the past

In a simple project with only four tasks, such as the one shown previously in Figure 10 - 26, it is very easy to spot tasks with incomplete work in the past. In a large project with hundreds or even thousands of tasks, it is very difficult to spot tasks with incomplete work in the past. There are several tricks you can use to make this process much easier. The first trick is to display the *Status* date line as a red dashed line in the Gantt Chart view by completing the following steps:

1. Apply the *Gantt Chart* view, then click the **Format** tab to display the *Format* ribbon with the *Gantt Chart Tools* applied.
2. In the *Format* section of the *Format* ribbon, click the **Gridlines** pick list button and select the **Gridlines** item, such as shown in Figure 10 - 27. Microsoft Project displays the *Gridlines* dialog.

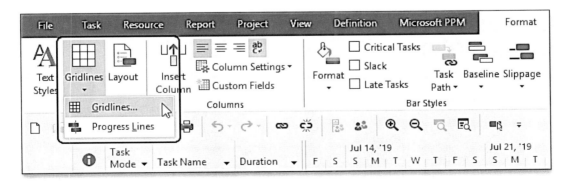

Figure 10 - 27: Click the Gridlines pick list button

Module 10

3. In the *Gridlines* dialog, scroll to the bottom of the *Line to change* list and select the **Status Date** item.

4. Click the **Type** pick list and select the *last item* on the pick list, such as shown in Figure 10 - 28.

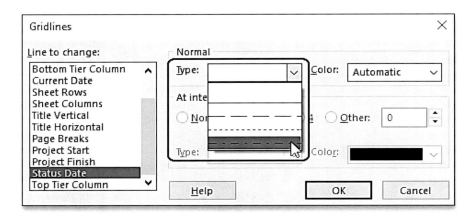

Figure 10 - 28: Select a Type value

5. Click the **Color** pick list and select the **Red** color in the *Standard Colors* section of the pick list.

6. In the *Gridlines* dialog, click the **OK** button to display the *Status Date* gridline in the *Gantt Chart* view.

 Information: When you display the *Status Date* gridline in the *Gantt Chart* view of your project, you only need to perform the steps once to display it, as Microsoft Project adds the *Status Date* gridline as a permanent part of the view for that particular project. You must still perform this step in each project you manage. In addition, Projility recommends that your organization display the *Status Date* gridline in every project template used to create projects in your organization.

The second step in the process of locating tasks with incomplete work in the past is to set the *Status* date for your project. You must do this in *every project* you manage, and you must do this *every week* as well. The *Status* date represents the last day of the previous reporting period, which is usually Friday of past week. Microsoft Project does not set the *Status* date for you automatically; this is a process you must perform manually every week. The red dashed *Status Date* line can serve as a nice reminder for you to perform this step every week. To set the *Status* date, complete the following steps:

1. Click the **Project** tab to display the *Project* ribbon.

2. In the *Status* section of the *Project* ribbon, click the **Status Date** button, as shown in Figure 10 - 29.

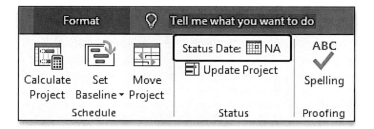

Figure 10 - 29: Click the Status Date button

3. In the *Status Date* dialog shown in Figure 10 - 30, click the **Select Date** pick list and select the *Status* date for your project, and then click the **OK** button.

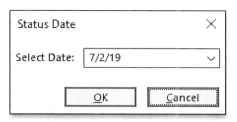

Figure 10 - 30: Select a
Status date value

The third and final step in locating incomplete work in the past is to filter the project for late tasks by completing the following steps:

1. Click the **View** tab to display the *View* ribbon.
2. In the *Data* section of the *View* ribbon, click the **Filter** pick list and select the **Late Tasks** item.
3. Study the late tasks shown by the *Late Tasks* filter to see which tasks have incomplete work in the past.
4. Press the **F3** function key on your computer keyboard to clear the filter.

To reschedule incomplete work from the past reporting period into the current reporting period, complete the following steps:

1. Click the **Project** tab to display the *Project* ribbon.
2. In the *Status* section of the *Project* ribbon, click the **Update Project** button, as shown in Figure 10 - 31.

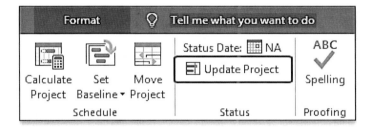

Figure 10 - 31: Click the
Update Project button

3. In the *Update Project* dialog shown in Figure 10 - 32, select the **Reschedule incomplete work to start after** option. Notice that Microsoft Project automatically enters your selected *Status* date in the accompanying field.
4. Leave the **Entire project** option selected, and then click the **OK** button.

 Information: If you select the *Entire project* option in the *Update Project* dialog, Microsoft Project automatically reschedules every task with incomplete work in the past. This option is very useful for large projects with hundreds or thousands of tasks.

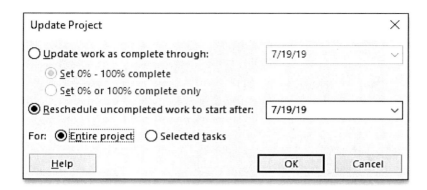

Figure 10 - 32: Update Project dialog set
to reschedule incomplete work

5. Drag the split bar to the right edge of the *Finish* column.

6. Look for the light blue cell background color (the change highlighting) in the *Start* and/or *Finish* column of any task, as this reveals which tasks the software rescheduled as a result of this operation.

When you reschedule incomplete work from the past into the current reporting period, Microsoft Project does the following in response:

- The software *does not* touch the actual progress on any task.

- If a task with incomplete work in the past is an in-progress task, the software leaves the actual progress untouched, but creates a task split to reschedule the remaining work from the past into the current reporting period.

- If a task with incomplete work in the past is an unstarted task, the software applies a Start No Earlier Than constraint to delay the *Start* date of the task to the first day of the current reporting period.

Figure 10 - 33 shows the same set of four Gantt bars shown previously in Figure 10 - 26, but after rescheduling incomplete work from the past to the current reporting period. Notice that Microsoft Project did not touch the actual progress for the first two tasks. Notice that the software created a task split (the … pattern) in the third Gantt bar, which represents an in-progress task. Notice finally that the software delayed the fourth task in its entirety to the next week because this task is an unstarted task.

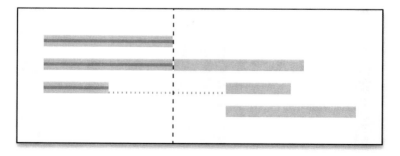

Figure 10 - 33: Incomplete work rescheduled

Approving Time and Task Progress

Hands On Exercise

Exercise 10 - 3

Locate tasks with incomplete work in the past and reschedule them into the current reporting period.

1. Launch Microsoft Project and connect to Project Web App in the *Login* dialog.
2. Reopen the enterprise project you created for use with this class and then check out the project for editing.
3. Apply the *Gantt Chart* view, if necessary.
4. Click the **Format** tab to display the *Format* ribbon.
5. In the *Format* section of the *Format* ribbon, click the **Gridlines** pick list button and select the **Gridlines** item.
6. In the *Gridlines* dialog, scroll to the bottom of the *Line to change* list and select the **Status Date** item.
7. Click the **Type** pick list and select the **last item** on the list.
8. Click the **Color** pick list and select the **Red** color in the *Standard Colors* section of the pick list.
9. In the *Gridlines* dialog, click the **OK** button.
10. Click the **Project** tab to display the *Project* ribbon.
11. In the *Status* section of the *Project* ribbon, click the **Status Date** button.
12. In the *Status Date* dialog, click the **Select Date** pick list and set the *Status* date to *Friday of the current week*.
13. In the *Status Date* dialog, click the **OK** button.

Information: If you are working this exercise using self-study, set the *Status* date value to the Friday of the week for the time reporting period during which you entered task progress in the *Timesheet* page in PWA.

14. In the *Status* section of the *Project* ribbon, click the **Update Project** button.
15. In the *Update Project* dialog, select the **Reschedule incomplete work to start after** option.

Notice that Microsoft Project automatically enters your selected *Status* date in the field to the right of the *Reschedule incomplete work to start after* option.

16. Leave the **Entire project option** selected, and then click the **OK** button.
17. In the *Planning Wizard* dialog, click the **OK** button.

Module 10

18. Study the impact to your project schedule after rescheduling incomplete work from the past to the current reporting period.

19. Save the latest changes to your enterprise project.

20. Publish the latest changes to your enterprise project.

Updating Expense Cost Resource Information

After you reschedule incomplete work in past time periods, you are ready to update actual project costs for *Expense Cost* resources. To update the *Expense Cost* resource information, complete the following steps:

1. Click the **View** tab to display the *View* ribbon.

2. In the *Task Views* section of the *Task* ribbon, click the **Task Usage** button to apply the *Task Usage* view.

3. In the *Data* section of the *View* ribbon, click the **Tables** pick list and select the **Cost** item to apply the *Cost* table.

4. Drag the split bar to the right edge of the *Remaining Cost* column.

5. Enter the actual expenditure in the **Actual Cost** column for the *Expense Cost* resource assignment.

6. Optionally reduce the **Remaining Cost** value to **$0.00**, if necessary.

Figure 10 - 34 shows that I entered *$4,800* in the *Actual Cost* column for the *Training Expense* cost resource assigned to the *Conduct needs analysis* task. Notice that Microsoft Project calculated a *Remaining Cost* value of *$200* for the cost resource. Because the actual training expenditure was only $4,800, I must reduce the *Remaining Cost* value to *$0*.

	Task Name	Fixed Cost	Fixed Cost Accrual	Total Cost	Baseline	Variance	Actual	Remaining
7	▲ Analysis/Software Requirements	$0.00	Prorated	$26,040.00	$26,040.00	$0.00	$12,000.00	$14,040.00
8	▲ Conduct needs analysis	$0.00	Prorated	$12,200.00	$12,200.00	$0.00	$12,000.00	$200.00
	Sharon Brignole			$4,000.00	$4,000.00	$0.00	$4,000.00	$0.00
	Gary Manche			$3,200.00	$3,200.00	$0.00	$3,200.00	$0.00
	Training Expense			$5,000.00	$5,000.00	$0.00	$4,800.00	$200.00
9	▲ Draft preliminary software specifications	$0.00	Prorated	$4,320.00	$4,320.00	$0.00	$0.00	$4,320.00

Figure 10 - 34: Enter Actual Cost for an Expense Cost resource

Microsoft Project distributes the *Actual Cost* value evenly across the duration of the task. In reality, the expense likely accrued on a specific day due to payment of an invoice or the filing of an expense report. To reallocate the *Actual Cost* value to a specific day or days, complete the following additional steps:

1. Drag the split bar to the right edge of the *Task Name* column to display the timephased grid on the right side of the *Task Usage* view.

2. Right-click anywhere in the timephased grid and select the **Actual Cost** item on the shortcut menu, such as shown in Figure 10 - 35.

Approving Time and Task Progress

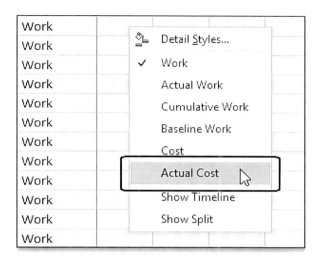

Figure 10 - 35: Display the Actual Cost
row in the timephased grid

3. Right-click again in the timephased grid and *deselect* the **Work** item on the shortcut menu.

Figure 10 - 36 shows the timephased *Actual Cost* values for the Training Expense cost resource. Notice that the software applied $1,000/day of actual expense for the first four days of the task, and applied $800 of expense on the last day of the task.

	Task Name	Details	W	T	F	S	Oct 27, '19 S	M	T
7	▲ Analysis/Software Requirements	Act. Cost	$2,440.00	$2,440.00	$2,440.00	$0.00	$0.00	$2,440.00	$2,240.00
8	▲ Conduct needs analysis	Act. Cost	$2,440.00	$2,440.00	$2,440.00	$0.00	$0.00	$2,440.00	$2,240.00
	Sharon Brignole	Act. Cost	$800.00	$800.00	$800.00			$800.00	$800.00
	Gary Manche	Act. Cost	$640.00	$640.00	$640.00			$640.00	$640.00
	Training Expense	Act. Cost	$1,000.00	$1,000.00	$1,000.00	$0.00	$0.00	$1,000.00	$800.00
9	▲ Draft preliminary software specifications	Act. Cost							
	Sharon Brignole	Act. Cost							
	Gary Manche	Act. Cost							

Figure 10 - 36: Timephased Actual Cost for an Expense Cost resource

4. Edit the *Actual Cost* information in the timephased grid on the day(s) the expense occurred for the *Expense Cost* resource.

5. Manually type **$0** in the **Actual Cost** row for every other day of the task duration.

Because the actual expenditure for the Training Expense cost resource occurred on the first day of the task, notice in Figure 10 - 37 that I entered $4,800 on the first day of the task in the *Actual Cost* row for the Training Expense cost resource. Notice also that I entered a $0 value for the other four days of the task.

327

	Task Name	Details	W	T	F	S	Oct 27, '19 S	M	T
7	▲ Analysis/Software Requirements	Act. Cost	$6,240.00	$1,440.00	$1,440.00	$0.00	$0.00	$1,440.00	$1,440.00
8	▲ Conduct needs analysis	Act. Cost	$6,240.00	$1,440.00	$1,440.00	$0.00	$0.00	$1,440.00	$1,440.00
	Sharon Brignole	Act. Cost	$800.00	$800.00	$800.00			$800.00	$800.00
	Gary Manche	Act. Cost	$640.00	$640.00	$640.00			$640.00	$640.00
	Training Expense	Act. Cost	$4,800.00	$0.00	$0.00	$0.00	$0.00	$0.00	$0.00
9	▲ Draft preliminary software specifications	Act. Cost							
	Sharon Brignole	Act. Cost							
	Gary Manche	Act. Cost							

Figure 10 - 37: Actual Cost occurred on the first day of the task

Warning: When you enter an *Actual Cost* value in the timephased grid on the specific day that the expenditure actually occurred, you **must** enter a **$0** value for each of the remaining days of the task. You must do this so that Microsoft Project correctly calculates the cost of the actual expenditure. You can then compare the *Actual Cost* values you entered in the timephased grid with the value in the *Actual Cost* column to confirm that your entries are correct.

Publishing the Latest Schedule Changes

After you update task progress into your project, reschedule incomplete work, and update *Expense Cost* resource information, you must always save the project and then publish the latest schedule changes. This step is extremely important, as it updates the latest schedule information on the *Timesheet* page and the *Tasks* page of each team member in the project. To publish the latest schedule changes, click the **File** tab and then click the **Publish** button on the *Info* page of the *Backstage*. The system updates the Project Online database with the current information for the project, making the current schedule visible everywhere in Project Web App.

Warning: If you fail to publish the latest schedule changes after updating a project, your project becomes "out of sync" between the Microsoft Project schedule and Project Web App. "Out of sync" project information adversely impacts your project team members, as they cannot see the current task schedule on their *Timesheet* and *Tasks* pages in PWA. Remember that you **must** always publish your enterprise project after updating it with task progress.

Best Practice: Projility recommends that your organization make project publishing a "training and performance" issue. Teach your project managers to publish their enterprise project after every session of approving updates, and then hold them accountable for their performance in this area.

Axioms for Success with Tracking Progress

To be successful in tracking project progress using Project Web App, your organization should keep the following in mind:

- Everyone in your organization who is responsible for reporting progress should submit their updates on a standard day and time, whether this is daily, weekly, semi-weekly, or another predictable and appropriate reporting standard.

- Project managers should process task updates on a standard day each reporting period.

- Your organization should deal appropriately with team members who are responsible for reporting progress, but who fail to cooperate or participate fully in the process. Your organization should also deal appropriately with project manager who are responsible for updating progress, but who fail to cooperate or participate fully in the process You must take all necessary steps to ensure the full participation of everyone in your organization in order to validate the project data in the system.

- Your organization should track and manage the absence of resources during each update cycle. If necessary, you can create a Delegation session to manage timesheets for absent resources.

- Your organization should also track and manage the absence of project managers during each update cycle. For updating purposes, other project managers can create a Delegation session for an absent manager and then process the pending task updates.

- Stay current with progress reporting and updates to make sure that you are managing your projects with current data.

Module 11

Variance Analysis, Plan Revision, and Reporting

Learning Objectives

After completing this module, you will be able to:

- Understand and analyze variance in an enterprise project
- Revise a project schedule to bring it back on track
- Use a change control process to add new tasks to a project schedule
- Baseline new tasks added to a project schedule
- Report about an enterprise project using the Timeline view and Dashboard Reports
- Report about enterprise data using Power BI reports

Inside Module 11

Understanding Variance	333
Understanding Variance Types	*333*
Calculating Variance	*333*
Understanding Actual vs. Estimated Variance	*334*
Analyzing Project Variance	335
Analyzing Date Variance	*335*
Analyzing Work Variance	*337*
Analyzing Cost Variance	*338*
Revising a Project Plan	341
Potential Problems with Revising a Plan	*341*
Using a Change Control Process	343
Inserting New Tasks in a Project	*343*
Updating the Baseline in Your Project	345
Baselining Only Selected Tasks	*345*
Backing up the Current Baseline Data	*347*
Project Reporting Overview	349
Project Reporting Using the Timeline View	349
Adding Tasks to the Timeline View	*350*

 Arranging Tasks in the Timeline View *350*
 Formatting the Timeline View *351*
Project Reporting Using the Dashboard Reports **353**
 Customizing a Chart *354*
 Customizing a Table *356*
Project Reporting Using Power BI Reports **359**
 Using Ad Hoc Filtering *368*
 Using Natural Language Queries *370*

Understanding Variance

At the end of every reporting period, you should analyze project variance by comparing actual progress and remaining estimates against the original project baseline. This is the way you determine schedule slippage and overruns, as well as identifying existing and/or potential problems with your project schedule. Analyzing variance is the first step in revising the project plan to bring it back on track with its original goals and objectives.

Understanding Variance Types

In Module 08, *Project Execution*, I documented that when you save a baseline in Microsoft Project, the software saves the baseline for the current values in five major task fields. These fields include the *Duration, Start, Finish, Work*, and *Cost*. Because the software saves these five task values in the project baseline, the software can calculate five types of task variance:

- Duration variance
- Start variance
- Finish variance
- Work variance
- Cost variance

> **About Those Extra Task Baseline Fields**
>
> In addition to the five major task fields captured in the baseline, Microsoft Project also captures additional baseline information in several minor task fields. The software captures this minor baseline information in the following fields: *Baseline Fixed Cost, Baseline Fixed Cost Accrual, Baseline Estimated Duration, Baseline Estimated Start, Baseline Estimated Finish, Baseline Budget Cost* and *Baseline Budget Work* fields.
>
> Even though Microsoft Project captures the extra task baseline information in these seven fields, the software **does not** include any corresponding variance fields for them. This means that if you want to analyze *Budget Cost* variance, for example, there is no default field called *Budget Cost Variance*. If you want to analyze *Budget Cost* variance, you must create a custom task field containing a formula to calculate this variance. The same is true for the other six extra baseline fields as well.

Calculating Variance

To calculate variance, Microsoft Project uses the following formula:

Variance = (Actual Progress + Remaining Estimates) - Baseline

In Microsoft Project, a positive variance value is unfavorable to the project, and means that the project schedule is late, or that work and/or cost are over budget. A negative variance value is favorable to the project, and means that the project is ahead of schedule, or that work and/or cost are under budget.

For example, suppose that the *Actual Work* for a task is *60 hours*, the *Remaining Work* estimate is *40 hours*, and the *Baseline Work* for the task is *80 hours*. Using the formula above, Microsoft Project calculates the *Work Variance* as:

Work Variance = (Actual Work + Remaining Work) – Baseline Work

Work Variance = (60 hours + 40 hours) – 80 hours

Work Variance = 100 hours - 80 hours

Work Variance = 20 hours

The resulting *Work Variance* value of *20 hours* is **unfavorable** to the project because the total *Work* hours exceed the original *Baseline Work* budget. Using another example, suppose that the *Actual Work* for a task is *32 hours*, the *Remaining Work* estimate is *0 hours* (the task finished early), and the *Baseline Work* for the task is *40 hours*. Using the formula above, Microsoft Project calculates the *Work Variance* as:

Work Variance = (Actual Work + Remaining Work) – Baseline Work

Work Variance = (32 hours + 0 hours) – 40 hours

Work Variance = 32 hours - 40 hours

Work Variance = -8 hours

The resulting *Work Variance* value of *-8 hours* is **favorable** to the project because the total *Work* hours for the project are now less than the original *Baseline Work* budget.

Understanding Actual vs. Estimated Variance

Microsoft Project measures two types of variance in any project: **Actual Variance** and **Estimated Variance**. It is important that you understand the distinction between the two. Actual variance occurs when an actual value, such as *Actual Work*, exceeds its original baseline value. For example, suppose that a task has a *Baseline Work* of *40 hours*, but the task is completed, and the *Actual Work* on the task is *50 hours*. Using the formula for variance, Microsoft Project calculates the variance as follows:

Work Variance = (Actual Work + Remaining Work) – Baseline Work

Work Variance = (50 hours + 0 hours) – 40 hours

Work Variance = 10 hours

Because the task is complete and the *Actual Work* exceeds the *Baseline Work* by 10 hours, this type of variance is actual variance. In other words, the task went over its baseline budget on work and it is now too late for the project manager to do anything about it.

On the other hand, estimated variance is variance that "might" occur, based on the *Remaining Work* estimates submitted by the project team members. Estimated variance occurs when actual progress plus remaining estimates exceed the baseline. For example, a task has a *Baseline Work* of *40 hours*. At the end of the first week of work on the task, the resource reports *Actual Work* of *25 hours*, plus a *Remaining Work* estimate of *30 hours*. Using the formula for variance, Microsoft Project calculates the variance as follows:

Work Variance = (Actual Work + Remaining Work) – Baseline Work

Work Variance = (25 hours + 30 hours) – 40 hours

Work Variance = 55 hours – 40 hours

Work Variance = 15 hours

The 15 hours of work variance is only an "estimate" at this point, which is caused by the resource "estimating" 15 hours more work than originally scheduled. Estimated variance is very important to you because it is variance that "might" occur, and which gives you time to plan for the possible slippage or overrun.

Analyzing Project Variance

Microsoft Project offers you the following locations from which to analyze project variance:

- *Tracking Gantt* view
- *Variance* table
- *Work* table
- *Cost* table

The *Tracking Gantt* view and the *Variance* table allow you to analyze start and finish variance for tasks. The *Work* and *Cost* tables allow you to analyze work and cost variance respectively.

Information: Microsoft Project does not offer a default table in which to analyze *Duration* variance. If you want to see *Duration* variance, you must create your own custom table for this purpose.

Analyzing Date Variance

Date variance is a major concern for every project manager because many projects have an inflexible project finish date. You can analyze date variance graphically by applying the *Tracking Gantt* view. To apply the *Tracking Gantt* view, use one of the following methods:

- In the *View* section of the *Task* ribbon, click the **Gantt Chart** pick list button, and then select the **Tracking Gantt** view.

- In the *View* section of the *Resource* ribbon, click the **Team Planner** pick list button, and then select the **Tracking Gantt** view.

- In the *Tasks Views* section of the *View* ribbon, click the **Gantt Chart** pick list button, and then select the **Tracking Gantt** view.

Using any of the preceding methods, Microsoft Project displays the *Tracking Gantt* view, such as the one shown in Figure 11 - 1. In the *Tracking Gantt* view, following are the Gantt bar symbols of the most importance to variance analysis:

- **Red Gantt bars** represent tasks on the Critical Path (Critical tasks).
- **Blue Gantt bars** represent tasks not on the Critical Path (non-Critical tasks).
- **Gray Gantt bars** represent the original baseline schedule for each task.
- **Black solid diamonds** represent the current schedule for each milestone task.
- **Hollow diamonds** represent the baseline schedule for each milestone task.

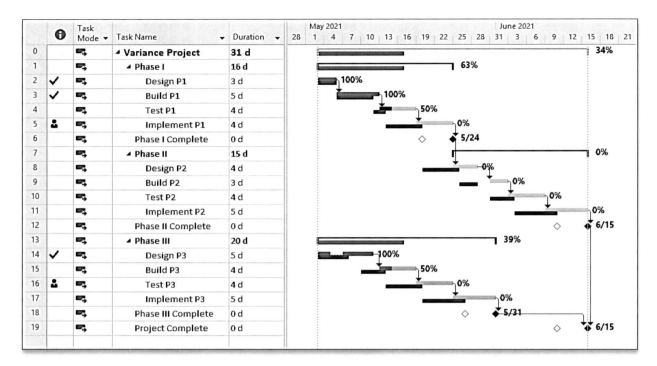

Figure 11 - 1: Tracking Gantt view

The *Tracking Gantt* view allows you to see schedule slippage presented graphically. Using this view, you can compare red and blue Gantt bars (the current schedule of each task) with their accompanying gray Gantt bars (the baseline schedule for each task). If a red or blue Gantt bar slips to the right of its gray Gantt bar, then the task is slipping. Using the *Tracking Gantt* view to analyze schedule variance, it is easy to see the slippage for all tasks of the project.

While in the *Tracking Gantt* view, you can use the *Variance* table to view the date variance measured in days. To apply the *Variance* table, use either of the following methods:

- Right-click on the **Select All** button and select the **Variance** table in the shortcut menu.
- In the *Tables* section of the *View* ribbon, click the **Tables** pick list, and then select the **Variance** table.

Microsoft Project displays the *Variance* table, such as shown in Figure 11 - 2. To analyze date variance, examine each value in the *Start Variance* and *Finish Variance* columns. As you examine the data shown in these two columns, keep the following information in mind:

- A positive value in the *Start Variance* column means that the task started late against its original baseline, while a negative value in the *Start Variance* column means that the task started early against its original baseline.

- A positive value in the *Finish Variance* column means that the task finished late against its original baseline, while a negative value in the *Finish Variance* column means that the task finished early against its original baseline.

- The *Finish Variance* value for the Project Summary Task (Row 0 or Task 0) shows you the finish variance for the entire project. Notice in Figure 11 - 2 that the project is currently 3 days late, indicated by the *Finish Variance* value of *3 days* shown for the Project Summary Task.

Variance Analysis, Plan Revision, and Reporting

	Task Mode	Task Name	Start	Finish	Baseline Start	Baseline Finish	Start Var.	Finish Var.
0		▲ Variance Project	5/3/21	6/15/21	5/3/21	6/10/21	0 d	3 d
1		▲ Phase I	5/3/21	5/24/21	5/3/21	5/19/21	0 d	3 d
2		Design P1	5/3/21	5/5/21	5/3/21	5/5/21	0 d	0 d
3		Build P1	5/6/21	5/12/21	5/6/21	5/11/21	0 d	1 d
4		Test P1	5/13/21	5/18/21	5/12/21	5/13/21	1 d	3 d
5		Implement P1	5/19/21	5/24/21	5/14/21	5/19/21	3 d	3 d
6		Phase I Complete	5/24/21	5/24/21	5/19/21	5/19/21	3 d	3 d
7		▲ Phase II	5/25/21	6/15/21	5/20/21	6/10/21	3 d	3 d
8		Design P2	5/25/21	5/28/21	5/20/21	5/25/21	3 d	3 d
9		Build P2	5/31/21	6/2/21	5/26/21	5/28/21	3 d	2 d
10		Test P2	6/3/21	6/8/21	5/31/21	6/3/21	2 d	3 d
11		Implement P2	6/9/21	6/15/21	6/4/21	6/10/21	3 d	3 d
12		Phase II Complete	6/15/21	6/15/21	6/10/21	6/10/21	3 d	3 d
13		▲ Phase III	5/3/21	5/31/21	5/3/21	5/26/21	0 d	2 d
14		Design P3	5/3/21	5/11/21	5/3/21	5/7/21	0 d	2 d
15		Build P3	5/13/21	5/18/21	5/10/21	5/13/21	3 d	3 d
16		Test P3	5/19/21	5/24/21	5/14/21	5/19/21	3 d	3 d
17		Implement P3	5/25/21	5/31/21	5/20/21	5/26/21	3 d	2 d
18		Phase III Complete	5/31/21	5/31/21	5/26/21	5/26/21	2 d	2 d
19		Project Complete	6/15/21	6/15/21	6/10/21	6/10/21	3 d	3 d

Figure 11 - 2: Variance table

Analyzing Work Variance

Use the *Work* table to analyze work variance and to determine when project work exceeds its original planned work budget. To apply the *Work* table, use either of the following methods:

- Right-click on the **Select All** button and select the **Work** table in the shortcut menu.
- In the *Tables* section of the *View* ribbon, click the **Tables** pick list, and then select the **Work** table.

Microsoft Project displays the *Work* table, such as shown in Figure 11 - 3. To analyze work variance, examine each value in the *Variance* column. Remember that a positive value indicates that the task is over budget on work, while a negative value indicates that the task is under budget on work. Notice in Figure 11 - 3 that the project is currently 32 hours over its budget on work, as indicated by the *Variance* value of *32 hours* for the Project Summary Task.

Information: In the *Work* table, the real name of the *Variance* column is *Work Variance*. Microsoft uses the shorter name as the title of this column for display purposes only. Remember that you can see the real name of any column by floating your mouse pointer over its column header. The software displays a tool tip that shows the title of the column, followed by the real name of the column in parentheses.

	Task Name	Work	Baseline	Variance	Actual	Remaining	% W. Comp.
0	⊿ Variance Project	496 h	464 h	32 h	176 h	320 h	35%
1	⊿ Phase I	168 h	136 h	32 h	120 h	48 h	71%
2	Design P1	24 h	24 h	0 h	24 h	0 h	100%
3	Build P1	80 h	64 h	16 h	80 h	0 h	100%
4	Test P1	32 h	16 h	16 h	16 h	16 h	50%
5	Implement P1	32 h	32 h	0 h	0 h	32 h	0%
6	Phase I Complete	0 h	0 h	0 h	0 h	0 h	0%
7	⊿ Phase II	152 h	152 h	0 h	0 h	152 h	0%
8	Design P2	32 h	32 h	0 h	0 h	32 h	0%
9	Build P2	48 h	48 h	0 h	0 h	48 h	0%
10	Test P2	32 h	32 h	0 h	0 h	32 h	0%
11	Implement P2	40 h	40 h	0 h	0 h	40 h	0%
12	Phase II Complete	0 h	0 h	0 h	0 h	0 h	0%
13	⊿ Phase III	176 h	176 h	0 h	56 h	120 h	32%
14	Design P3	40 h	40 h	0 h	40 h	0 h	100%
15	Build P3	32 h	32 h	0 h	16 h	16 h	50%
16	Test P3	64 h	64 h	0 h	0 h	64 h	0%
17	Implement P3	40 h	40 h	0 h	0 h	40 h	0%
18	Phase III Complete	0 h	0 h	0 h	0 h	0 h	0%
19	Project Complete	0 h	0 h	0 h	0 h	0 h	0%

Figure 11 - 3: Work table

Analyzing Cost Variance

Use the *Cost* table to analyze cost variance and to determine when project costs exceed its original planned cost budget. To apply the *Cost* table, use either of the following methods:

- Right-click on the **Select All** button and select the **Cost** table in the shortcut menu.
- In the *Tables* section of the *View* ribbon, click the **Tables** pick list, and then select the **Cost** table.

Microsoft Project displays the *Cost* table, such as shown in Figure 11 - 4. To analyze cost variance, examine each value in the *Variance* column. A positive value indicates that the task is over budget on cost, while a negative value indicates that the task is under budget on cost. Notice in Figure 11 - 4 that the project is currently $1,600 over its budget on work, as indicated by the *Variance* value of $1,600 for the Project Summary Task

Information: In the *Cost* table, the real name of the *Variance* column is *Cost Variance*. Microsoft uses the shorter name as the title of this column for display purposes only. Remember that you can see the real name of any column by floating your mouse pointer over its column header. The software displays a tool tip that shows the title of the column, followed by the real name of the column in parentheses.

Variance Analysis, Plan Revision, and Reporting

	Task Name	Fixed Cost	Fixed Cost Accrual	Total Cost	Baseline	Variance	Actual	Remaining
0	⊿ Variance Project	$0.00	Prorated	$27,300.00	$25,700.00	$1,600.00	$11,300.00	$16,000.00
1	⊿ Phase I	$0.00	Prorated	$10,900.00	$9,300.00	$1,600.00	$8,500.00	$2,400.00
2	Design P1	$0.00	Prorated	$1,200.00	$1,200.00	$0.00	$1,200.00	$0.00
3	Build P1	$2,500.00	Prorated	$6,500.00	$5,700.00	$800.00	$6,500.00	$0.00
4	Test P1	$0.00	Prorated	$1,600.00	$800.00	$800.00	$800.00	$800.00
5	Implement P1	$0.00	Prorated	$1,600.00	$1,600.00	$0.00	$0.00	$1,600.00
6	Phase I Complete	$0.00	Prorated	$0.00	$0.00	$0.00	$0.00	$0.00
7	⊿ Phase II	$0.00	Prorated	$7,600.00	$7,600.00	$0.00	$0.00	$7,600.00
8	Design P2	$0.00	Prorated	$1,600.00	$1,600.00	$0.00	$0.00	$1,600.00
9	Build P2	$0.00	Prorated	$2,400.00	$2,400.00	$0.00	$0.00	$2,400.00
10	Test P2	$0.00	Prorated	$1,600.00	$1,600.00	$0.00	$0.00	$1,600.00
11	Implement P2	$0.00	Prorated	$2,000.00	$2,000.00	$0.00	$0.00	$2,000.00
12	Phase II Complete	$0.00	Prorated	$0.00	$0.00	$0.00	$0.00	$0.00
13	⊿ Phase III	$0.00	Prorated	$8,800.00	$8,800.00	$0.00	$2,800.00	$6,000.00
14	Design P3	$0.00	Prorated	$2,000.00	$2,000.00	$0.00	$2,000.00	$0.00
15	Build P3	$0.00	Prorated	$1,600.00	$1,600.00	$0.00	$800.00	$800.00
16	Test P3	$0.00	Prorated	$3,200.00	$3,200.00	$0.00	$0.00	$3,200.00
17	Implement P3	$0.00	Prorated	$2,000.00	$2,000.00	$0.00	$0.00	$2,000.00
18	Phase III Complete	$0.00	Prorated	$0.00	$0.00	$0.00	$0.00	$0.00
19	Project Complete	$0.00	Prorated	$0.00	$0.00	$0.00	$0.00	$0.00

Figure 11 - 4: Cost table

Hands On Exercise

Exercise 11 - 1

Information: Because the enterprise project you created for this class does not contain much variance, this exercise uses a Microsoft Project sample file containing lots of variance for you to analyze.

Analyze schedule variance, along with date, work, and cost variance in your enterprise project.

1. Close and check in the enterprise project you created for this class.
2. Open the **Training Advisor Rollout 2019.mpp** sample file from your student sample files folder.

Actual progress is current through Friday, February 12, 2021 in your Training Advisor Rollout project.

3. Click the **Task** tab to display the *Task* ribbon.

339

Module 11

4. In the *View* section of the *Task* ribbon, click the **Gantt Chart** pick list button, and then select the **Tracking Gantt** view.

5. Scroll down to the last task in the project schedule.

Notice the "missed Deadline" indicator in the *Indicators* column for task ID #23, the *Project Complete* milestone task.

6. Select task ID #23, the *Project Complete* milestone task and then click the **Scroll to Task** button in your Quick Access Toolbar to scroll the task's Gantt bar into view.

Notice that this milestone task is currently 5 days late compared with its deadline date.

7. Scroll through the *Tracking Gantt* pane to analyze schedule slippage for each task in the project.

8. Right-click on the **Select All** button and select the **Variance** table in the shortcut menu.

9. Widen the *Task Name* column and then drag the split bar to the right edge of the *Finish Variance* column.

10. In the *Start Variance* and *Finish Variance* columns, analyze the schedule slippage data for every task in the project.

For the Project Summary Task (Row 0), notice in the *Finish Variance* column that the project is currently 8 days late. Notice the schedule slippage in the INSTALLATION, TESTING, and TRAINING phases of the project. Notice finally that the *Create Training Materials* deliverable section is 9 days late.

11. Right-click on the **Select All** button and select the **Work** table in the shortcut menu.

12. Widen the *Task Name* column and then drag the split bar to the right edge of the *Variance* column.

13. Examine the data in the *Variance* column and analyze work variance for every task in the project.

For the Project Summary Task (Row 0), notice in the *Variance* column that the project is currently 60 hours over budget on work. Notice the work overruns in the INSTALLATION, TESTING, and TRAINING phases of the project.

14. Right-click on the **Select All** button and select the **Cost** table in the shortcut menu.

15. Widen the *Task Name* column and then drag the split bar to the right edge of the *Variance* column.

16. Examine the data in the *Variance* column and analyze cost variance for every task in the project.

For the Project Summary Task (Row 0), notice in the *Variance* column that the project is nearly $30,000 over budget on cost. Notice the work overruns in the INSTALLATION, TESTING, and TRAINING phases of the project.

17. Double-click task ID #8, the *Install Training Advisor Clients* task, and then click the **Notes** tab in the *Task Information* dialog.

Notice that the task note explains why this task is more than $15,000 over budget, due to a need for more desktop software licenses for the new regional office in Irvine, CA.

18. Click the **Cancel** button to close the *Task Information* dialog.
19. Right-click on the **Select All** button and select the **Entry** table in the shortcut menu.
20. Drag the split bar back to the right edge of the *Duration* column.
21. Save but *do not* close the **Training Advisor Rollout 2019.mpp** sample file.

Revising a Project Plan

After completing variance analysis, you may need to revise your project plan to bring it "back on track" against its original goals, objectives, and schedule. There are a number of strategies for revising a project plan, but each one requires careful consideration before you make the revision. You should perform a "what-if" analysis before making plan revisions, especially if you need formal approval to make the revisions.

Microsoft Project offers a number of methods for revising a project plan. These methods include:

- Add resources to *Effort Driven* tasks to shorten the duration of the task.
- Ask project team members to work overtime or on weekends.
- Increase project team availability for your project.
- Modify dependencies, including reducing or removing *Lag* time, or adding *Lead* time.
- Reduce the scope of the project.
- Renegotiate the project *Finish* date.

Potential Problems with Revising a Plan

Prior to using any of the preceding techniques, you should be aware of potential problems that may arise when you implement your revisions. Some of the potential problems include:

- Adding resources to an *Effort Driven* task can increase the total work on the task due to increased communication needs between the team members.
- Asking team members to work overtime on a regular basis can create morale problems and increase your employee turnover rate.
- Increasing team member availability for your project reduces their availability for other projects, potentially causing those projects to slip.
- Reducing *Lag* time on task dependencies can create an overly optimistic project schedule.
- Adding *Lead* time on task dependencies can create a scheduling crisis when the predecessor task must finish completely, thus negating the intent of adding the *Lead* time.
- The scope of your project may be non-negotiable.
- The finish date of your project may be non-negotiable.

Module 11

Hands On Exercise

Exercise 11 - 2

Revise the Training Advisor Rollout project by adding resources to *Effort Driven* tasks and by adjusting resource availability to bring it "back on track" against its original baseline schedule.

1. In your Quick Access Toolbar, click the **Zoom Out** button until you zoom the timescale to the **Months Over Weeks** level of zoom.
2. Click the **View** tab to display the *View* ribbon.
3. In the *Split View* section of the *View* ribbon, select the **Details** checkbox.

> **Information**: The temporary combination view you just created (*Tracking Gantt* view in the top pane and *Task Form* view in the bottom pane) is an excellent view to use when revising your project. This view shows you the immediate result of each revision compared against the original project baseline schedule.

Due to commitments to his Help Desk work, *Mike Andrews* is only able to work half-time on the *Verify Connectivity* task. You negotiate with his functional manager to "borrow" *Mike Andrews* to work three-quarter time on this task to complete it sooner. You now owe *Mike Andrew's* functional manager a **big favor** in return!

4. In the *Gantt Chart* pane, select task ID #9, the *Verify Connectivity* task, and then click the **Scroll to Task** button in your Quick Access Toolbar to scroll the task's Gantt bar into view.
5. In the *Task Form* pane, increase the **Units** value for **Mike Andrews** to **75%**, and then click the **OK** button.

Notice that Microsoft Project shortens the *Duration* of this task to *6.67 days*. Remember that on a *Fixed Units* task, when you change the *Units* value, the software always recalculates the *Duration* of the task.

6. In the *Task Form* pane, click the **Next** button to select task ID #10, the *Resolve Connectivity Errors* task.
7. In the *Task Form* pane, click the first blank row below *Bob Jared*.
8. In the *Task Form* pane, add **Jeff Holly** with a **Units** value of **25%** (**do not** enter a value in the *Work* field) and click the **OK** button.
9. To account for the increased communication needs, change the **Work** value to **24h** for each resource and then click the **OK** button.
10. In the *Split View* section of the *View* ribbon, **deselect** the **Details** checkbox to close the *Task Form* pane and return to a single-pane *Tracking Gantt* view.

Variance Analysis, Plan Revision, and Reporting

If you compare the project's current schedule (red Gantt bar) to its original baseline schedule (dark gray Gantt bar) for the *Resolve Connectivity Errors* task, your project should appear **ahead** of schedule.

11. In the *Task Views* section of the *View* ribbon, click the **Gantt Chart** pick list button, and then select the **Gantt Chart** view.
12. Save but *do not* close the **Training Advisor Rollout 2019.mpp** sample file.

Using a Change Control Process

Change control is the process of managing requested changes in your project. Change requests can arise from a variety of sources, including your customer, your project sponsor, your project stakeholders, your company's executives, your fellow project managers, and even from your project team members. Because each change can result in schedule slippage and cost overruns, it is important that you manage all changes in your project. Remember the old project management saying, "Either you manage change, or change manages you!"

Your change control process should identify and maximize the benefits of change, and should avoid all changes that offer no benefit to the project or that impact the project negatively. Document your change management process in both the Statement of Work document and in the "rules of engagement" with your project sponsor and/or client.

Inserting New Tasks in a Project

Perhaps the most common change request is to add new tasks to a project. When you insert a new task between two dependent tasks, the *Autolink* feature of Microsoft Project determines whether the software automatically adds dependency links to the new task. If you **disabled** the *Autolink* feature in the *Project Options* dialog, the software does not automatically link the new task to the existing tasks in the project. However, if you **enabled** the *Autolink* feature, then Microsoft Project handles the task linking operation as follows when you insert a new task between two dependent tasks:

- If the dependent tasks have a Finish-to-Start (FS) dependency, the software automatically links the new task to the existing tasks using the Finish-to-Start dependency.

- If the dependent tasks have any other type of dependency (SS, FF, or SF), then Microsoft Project **does not** automatically link the new task to the existing tasks. Instead, the software leaves the new task unlinked.

Best Practice: Because you should always make task dependency decisions, and not the software, Projility recommends that you **disable** the *Autolink* feature. If this is not possible due to corporate policy, then break the task dependency links on the tasks in the section where you intend to insert new task. After inserting the new task, establish appropriate task dependencies for the tasks in that section of your project.

Best Practice: When you add new tasks to a project through a change control process, Projility recommends that you format the new tasks with a unique color. You can format the font, the cell background color, and/or the Gantt bar color, as needed. Keep in mind that these formatting changes are visible only in the view in which you apply the formatting.

343

Module 11

 Hands On Exercise

Exercise 11 - 3

While working on the *Create Training Module 03* task, *Ruth Andrews* believes the course training materials need one additional module. She requests that you add a new task named *Create Training Module 04* in the *Create Training Materials* section of the project. After reviewing the change request and its impact to the project, you approve the change request.

1. Click the **Task** tab to display the *Task* ribbon.

2. Select task ID #18, the *Training Materials Created* milestone task, and then press the **Insert** key on your computer keyboard.

3. In the new blank row, enter the name **Create Training Module 04** and enter a **Duration** value of **5 days** for the new task.

Because the *Autolink* feature is disabled in this project, notice that Microsoft Project *does not* automatically link the new task with the other tasks in the *Create Training Materials* section. This means that you have full control over the task linking process.

4. Select task IDs #17-19, from the *Create Training Module 03* task to the *Training Materials Created* milestone task, and then click the **Unlink Tasks** button in your customized Quick Access Toolbar.

5. With task IDs #17-19 still selected, click the **Link the Selected Tasks** button to link the three selected tasks with a default Finish-to-Start (FS) dependency.

6. Right-click on task ID #18, the new *Create Training Module 04* task, and select the **Assign Resources** item on the shortcut menu.

7. In the *Assign Resources* dialog, select **Ruth Andrews**, enter a **Units** value of **100%**, and then click the **OK** button to assign her to the new task.

8. In the *Assign Resources* dialog, click the **Close** button.

9. Double-click task ID #18, the new **Create Training Module 04** task, and then click the **Notes** tab in the *Task Information* dialog.

10. Enter the following note in the *Notes* page of the dialog:

    2/8/21 – New task added through change control at request of Ruth Andrews.

11. Click the **OK** button to close the *Task Information* dialog.

12. Click the row header for task ID #18 to select the entire task row for the new task.

13. In the *Font* section of the *Task* ribbon, click the **Background Color** pick list and select the **Light Green** item in the *Standard Colors* section of the pick list.

14. Select any other task in the project schedule and then examine the cell background formatting applied to the new task added through change control.

15. Save but *do not* close the **Training Advisor Rollout 2019.mpp** sample file.

Updating the Baseline in Your Project

After you add new tasks to your project through change control, you must update the baseline for your project. Although there are a number of methodologies you can use for updating a baseline, Projility recommends the following process:

1. Baseline only the new tasks you added to the project and then optionally "roll up" the baseline information to the respective summary task(s) of the new tasks.

2. Back up your new baseline in the next available set of numbered *Baseline* fields.

Warning: If you rebaseline the entire project using the default Baseline set of fields, Microsoft Project destroys all variance that existed in the project before you added the new tasks, and makes the project appear perfectly on schedule. Because of this, Projility recommends that you never rebaseline the entire project unless your organization's methodologies allow or mandate this.

Baselining Only Selected Tasks

After adding new tasks to the project through a change control procedure, an ideal method for updating the baseline is to baseline **only** the new tasks. Microsoft Project offers you two methods for baselining only selected tasks, which are:

- Baseline only the selected tasks, but do not roll up the baseline values to any summary tasks in the project. Using this technique, the data for the new tasks shows as variance against the original project baseline.

- Baseline only the selected tasks, but roll up the baseline values to all summary tasks in the project. When you choose this option, the baseline data rolls up to all summary tasks for which the selected tasks are subtasks, including the Project Summary Task (Row 0). Using this technique, the data for the new tasks does not show as variance against the original project baseline.

To baseline only selected tasks using either of these options, complete the following steps:

1. Select only the new tasks added to the project through the change control procedure.

2. Click the **Project** tab to display the *Project* ribbon.

3. In the *Schedule* section of the *Project* ribbon, click the **Set Baseline** pick list button, and then click the **Set Baseline** item. Microsoft Project displays the *Set Baseline* dialog.

4. Leave the **Set baseline** option selected and leave the **Baseline** value selected in the **Set baseline** pick list.

5. In the *For:* section of the dialog, choose the **Selected tasks** option, as shown in Figure 11 - 5.

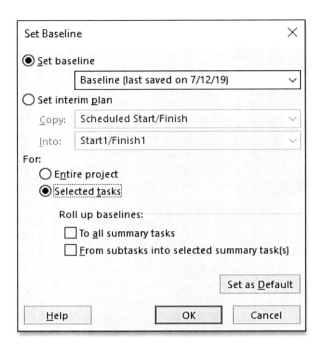

Figure 11 - 5: Choose the Selected tasks option

6. If you want to roll up the baseline values to summary tasks, choose one of the following options in the *Roll up baselines* section:

    - Select the **To all summary tasks** option if you want the software to roll up the baseline values to all summary tasks for which the selected tasks are subtasks and to the Project Summary Task as well.

    - Select the **From subtasks into selected summary tasks** option if you want the software to roll up the baseline values to only the summary tasks currently selected (you must select these summary tasks **before** you begin the process of updating the baseline).

 Information: If you do not want to roll up the baseline values to any summary tasks, do not select either of the checkboxes in the *Roll up baselines* section of the dialog. This means that data from the selected tasks continues to show as variance against the current project baseline.

7. Click the **OK** button. Microsoft Project warns you about overwriting the baseline data in the confirmation dialog shown in Figure 11 - 6.

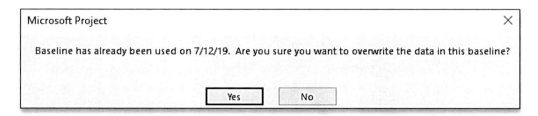

Figure 11 - 6: Overwrite baseline confirmation dialog

8. Click the **Yes** button in the confirmation dialog.

Information: In spite of the severity of the warning in the dialog, using this procedure does not actually "overwrite" the data in your original baseline. Instead, this procedure "appends" the baseline data from the new tasks to the current project baseline.

Backing up the Current Baseline Data

After you update your baseline for the new tasks added through the change control procedure, it is wise to back up the current baseline data stored in the default *Baseline* set of fields. As you know by now, Microsoft Project offers you 11 sets of fields in which to save baseline data. These sets of fields include the default *Baseline* set of fields, plus the *Baseline 1* through *Baseline 10* sets. You can use any of these ten sets of alternate baseline fields to back up the current baseline in your project. To back up your current baseline values, complete the following steps:

1. Click the **Project** tab to display the *Project* ribbon.

2. In the *Schedule* section of the *Project* ribbon, click the **Set Baseline** pick list button, and then select the **Set Baseline** item.

3. Click the **Set baseline** pick list and select next *unused* numbered baseline field, such as the **Baseline 2** field, as shown in Figure 11 - 7.

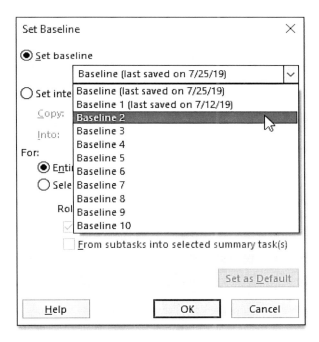

Figure 11 - 7: Select a numbered Baseline field

Information: Remember that in the *Set baseline* pick list, the *(last saved on _____)* text appended to the right of the *Baseline* field reveals whether you saved a baseline in a particular *Baseline* field and when you saved the baseline information. If a baseline does not include the *(last saved on _____)* text to the right of its name, this indicates that the baseline is currently unused, and is available for backing up the current baseline information.

4. In the *For:* section of the dialog, leave the **Entire project** option selected, and then click the **OK** button.

5. Save the latest changes to your project.

When you use this procedure, Microsoft Project copies all baseline information from the *Baseline* set of fields to the set of fields for the alternate baseline. You can use this process for up to nine change control procedures, at which point you run out of alternate sets of baseline fields.

 Hands On Exercise

Exercise 11 - 4

Baseline only selected tasks in the Training Advisor Rollout project and then back up the baseline into one of the unused baseline fields.

1. Right-click on the **Select All** button and select the **More Tables** item on the shortcut menu.
2. In the *More Tables* dialog, select the **Baseline** table, and then click the **Apply** button.
3. Drag the split bar to the right edge of the *Baseline Cost* column.
4. Select task ID #18, the new *Create Training Module 04* task, then click the **Scroll to Task** button in your Quick Access Toolbar to click to bring the Gantt bar into view.

Notice that Microsoft Project displays no baseline data for the new *Create Training Module 04* task. This is because the new task lacks baseline information.

5. Click the **Project** tab to display the *Project* ribbon.
6. In the *Schedule* section of the *Project* ribbon, click the **Set Baseline** pick list button, and then select the **Set Baseline** item.
7. In the *Set Baseline* dialog, leave the **Set baseline** option selected and leave the **Baseline** value selected in the *Set baseline* pick list.
8. In the *For:* section of the dialog, choose the **Selected tasks** option.
9. In the *Roll up baselines* section of the dialog, select the **To all summary tasks** option.
10. Click the **OK** button.
11. When the software warns you about overwriting the current baseline, click the **Yes** button in the confirmation dialog.

Look for the cells with change highlighting applied (light blue cell background color), which indicates which tasks contain updated baseline information. You should see change highlighting for the new task, for its direct summary tasks, and for the Project Summary Task as well.

12. In the *Schedule* section of the *Project* ribbon, click the **Set Baseline** pick list button, and then select the **Set Baseline** item again.

Variance Analysis, Plan Revision, and Reporting

13. In the *Set Baseline* dialog, click the **Set baseline** pick list, and select the **Baseline 2** item.

Notice that the *Baseline 2* is the next set of unused baseline fields.

14. In the *For:* section of the dialog, leave the **Entire project** option selected, and then click the **OK** button.
15. Right-click on the **Select All** button and select the **Entry** table.
16. Drag the split bar to the right edge of the *Duration* column.
17. *Save* but *do not* close the **Training Advisor Rollout 2019.mpp** sample file.

Project Reporting Overview

Microsoft Project and Project Online offer multiple methods for reporting about projects, resources, and tasks. The primary methods for reporting are as follows:

- The **Timeline view** allows you to report about significant tasks in a single project schedule using a timeline presentation similar to what you see in Microsoft Visio. When you publish your enterprise project, the *Timeline* view appears automatically at the top of the *Schedule* PDP (Project Detail Page) and at the top of the Project Site.

- **Dashboard reports** allow you to report about tasks and resources in a single project schedule using graphical reports based on Microsoft Word and Microsoft Excel functionality. You can view or print dashboard reports, and you can even export them to other Office applications such as Microsoft PowerPoint.

- **Power BI reports** offer a robust reporting platform about the status of all projects in your portfolio, all resources assigned to project work, and project artifacts in the Project Site, such as risks or issues. Although Project Online does not offer any Power BI reports by default, Microsoft offers a free Power BI report pack which can be easily installed in your Project Online system, with reports that are ready to use "out of the box" for Project Online reporting.

I discuss each of these reporting methods individually.

Project Reporting Using the Timeline View

During project execution, you must report project progress to one or more stakeholder groups. These typically include your project sponsor, your customer, your company executives, and even your project team. Although Microsoft Project offers you a number of ways to report about your project, the *Timeline* view offers you a powerful means of reporting on your enterprise projects.

Microsoft Project includes the default *Timeline* view that displays the current project schedule using a timeline presentation similar to what you see in Microsoft Visio or in any other timeline software application. You can modify the default *Timeline* view to show your current project schedule according to your reporting requirements and you can create additional *Timeline* views as needed. You can also export any *Timeline* view to other Office applications, such as to PowerPoint. One special feature of the *Timeline* view in an enterprise project is when you publish the project, Project Online automatically displays the modified *Timeline* view at the top of the *Schedule* PDP (Project

Detail Page) and at the top of the Project Site in SharePoint. Figure 11 - 8 shows an example of a completed *Timeline* view.

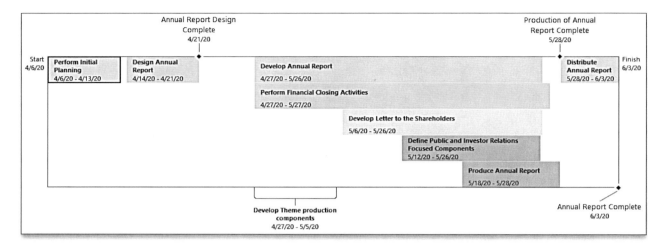

Figure 11 - 8: Timeline view

To display the *Timeline* view, use one of the following methods:

- In the *Split View* section of the *View* ribbon, select the **Timeline** checkbox.
- Right-click anywhere in the *Gantt Chart* pane and select the **Show Timeline** item on the shortcut menu.

Figure 11 - 9 displays a blank *Timeline* view, ready for modification for reporting purposes.

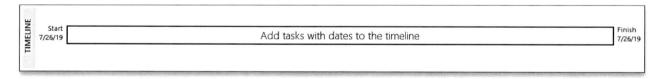

Figure 11 - 9: Blank Timeline view

Adding Tasks to the Timeline View

To add any task to the *Timeline* view, right-click on the name of the task in task list on the left side of the *Gantt Chart* view and then select the **Add to Timeline** item on the shortcut menu. To add multiple tasks to the *Timeline* view, select a block of tasks, right-click anywhere in the selected block of tasks, and then select the **Add to Timeline** item on the shortcut menu.

Arranging Tasks in the Timeline View

After you add tasks to the *Timeline* view, you can rearrange the tasks on the timeline bar using any of the following techniques:

- Drag a task to a new row above or below its current position in the timeline bar.
- Drag a task above or below the timeline bar to display the task as a callout.

Variance Analysis, Plan Revision, and Reporting

- Drag a block of tasks by selecting them while pressing and holding the **Control** key on your keyboard, and then dragging the block of the selected tasks to a new position.

- Right-click on any task in the timeline bar and select the **Display as Callout** item on the shortcut menu.

- Drag a new callout from the top of the timeline bar to a position below the timeline bar.

- Convert a callout to a task bar by right-clicking on the callout and then clicking the **Display as Bar** item on the shortcut menu.

Formatting the Timeline View

To format the *Timeline* view, click anywhere in the *Timeline* pane to activate the pane, and then click the **Format** tab. Microsoft Project displays the contextual *Format* ribbon with the *Timeline Tools* applied, as shown in Figure 11 - 10.

Figure 11 - 10: Format ribbon with the Timeline Tools applied

Using the features on the *Format* ribbon, the software offers you many ways to format the *Timeline* view:

- To format the text for specific set of objects shown in the *Timeline* view, such as all milestone dates, click the **Text Styles** button in the *Text* section of the contextual *Format* ribbon.

- To change the font or the cell background color of an individual object in the *Timeline* view, select the object and then change the formatting using the **Background Color** and **Font Color** buttons in the *Font* section of the contextual *Format* ribbon.

- To hide date items from the timeline, click the **Date Format** pick list button in the *Show/Hide* section of the contextual *Format* ribbon, and then deselect the date item at the bottom of the pick list, such as the **Today** item, for example.

- To hide the *Pan & Zoom* section of the timeline (the two green vertical lines), *deselect* the **Pan & Zoom** checkbox in the *Show/Hide* section of the contextual *Format* ribbon.

- To add an additional timeline to the *Timeline* view, click the **Timeline Bar** button in the *Insert* section of the contextual *Format* ribbon.

- To change the type of object displayed in the *Timeline* view, or to remove an object from the *Timeline* view, select the object and then use the buttons in the *Current Selection* section of the contextual *Format* ribbon.

- To add additional bars to the *Timeline* view, click the **Existing Tasks** button in the *Insert* section of the *Format* ribbon. In the *Add Tasks to Timeline* dialog, select the checkboxes for the tasks you want to add to the *Timeline* view, and then click the **OK** button.

Microsoft Project allows you to export the entire *Timeline* view to any Office application, such as PowerPoint or Word. To copy the *Timeline* view, click the **Copy Timeline** pick list button in the *Copy* section of the contextual *Format* ribbon, and then select one of the three items on the pick list, as shown in Figure 11 - 11.

Module 11

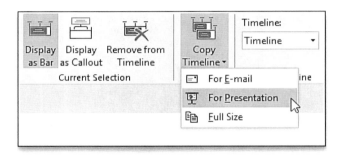

Figure 11 - 11: Copy Timeline pick list

If you select the **Full Size** item on the *Copy Timeline* pick list, Microsoft Project copies the full-size image of the current *Timeline* view to your Windows clipboard. If you select the **For Presentation** item, the software optimizes the image for use in PowerPoint by reducing the image size to approximately 90% of full size. If you select the **For E-Mail** item, the software optimizes the image for use in Microsoft Outlook by reducing the image size to approximately 60% of full size.

After copying the *Timeline* view to your clipboard, paste the image in one of the Office applications. If you use an application that has image editing capabilities, such as Microsoft PowerPoint, you can continue editing the image to further refine your *Timeline* view presentation.

 Hands On Exercise

Exercise 11 - 5

Add tasks to the *Timeline* view and then modify the view.

1. Right-click anywhere in the *Gantt Chart* pane and then select the **Show Timeline** item on the shortcut menu.

2. Expand the height of the *Timeline* pane to approximately double its default size.

3. In the *Gantt Chart* pane, right-click on task ID #1, the *INSTALLATION* summary task, and then select the **Add to Timeline** item on the shortcut menu.

4. Using the **Control** key on your computer keyboard, select the following tasks as a group and then select the **Add to Timeline** item on the shortcut menu:

    - Task ID #7, the *TESTING* summary task
    - Task ID #12, the *TRAINING* summary task
    - Task ID #14, the *Create Training Materials* summary task
    - Task ID #22, the *Provide End User* Training task
    - Task ID #24, the *Project Complete* milestone task

352

5. Click anywhere in the *Timeline* pane to select the pane.
6. Click the **Format** tab to display the contextual *Format* ribbon with the *Timeline Tools* applied.
7. In the *Show/Hide* section of the *Format* ribbon, ***deselect*** the **Pan & Zoom** checkbox.
8. In the *Show/Hide* section of the *Format* ribbon, click the **Date Format** pick list button and ***deselect*** the **Timescale** item in the *Show/Hide* section of the pick list.
9. Click and hold the *Project Complete* callout and drag it ***above*** the timeline bar.
10. Click and hold the *Provide End User Training* bar and drag it ***below*** the timeline bar to convert it to a callout.
11. Individually click the INSTALLATION, TESTING, TRAINING and *Create Training Materials* bars, then click the **Background Color** pick list button in the *Font* section of the *Format* ribbon, and select a background color of your choice.
12. ***Save and close*** the **Training Advisor Rollout 2019.mpp** sample file.

Project Reporting Using the Dashboard Reports

Microsoft Project includes a selection of graphical dashboard reports which you can view, print, or export to another Office application. To use any of these dashboard reports, click the **Report** tab to display the *Report* ribbon. The *View Reports* section of the *Report* ribbon contains eight pick list buttons that allow you to work with dashboard reports, as shown in Figure 11 - 12.

Figure 11 - 12: Categories of dashboard reports

Microsoft Project organizes the default reports into five categories: *Dashboards*, *Resources*, *Costs*, *In Progress*, and *Getting Started*. To view a report, click one of these five pick list buttons and select a report. For example, if you click the **Dashboards** pick list button and select the **Project Overview** report, Microsoft Project displays the *Project Overview* report, such as the one shown in Figure 11 - 13. Every report in Microsoft Project can contain any of the following types of objects:

- **Text boxes** – The software uses text boxes to show the title of the report, plus additional information about other objects in the report. In the *Project Overview* report shown in Figure 11 - 13, the title of the report is a text box.
- **Tables** – The software uses tables to display data from project fields in a table format based on Microsoft Word table functionality. In the *Project Overview* report shown in Figure 11 - 13, the tables include the two date fields in the upper left corner, plus the *% Complete* table, the *Milestones Due* table, and the *Late Tasks* table.

- **Charts** – The software uses charts based on Microsoft Excel charting functionality to graphically display data from fields in your project. In the *Project Overview* report shown in Figure 11 - 13, the *% Complete* chart in the upper right corner of the page is the only chart in the report.

- **Hyperlinks** – The software uses hyperlinks to direct you to a *Help* article in Office.com that is relevant to the type of report currently displayed. The *Project Overview* report shown in Figure 11 - 13 does not include any hyperlinks.

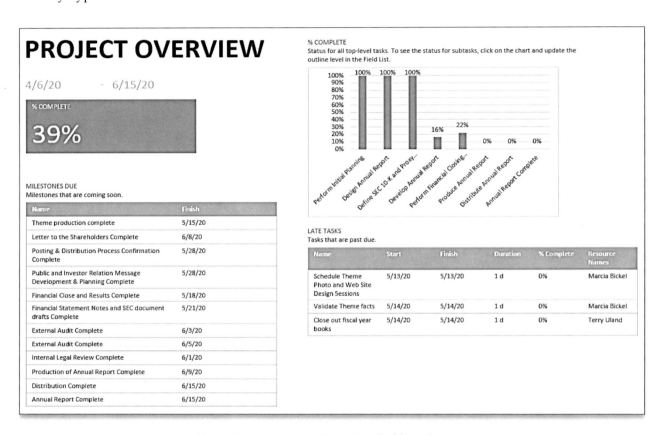

Figure 11 - 13: Project Overview dashboard report

Customizing a Chart

To customize a chart in your report, click once in the chart you want to customize. The software places the chart into editing mode, such as for the *% Complete* chart shown in Figure 11 - 14. When you click a chart to put it into editing mode, Microsoft Project displays the following elements:

- On the far right side of the report, the software displays the *Field List* sidepane.

- To the right of the selected chart, the software displays three formatting buttons: the *Chart Elements*, *Chart Styles*, and *Chart Filters* buttons. For the purpose of clarity, I placed a lasso around the buttons shown in Figure 11 - 14.

- At the top of the user interface, the software displays the *Design* and *Format* ribbon tabs with the *Chart Tools* applied. Figure 11 - 15 shows the *Design* ribbon and Figure 11 - 16 shows the *Format* ribbon with the *Chart Tools* applied to each ribbon.

Variance Analysis, Plan Revision, and Reporting

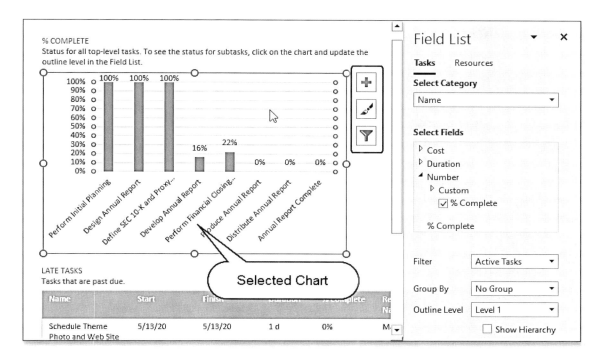

Figure 11 - 14: % Complete chart in editing mode

Figure 11 - 15: Design ribbon with the Chart Tools applied

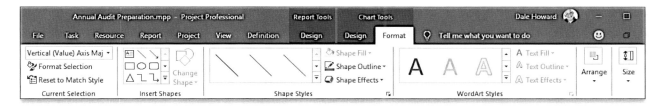

Figure 11 - 16: Format ribbon with the Chart Tools applied

You can use any of these three options to customize a chart, plus you can double-click any chart element to edit that individual element.

 Information: Customizing a chart in Microsoft Project is nearly identical to customizing a chart in Microsoft Excel. The more experience you have with formatting an Excel chart, the better you will be at formatting a Project chart.

355

Customizing a Table

To customize a table in your report, click anywhere in the table you want to customize. The software places the table into editing mode, such as for the *% Complete* table shown in Figure 11 - 17. When you click a table to put it into editing mode, Microsoft Project displays the following elements:

- On the far right side of the report, the software displays the *Field List* sidepane.
- At the top of the user interface, the software displays the *Design* and *Layout* ribbon tabs with the *Table Tools* applied. Figure 11 - 18 shows the *Design* ribbon and Figure 11 - 19 shows the *Format* ribbon with the *Table Tools* applied to each ribbon.

You can use either of these two options to customize a table.

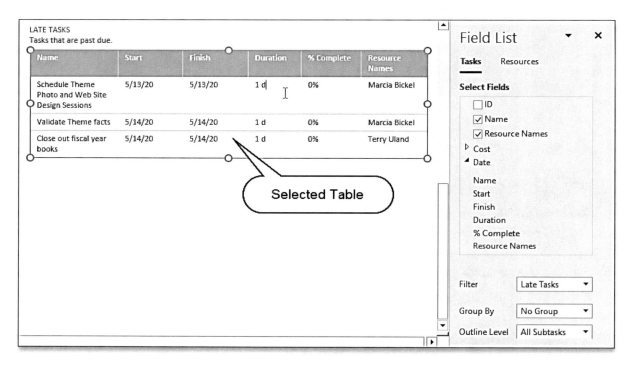

Figure 11 - 17: % Complete table in editing mode

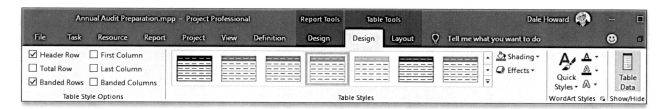

Figure 11 - 18: Design ribbon with the Table Tools applied

Variance Analysis, Plan Revision, and Reporting

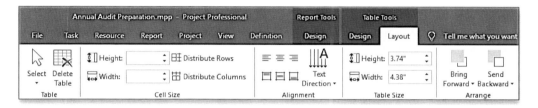

Figure 11 - 19: Format ribbon with the Table Tools applied

 Information: Customizing a table in Microsoft Project is very similar to customizing a table in Microsoft Word. The more experience you have with formatting Word tables, the better you will be at formatting a table in Microsoft Project.

 Hands On Exercise

Exercise 11 - 6

Explore the dashboard reports in Microsoft Project.

1. Open the **Project Navigation 2019.mpp** sample file from your sample files folder. The project opens Read-Only.

2. Click the **Report** tab to display the *Report* ribbon.

3. In the *View Reports* section of the *Report* ribbon, click the **Dashboards** pick list button and select the **Project Overview** item in the pick list.

Notice how Microsoft Project displays the *Design* ribbon with the *Report Tools* applied.

4. Click anywhere in the **Milestones Due** table and notice how the software displays the *Field List* sidepane on the right side of the page.

 Information: If you do not see the *Field List* sidepane when you click in the *Milestones Due* table, click the second **Design** tab to display the *Design* ribbon with the *Table Tools* applied. In the *Show/Hide* section of the *Design* ribbon, click the **Table Data** button to display the *Field List* sidepane on the right side of the report.

5. Examine the selected options in the *Field List* sidepane on the right side of the report, including the fields selected in the *Select Fields* section of the sidepane, along with the items selected on the *Filter*, *Group By*, *Outline Level*, and *Sort By* pick lists.

6. Click anywhere in the **% Complete** chart and then examine the selected options in the *Field List* sidepane.

357

Module 11

7. Click the **Report** tab to display the *Report* ribbon again.

8. In the *View Reports* section of the *Report* ribbon, click the **Resources** pick list button and select the **Resource Overview** item in the pick list.

9. Study the data shown in the *Resource Stats* and the *Work Status* charts, along with the data shown in the *Resource Status* table.

Exercise 11 - 7

Customize a table in a dashboard report in Microsoft Project.

1. Click the **Report** tab to display the *Report* ribbon.

2. In the *View Reports* section of the *Report* ribbon, click the **Dashboards** pick list button and select the **Project Overview** report.

3. Click anywhere in the **Milestones Due** table to select the table for editing and to display the *Field List* sidepane.

4. In the *Field List* sidepane, do each of the following:
    - Click the **Filter** pick list and select the **Summary Tasks** filter.
    - Click the **Outline Level** pick list and select the **Level 3** item.
    - Select the **Show Hierarchy** checkbox.

5. Click the **Milestones Due** title at the top of the table to select it.

6. Delete the *Milestone Due* text and replace it with **SUMMARY TASK DUE DATES**.

7. Click the caption below the *SUMMARY TASK DUE DATES* title to select it.

8. Delete the existing caption text and replace it with **Shows Due Dates for All Summary Tasks**.

Exercise 11 - 8

Customize a chart in a dashboard report in Microsoft Project.

1. Click the **Report** tab to display the *Report* ribbon.

2. In the *View Reports* section of the *Report* ribbon, click the **Dashboards** pick list button and select the **Work Overview** report.

3. Click anywhere in the **Work Stats** chart to select the chart for editing and to display the *Field List* sidepane.

4. Click the **Chart Elements** button (+ button) on the right side of the chart and select the checkbox for the **Chart Title** element in the flyout menu.

5. Click anywhere in the new **Chart Title** element in the chart, delete the default text, and enter **Work Comparison by Phase** as the title of the chart.

6. With the *Work Stats* chart still selected, click the **Filter** pick list in the *Field List* sidepane and select the **Summary Tasks** filter.

7. Click the **Chart Styles** button on the right side of the chart and select the **Style 8** item on the flyout menu.

8. Click the **Chart Styles** button again, click the **Color** tab at the top of the flyout menu, and select the **Colorful Palette 4** color scheme in the *Colorful* section of the flyout menu.

9. Scroll down the report and then click anywhere in the **Resource Stats** chart to select the chart.

10. Click the second **Design** tab to display the *Design* ribbon with the *Chart Tools* applied.

11. In the *Type* section of the *Design* ribbon, click the **Change Chart Type** button.

12. In the *Change Chart Type* dialog, select the **Column** type of chart in the list on the left side.

13. At the top of the dialog, select the **3-D Stacked Column** chart type, and then click the **OK** button.

14. Close but *do not* save the **Project Navigation 2019.mpp** sample file.

Project Reporting Using Power BI Reports

Warning: In this topical section, I provide you with only a high-level overview of using Power BI reports with Project Online. If you need in-depth information about Power BI reporting with Project Online, Microsoft provides abundant learning resources on the Internet for this purpose.

Introduced by Microsoft in 2015, Power BI represents the future of enterprise reporting with Project Online. The "power" behind Power BI includes the following:

- Power BI reports can pull data from **multiple enterprise applications** into a single report. For example, suppose that your organization uses Project Online for project portfolio management, and uses Microsoft Dynamics for your organization's financial management. Power BI makes it possible to view a dashboard with reports that pull project data from Project Online and financial data from Microsoft Dynamics.

- Power BI reports use attractive data visualizations that are **easy to understand**. For example, suppose that a Power BI report contains a column named *Cost Variance* that displays a red, yellow, or green stoplight indicator to show the severity of the cost variance. If the report displays all projects in your organization's portfolio, it is pretty easy to assume that projects with a red stoplight indicator are seriously over budget on cost.

- Power BI reports are **interactive**. Each report offers multiple tiles in which you can click a data element to either filter the report or to "drill down" to the underlying data. For example, if you click the *Software Development* slice of a donut chart visualization, the Power BI report refreshes to show the list of only software development projects.

Keep in mind that Project Online does not include any "out of the box" Power BI reports, which means that the report authors in your organization must create custom Power BI reports according to the reporting needs of your organization. To get a fast start with Power BI reporting, however, your Power BI administrator can install the free **Project Online Content Pack for Power BI** from Microsoft. The reports in this report pack target the reporting needs of project managers, program managers, portfolio managers, resource managers, and executives. Each user who needs to access the Power BI reports must have the necessary license to view the reports.

Once installed, and you access the free Power BI report pack for Project Online, the system displays the welcome page shown in Figure 11 - 20. In addition to the tiles that display the *Project* icon and the *Welcome* text, the home page includes of a series of tiles that display data visualizations from Project Online. For example, the *Project Count* visualization displays the number of projects currently in our organization's Project Online instance, while the *Portfolio Overview* visualization displays a donut chart that graphically displays the number of projects by Enterprise Project Type.

 Warning: Microsoft make updates from time to time to this report pack, so what you see may differ from the screenshots shown in this book. Please keep in mind that the information in this book about the report pack is only accurate as of the date of book publication.

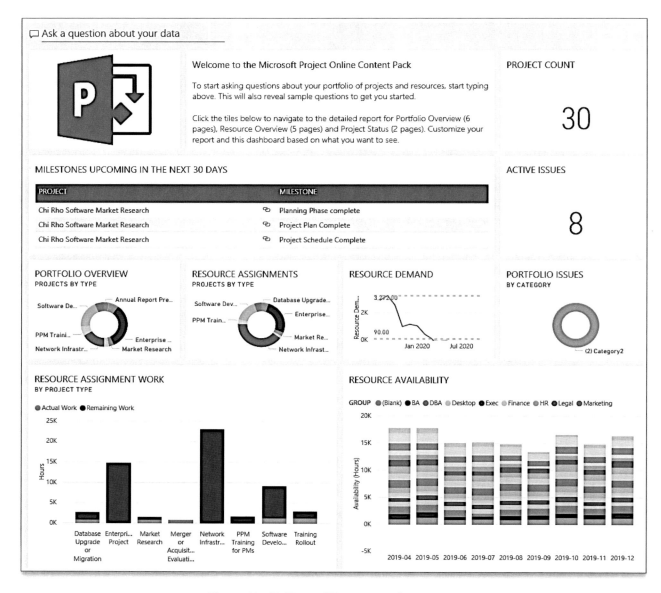

Figure 11 - 20: Power BI reports welcome page

Variance Analysis, Plan Revision, and Reporting

To begin exploring the reports in the Power BI report pack for Project Online, click the **Portfolio Overview** tile to view the reports beginning with the first report in the report pack. The system displays all the reports in the report pack, with each report displayed on its own worksheet tab. At the left end of the first worksheet tab, Power BI displays the left-arrow and right-arrow buttons to help you to navigate the worksheet tabs.

Figure 11 - 21 shows the *Portfolio Dashboard* report. Notice that the top of the report includes a series of data filters that allow you to filter the report by *Department*, by *Project Type*, by *Project Owner*, or by *Governance Phase*. Click the filter pick list in one or more of the filter fields to filter the data in the report according to your criteria. For example, I can click the *Department* filter pick list and select the *IT* checkbox to see only IT projects.

Information: Every report in the report pack includes a filtering section across the top of the report. All of the portfolio reports include the *Department*, *Project Type*, *Project Owner*, and *Governance Phase* filters. All of the resource reports except for the *Resource Details* report include the *Department*, *Group*, *Resource Calendar*, and *RBS* filters. The *Resource Details* report includes only a single filter, the *Resource Name* filter. The *Project Status* report includes only the *Project* filter. The *Project Risks & Issues* report includes the *Issue Status*, *Risk Status*, and *Project* filters.

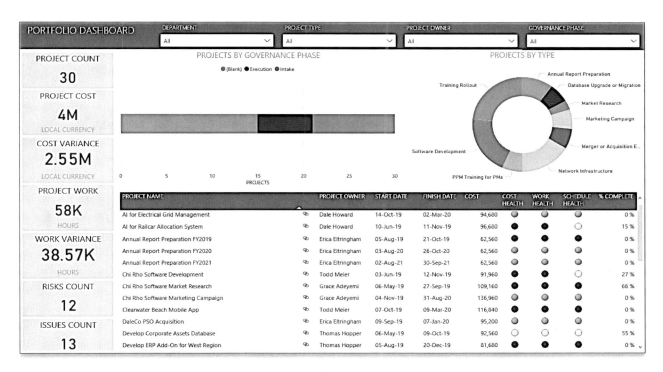

Figure 11 - 21: Portfolio Dashboard report

The **Portfolio Dashboard** report shown previously in Figure 11 - 21 provides a high-level overview of project data for all projects. The left side of the report contains a series of multi-row tiles that display totals data for all projects, including the number of projects, cost and cost variance, work and work variance, number of risks, and the number of issues. The top of the report includes the *Projects By Governance Phase* stacked bar chart and the *Projects By Type* donut chart. Click one of the data elements in either chart to filter the projects and the data shown in the report. The bottom half of the report includes a table that displays high-level information about each project, including the project owner, start date and finish date, cost, % complete, plus three columns that display stoplight indicators revealing the severity of cost, work, and schedule variance. To the right of the name of each project, the report

361

Module 11

displays a link icon (it looks like a double paper clip). Click this icon to navigate directly to the Project Site associated with the project.

> **Information**: The color criteria for the *Cost Health*, *Work Health*, and *Schedule Health* graphical indicators shown in the *Portfolio Dashboard* report are as follows:
> - Gray stoplight – No baseline
> - Green stoplight – Less than 10% variance
> - Yellow stoplight – Between 10% and 20% variance
> - Red stoplight – Greater than 20% variance

The **Portfolio Timeline** report shown in Figure 11 - 22 includes a timeline slicer at the top of the page that you can use to select the time span containing the *Start* dates of the projects you want to see in the report. In the upper left corner of the timeline slicer, the report displays a granularity indicator that allows you to select the granularity of time periods shown in the timeline slicer. You can specify a granularity level in either days, weeks, months, quarters, or years. After setting your desired granularity setting, click and drag the left end of the timeline slicer to select the start date of the time span, then click and drag the right end of the timeline slicer to select the finish date of the time span. The bottom half of the report updates to show a list of projects with a *Start* date during the time span you selected in the timeline slicer, along with a Gantt bar representing the schedule of each project. The report automatically color codes the Gantt bars according to the *Enterprise Project Type* value for each project, and indicates completed work using a darker color in the Gantt bar. The report displays the name of the project manager to the right of each Gantt bar.

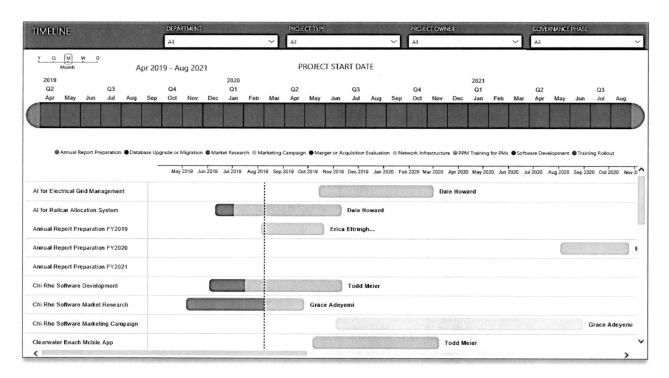

Figure 11 - 22: Portfolio Timeline report

The **Portfolio Costs** report provides high-level cost data for all projects in your portfolio. The upper left corner of the report includes the *Cost KPI* stoplight filter. Click the red, yellow, and/or green stoplight indicators to display only the projects with the variance you want to see. The left side of the report contains a series of multi-row tiles that display cost totals for all projects currently displayed in the report, including project cost, baseline cost, actual cost, and cost variance. The top of the report includes the *Top Projects by Cost* list. Float your mouse pointer over one of the items in the list to reveal the hidden scrollbar that you can use to scroll through the list of projects. The upper right corner of the report includes the *Portfolio Costs* gauge chart which displays actual cost, baseline cost, and total cost data for all projects currently displayed in the report. The bottom half of the report includes a table that displays high-level cost information about each project, including the budget cost, baseline cost, total cost (EAC), actual cost, remaining cost, and cost variance. The *Cost Variance* column includes red shading to indicate projects currently over budget on cost.

The **Portfolio Milestones** report displays milestone data for all projects in your portfolio. The left side of the report includes the *Milestones Completed in the Last 30 Days* table and the *Milestones Upcoming in the Next 30 Days* table. Each table displays the name of the project, the name of the milestone, and the finish date of each milestone. The right side of the report includes the *Milestones Completed By Project* donut chart and the *Milestones Planned By Project* donut chart. Click the name of a project in either donut chart to filter the report to show only the milestones in the selected project. To the right of the name of each project, the report displays a link icon (it looks like a double paper clip). Click this icon to navigate directly to the Project Site associated with the project

The **Portfolio Risks** report shown in Figure 11 - 23 displays the risks in the Project Sites associated with every project in your portfolio. The top of the report includes the *Risks by Category* donut chart, the *Risk Cost vs Risk Exposure* clustered column chart, and the *Risks by Project Owner* pie chart. The *Risks by Category* donut chart displays the number of risks by *Category* value, from *Category 1* through *Category 3*. The *Risk Cost vs Risk Exposure* clustered column chart shows the *Cost* and *Cost Exposure* values for all risks, grouped by their *Category* value. The *Risks by Project Owner* pie chart shows the total number of risks for the projects owned by each project manager. Click one of the data elements in any of these three charts to filter the data in the report. The bottom half of the report includes a table that displays high-level information about every risk, including the name of the risk, name of the project associated with the risk, user assigned to the risk, *Category* value for the risk, the due date for the risk, the *Cost* and *Cost Exposure* values for the risk, and the probability of the risk occurring. To the right of the name of each risk, the report displays a link icon (it looks like a double paper clip). Click this icon to navigate directly to the risk in its associated Project Site.

Information: The default values in the *Category* field (*Category 1*, *Category 2*, and *Category 3*) provide no useful data about risks. Rather, Microsoft intends for your organization to customize the *Category* field with values that are meaningful to how your organization manages project risks. For example, many years ago I worked for an organization that customized the *Category* field values to include *Internal*, *External*, and *Mixed* to reflect the source of the risk. Your Project Online application administrator can customize the *Category* field for both risks and issues in the Project Site template so that every Project Site automatically contains the custom *Category* values.

Information: Project Online calculates the *Exposure* value for each risk by multiplying its *Probability* value times its Impact value. SharePoint calculates the *Cost Exposure* value by multiplying the Exposure value times its *Cost* value.

Figure 11 - 23: Portfolio Risks report

The **Portfolio Issues** report displays the issues in the Project Sites associated with every project in your portfolio. The top of the report includes the *Issues by Category* donut chart, the *Issue Count by Status* column chart, and the *Issues by Project Owner* pie chart. The *Issues by Category* donut chart displays the number of issues by *Category* value, from *Category 1* through *Category 3*. The *Issue Count by Status* column chart shows the number of issues by their *Status* value, including the *Active*, *Postponed*, and *Closed* values. The *Issues by Project Owner* pie chart shows the total number of issues for the projects owned by each project manager. Click one of the data elements in any of these three charts to filter the data in the report. The bottom half of the report includes a table that displays high-level information about every issue, including the name of the issue, name of the project associated with the issue, user assigned to the issue, due date for the issue, *Category* value for the issue, *Priority* value for the issue, and status of the issue. To the right of the name of each issue, the report displays a link icon (it looks like a double paper clip). Click this icon to navigate directly to the issue in its associated Project Site.

 Information: The default values in the *Category* field (*Category 1*, *Category 2*, and *Category 3*) provide no useful data about issues. Rather, Microsoft intends for your organization to customize the *Category* field with values that are meaningful to how your organization manages project issues. For example, many years ago I worked for an organization that customized the *Category* field values to include *Internal*, *External*, and *Mixed* to reflect the source of the issue. Your Project Online application administrator can customize the *Category* field for both risks and issues in the Project Site template so that every Project Site automatically contains the custom *Category* values.

The **Resource Availability** report shown in Figure 11 - 24 displays resource availability information over time. The top of the report includes the *Demand and Capacity Over Time* combo chart and the *Availability Over Time* stacked column chart. The top of each chart displays the names of the *Group* values specified for the resources in your organization's Enterprise Resource Pool. Click one of the *Group* names to filter the report to display only those resources in the selected group. The bottom half of the report includes the *Availability Heatmap* table. This table is incredibly useful for project managers who need to determine whether a particular human resource is available for work in their projects. The *Availability Heatmap* table displays the name of each enterprise resource, sorted by availability from the least available resource (most utilized) to the most available resource (least utilized). The columns in the table display the monthly availability for each resource, formatted by their availability values. Cells shaded with the darkest green cell background color indicate periods of full-time availability, while lighter green, yellow, and orange cell background colors indicate periods with only partial availability. Red cell background formatting indicates periods in which a resource is overallocated. Click the name of any resource in the table to focus on the availability data for only the selected resource.

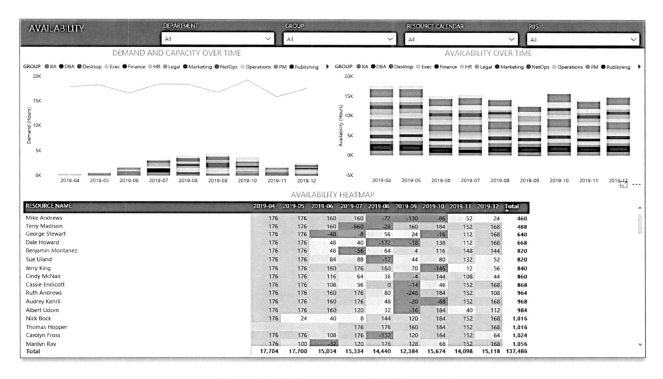

Figure 11 - 24: Resource Availability report

 Information: You can sort the names in the *Availability Heatmap* table several ways. To sort the names from the most available to the least available resources, click the *Total* column header to display a "down arrow" indicator in the column header. To sort the names in alphabetical order, click the *Resource Name* column header to display an "up arrow" indicator in the column header.

The **Resource Overview** report displays high-level information for all of the resources in your Enterprise Resource Pool. The top half of the report includes the *Resources by Group* donut chart, the *Resources by Department* bar chart, and the *Resources by RBS* pie chart. Click one of the data elements in any of these charts to filter the data in the report. The bottom half of the report includes a table that displays high-level information about each resource. This information includes their *Max. Units*, *Department*, and *Group* values, a Yes/No value to indicate whether the resource is a Generic resource, the base calendar assigned to the resource, and the *RBS* value for the resource.

The **Resource Assignments** report shown in Figure 11 - 25 displays high-level information about the project work assigned to resources in your Enterprise Resource Pool. The top half of the report includes the *Work by Group >*

Module 11

Resource donut chart, the *Work by Project Type > Project* donut chart, and the *Assignment Actual and Remaining Work by Project Type* stacked column chart. Click one of the data elements in any of these charts to filter the data in the report. The lower left corner of the report includes a table that displays the name of each resource, along with the names of the projects in which the resource is assigned task work. The table displays the start date and finish date of their work in each project, their current *% Work Complete* value for each project, and the total amount of work assigned to the resource in each project. The lower right corner of the report includes the *Resource Name* data slicer. Select the checkbox for a resource to filter the report data for only the selected resource.

Figure 11 - 25: Resource Assignments report

The **Resource Details** report displays summary information about the utilization of a single resource in your Project Online system. Before you can use this report, click the **Resource Name** pick list at the top of the report and then select the name of a resource. The report updates with all of the relevant information for the selected resource. The top of the report displays the name of the resource and includes an *Availability* heatmap chart for the resource. The upper left corner of the report includes single line tiles that display the e-mail address, the *Standard Rate* value, the cost type, and the *Max. Units* value for the selected resource. The upper right corner of the report includes a series of single line tiles in the *Current Year Assignment Metrics* section that display summary information for the resource, such as the total number of task assignments and projects, total amount of work and cost, etc. The bottom half of the report includes the *Work by Project* donut chart and the *Availability, Demand, and Capacity Over Time* combo chart. Click one of the data elements in either of these charts to filter the data in the report.

The **Resource Demand Forecast** report displays a combination line chart/area chart that shows resource demand phased over time. This unusual report uses predictive analytics to forecast future resource demand based on historical data. The green line in the report represents the resource demand based on resources assigned to tasks in published enterprise projects. The black line and corresponding gray shaded areas represent the forecast demand in the future, calculated by the predictive analytics engine in Power BI.

Variance Analysis, Plan Revision, and Reporting

The **Project Status** report shown in Figure 11 - 26 displays summary information about a single project in your Project Online system. Before you can use this report, click the **Project** pick list at the top of the report and then select the name of a project. The report updates with all of the relevant information for the selected project. The left side of the report contains a series of multi-row tiles that display totals data for the project. The top of the report displays stoplight indicators that show the severity of variance, along with high-level information about the project. The middle of the report includes the *Work Over Time* chart and the *Costs Over Time* chart. The bottom of the report includes the *Completed Milestones* table and *Upcoming Milestones* table. The lower right corner of the report includes the *% Complete* gauge chart, with a section immediately above it containing links that allow the user to navigate to the Project Site for the project, or to the project schedule in Project Web App.

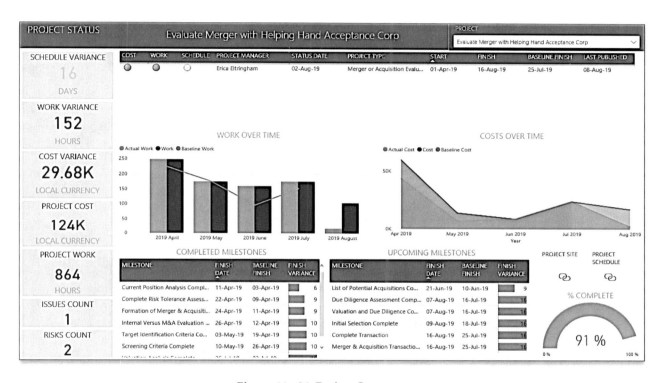

Figure 11 - 26: Project Status report

The **Project Risks & Issues** report provides summary information about all of the risks and issues associated with a single project or with all projects in your Project Online system. To use this report with a single project, click the **Project** pick list at the top of the report and then select the name of a project. The report updates with all of the relevant information for the selected project. To use this report with all projects, click the **Project** pick list and select the (Blank) item, then *deselect* the (Blank) item. The report updates with all of the relevant information for all projects. The top of the report displays the *Issue Count by Owner > Status* donut chart, the *Issue Count by Priority > Status* pie chart, the *Risk Matrix* bubble chart, and the *Risk Count by Owner > Status* donut chart. Click one of the data elements in any of these charts to filter the data in the report. The bottom half of the report includes the *Project Issues* table and the *Project Risks* table. The *Project Issues* table includes the relevant information about each issue, including the owner, priority, due date, and status of the issue. The *Project Risks* table includes relevant information about each risk, including the owner, due date, probability, impact, and status of the risk.

Using Ad Hoc Filtering

On the right side of every report, Power BI displays a collapsed *Filters* sidepane. Click the **Show/hide pane** indicator (the < indicator) at the top of the *Filters* sidepane to expand it, and then click one of the visualizations in the report to show the applied filters for the selected visualization. For example, Figure 11 - 27 shows the expanded *Filters* sidepane with the *Availability Heatmap* selected in the *Resource Availability* report. Notice that the *Visual level filters* section contains filters for *Availability, Resource Name, Time,* and *Year-Month*. Visual level filters impact only the selected visualization, by the way. Notice that the *Page level filters* section contains filters for *Resource Type, Show Local Generic Resources,* and *Time.* As you might surmise, *Page level filters* impact the data displayed in the entire report.

Figure 11 - 27: Filters sidepane

In the *Resource Availability* report used for authoring this book, the date range for the *Availability Heatmap* currently spans from April 2019 through December 2019. Instead, I would like to see a 12-month date range spanning from September 2019 through August 2020. To change the date range for the report, complete the following steps in the expanded *Filters* sidepane:

1. Expand the **Time** filtering options in the *Page level filters* section of the sidepane.
2. Leave the **Filter type** value set to **Relative date filtering**.
3. Click the first **Show items when the value** pick list and select the **is in the next** item.
4. In the field below the first pick list, manually enter **12**.

Variance Analysis, Plan Revision, and Reporting

5. Click the second **Show items when the value** pick list and select the **calendar months** item.
6. Click the **Apply filter** link at the bottom of the *Page level filters* section of the sidepane, such as shown in Figure 11 - 28.

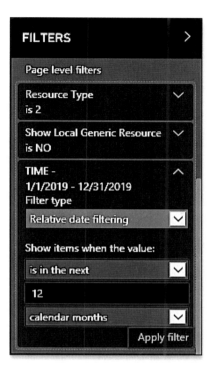

Figure 11 - 28: Updated Time filter

Power BI updates the *Availability Heatmap* with the new date range. In the *Filters* sidepane, continue changing the filter settings to apply your own ad hoc filtering to the report data.

When you finish a session of ad hoc filtering, you might want to reset the current report page back to its original default settings. To do this, click the **Reset to default** button in the toolbar at the top of the Power BI user interface. The system displays the *Reset to default* confirmation dialog shown in Figure 11 - 29. In the dialog, click the **Reset** button to complete the reset process.

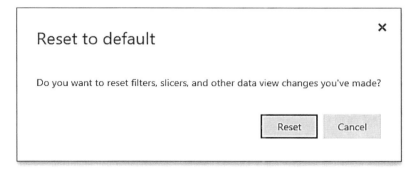

Figure 11 - 29: Reset to default dialog

369

Using Natural Language Queries

Another one of the powerful features of Power BI reporting provides users with the ability to write a natural language query of the Project Online data. You find this feature in the upper left corner of the welcome page shown in Figure 11 - 30.

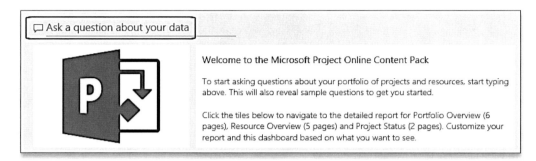

Figure 11 - 30: Natural language query feature

To create a natural language query, click anywhere in the **Ask a question about your data** field. Power BI displays the *Questions to get you started* page shown in Figure 11 - 31. The page includes a series of sample questions listed down the left side of the page.

To begin learning how to type a natural language query, I recommend that you click each of the sample queries and then study the results. You should also try modifying the text in the sample queries to see how Power BI modifies the results based on the changes you make. Once you feel comfortable with the process, you can type your own question in the **Ask a question about your data** field.

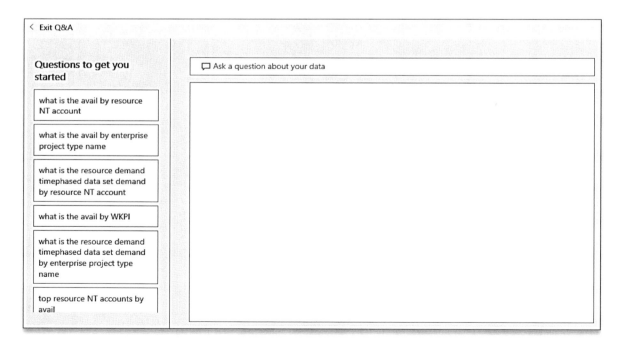

Figure 11 - 31: Natural language query – getting started page

For example, suppose that I want to know the total work for each resource group in our Project Online instance and display the results as a pie chart. To enter my query, I could simply type the text, **total work by resource group as pie**. Figure 11 - 32 shows the resulting pie chart with the requested data. To see the underlying data for any pie slice, simply float your mouse pointer over the slice, and Power BI displays the underlying data in a floating tooltip.

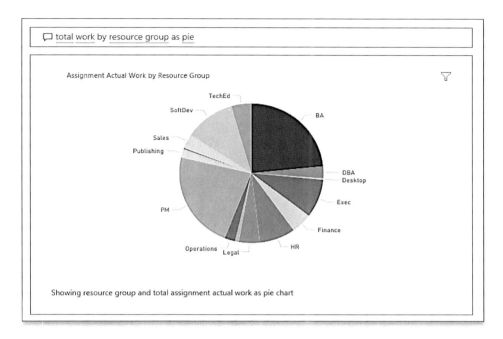

Figure 11 - 32: Results of a natural language query

To exit the natural language query editor, click the **Exit Q&A** link in the upper left corner of the page.

Module 11

Module 12

Working with the Project Site

Learning Objectives

After completing this module, you will be able to:

- Understand the Project Site
- Create and work with risks and issues
- Upload and work with project documents
- Associate a risk, issue, or document with a task
- Work with a shared OneNote notebook
- Work with the Tasks item

Inside Module 12

Understanding the Project Site **375**
Who Has Access to the Project Site? *375*
Navigating to the Project Site **375**
Navigating from the Microsoft Project Schedule *375*
Navigating from Project Web App *376*
Managing Project Risks **379**
Creating a New Risk *379*
Working with Existing Risks *383*
Managing Project Issues **389**
Creating a New Issue *390*
Viewing and Editing Existing Issues *392*
Viewing Your Assigned Issues and Risks **392**
Managing Project Documents **393**
Accessing the Documents Library *393*
Creating a New Document in a Document Library *394*
Creating a New Folder in a Documents Library *395*
Uploading Templates and Documents to a Document Library *396*
Renaming a Document *399*

Sharing a Document *401*
Checking Out a Document for Editing *402*
Editing a Document *403*
Working with the Version History of a Document *407*
Subscribing to E-Mail Alerts about a Document *411*

Attaching Risks, Issues, and Documents to Tasks **412**

Using Other Features in the Project Site **419**
Using OneNote Notebooks *419*
Using the Tasks Page *420*
Using the Newsfeed Web Part *422*

Understanding the Project Site

When you publish a new enterprise project, Project Online automatically provisions a new Project Site in SharePoint if your system options are set to automatic Project Site creation. The Project Site for an enterprise project serves as the central location for managing and sharing project-related information such as documents, risks, issues, and deliverables. When you navigate to a Project Site, you are working mostly with the native capabilities of SharePoint, supplemented with unique features from Project Online.

Who Has Access to the Project Site?

SharePoint dynamically provides access rights to a Project Site are based upon the roles of the people involved with the project. This assumes, however, that your application administrator selected the *Sync SharePoint Tasks Lists* option for the Enterprise Project Types (EPTs) from which you create your enterprise projects. Following is how SharePoint determines access rights:

- SharePoint grants **Full Control** access to the *Owner* (project manager) of the project. This means that the project manager has full site permissions, including the ability to delete content, add and edit items such as lists and libraries, etc.

- SharePoint grants **Contribute** access to all team members in the project, even to those project team members not assigned to any tasks in the project. This means that team members have Read/Write permissions to create and edit items such as risks, issues, and documents.

- SharePoint grants **Reader** access to members of the *Portfolio Views* group (typically the executives in your organization). Reader access provides executives with Read-Only access to items such as risks, issues, and documents. This means that executives can view these items, but cannot edit them.

Navigating to the Project Site

Project Online offers two methods for navigating to the Project Site. If you are a project manager, you can navigate to the Project Site directly from an open enterprise project. If you are a project team member or an executive, you can navigate to the Project Site from the Project Center page in Project Web App.

Navigating from the Microsoft Project Schedule

As a project manager, the fastest way to navigate to a Project Site is from Microsoft Project by completing the following steps:

1. Launch Microsoft Project and connect to Project Web App.

2. Open the enterprise project associated with the Project Site to which you want to navigate.

3. Click the **File** tab to display the *Backstage*.

4. In the *Related Items* section in the lower right corner of the *Backstage* shown in Figure 12 - 1, click either the **Documents**, **Issues**, or **Risks** link to navigate to the associated page in the Project Site.

Module 12

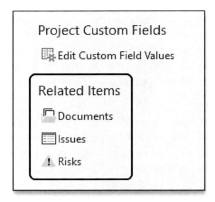

Figure 12 - 1: Related items
section of the Backstage

Navigating from Project Web App

Project team members and executives must navigate to the Project Site from Project Web App by completing the following steps:

1. Navigate to the *Project Center* page in Project Web App.

2. In the data grid of projects, click the name of an enterprise project.

3. In the "drill down" menu at the top of the *Quick Launch* menu, click the **Project Site** link, as shown in Figure 12 - 2.

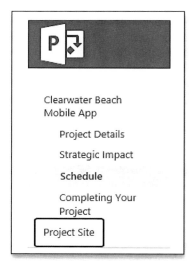

Figure 12 - 2: Click the
Project Site link

Project Web App displays the *Home* page of Project Site for the associated project, such as the own shown in Figure 12 - 3. The default layout of the *Home* page of the Project Site includes the following:

376

Working with the Project Site

- The page includes a *Quick Launch* menu on the left side of the page. The *Quick Launch* menu includes links for the pages you have permission to access in the Project Site, such as the *Risks*, *Issues*, *Deliverables*, or *Documents* pages.

- The page includes a *Project Summary* section at the top of the page. If you created a *Timeline* report in your enterprise project by customizing the default *Timeline* view, the Project Summary section initially displays your customized *Timeline* view. After a few seconds, the *Project Summary* section displays a list of late and upcoming tasks, with the due date for each task displayed. After a few seconds, the *Project Summary* section displays the *Timeline* view again. You can click the **Previous** and **Next** buttons in the upper right corner of the *Project Summary* section to navigate between the *Timeline* view and the list of late and upcoming tasks.

- The bottom left side of the page includes a *Newsfeed* section that you can use to start a conversation between you and your project team members.

- The bottom right side of the page includes the *Documents* section. In this section, you can see existing documents already uploaded to the *Documents* library, and you can use this section to upload additional documents as well.

 Information: The *Home* page of the Project Site shown in Figure 12 - 3 displays the default layout of the page. Your organization may use a customized *Home* page for the Project Site that includes additional sections of use to you and your project team members.

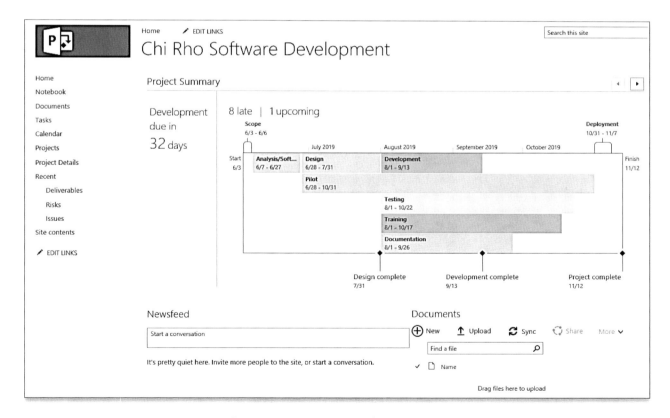

Figure 12 - 3: Home page of a Project Site

377

Module 12

 Hands On Exercise

Exercise 12 - 1

Navigate to the Project Site for your class project and explore the features of the Project Site.

1. Launch Microsoft Project and connect to Project Web App.
2. Open the enterprise project you created for this class, but *do not* check out the project for editing.
3. Click the **File** tab to display the *Backstage*.
4. In the *Related Items* section in the lower right corner of the *Backstage*, click the **Risks** link.

Notice how Project Online launches your preferred web browser, logs you into Project Web App, and then navigates you to the *Risks* page of the Project Site associated with your enterprise project. Continue with the following steps to learn how project team members and executives navigate to the Project Site from Project Web App.

5. On the *Risks* page of your Project Site, click the **Project Center** link in the *Quick Launch* menu to display the *Project Center* page in PWA.
6. In the *Project Center* data grid, click the name of your enterprise project.
7. In the "drill down" section at the top of the *Quick Launch* menu, click the **Project Site** link.

Notice how SharePoint navigates you to the *Home* page of your Project Site. Also notice the information about tasks in your project displayed in the *Project Summary* section at the top of the *Home* page.

8. In the *Recent* section of the *Quick Launch* menu, click the **Deliverables** link.

On the *Deliverables* page, notice the deliverables you created in your enterprise project earlier in this book.

9. In the *Quick Launch* menu, click the **Home** link to return to the *Home* page of your Project Site.
10. Return to your Microsoft Project application window, then *close but do not save* your enterprise project.
11. Exit Microsoft Project and return to your web browser that displays the *Home* page of your Project Site.

Working with the Project Site

Managing Project Risks

The Project Management Institute defines a risk as "an uncertain event or condition that, if it occurs, has a positive or negative effect on a project objective." Risks have both causes and consequences. If a risk occurs, the consequence can be either negative or positive. Your organization's risk management methodologies may dictate that you log and document anticipated risks to your project work.

Creating a New Risk

To create a new risk in the Project Site, click the **Risks** link on the *Quick Launch* menu. SharePoint displays the *Risks* page for the project, such as shown in Figure 12 - 4. You can use the *Risks* page to view all current risks for the project, such as the risk shown in Figure 12 - 4, and to create new risks as well.

Figure 12 - 4: Risks page in a Project Site

To create a new risk, click the **+ new item** link near the top of the *Risks* page. SharePoint displays the *Risks* page, ready for you to create a new risk. Figure 12 - 5 shows the top half of the *Risks* page, while Figure 12 - 6 shows the bottom half of the page. The *Risks* page for creating a new risk contains fields into which you can enter relevant information about the risk. SharePoint requires you to enter information into only three fields: *Title*, *Probability*, and *Impact*. Project Online requires you to enter *Probability* and *Impact* values to determine the *Exposure* value for the risk, which it calculates by multiplying these two numbers.

 Information: SharePoint displays the *Exposure* field on the *Risks* page. You can use the *Exposure* value to rank risks in descending order by severity. For example, your organization might establish a risk management methodology that requires you to create a risk management plan for the top four risks ranked by exposure.

379

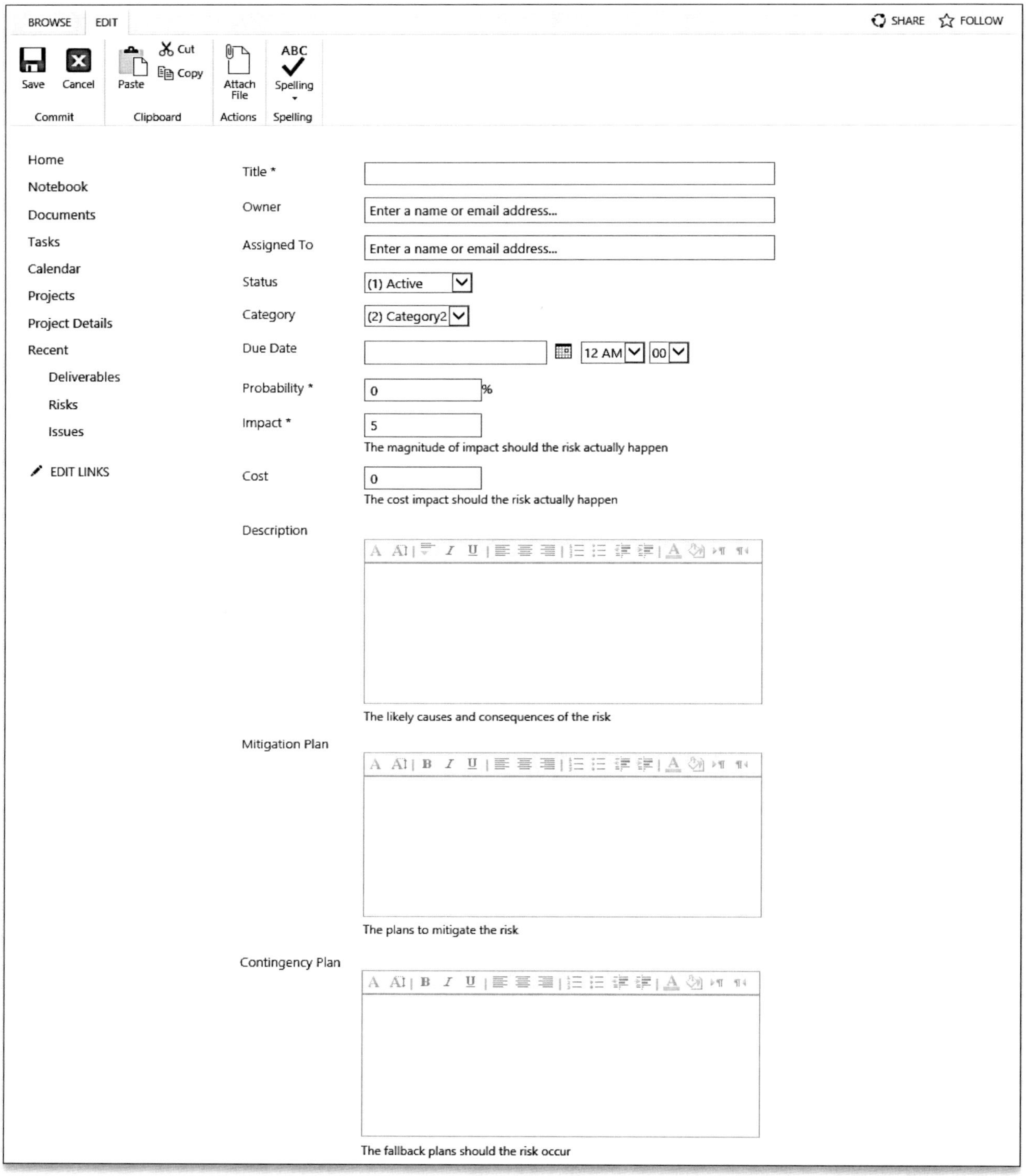

Module 12

Figure 12 - 5: Create a new risk – top of Risks page

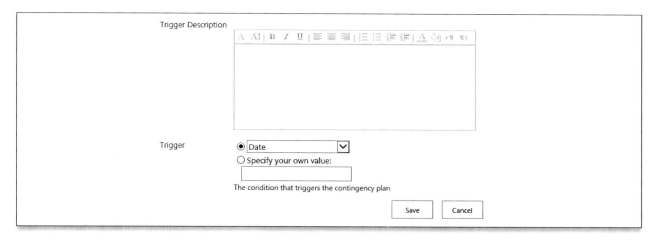

Figure 12 - 6: Create a new risk – bottom of Risks page

In the **Title** field, enter a brief description of the risk. In the **Owner** field, enter the name of the individual who owns the new risk. Start typing the name or email address and SharePoint automatically starts searching for names, such as shown in Figure 12 - 7. Use your mouse pointer to select the name of the owner from the list of suggested names.

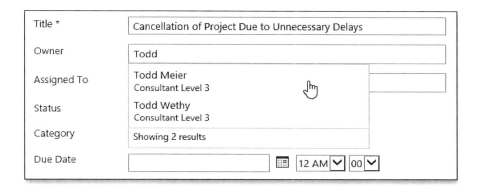

Figure 12 - 7: Select the name of the risk Owner

In the **Assigned To** field, enter the name of the person responsible for managing the risk. This is the person assigned to monitor risk conditions and enact a backup plan if the risk occurs.

Click the **Status** pick list and select the status of the risk. Options on the pick list include *Active*, *Postponed*, and *Closed*. The *Active* status is the default for all new risks unless you select an alternate value.

 Warning: Do not change the default values on the *Status* pick list and replace them with your own values, as doing so can break some of the integration between your Project Site and Project Online.

The **Category** field is a field your organization should modify to display data relevant to your organization's risk management methodologies. The default values of *Category 1*, *Category 2*, and *Category 3* do not provide any useful data, so unless you or your application administrator modifies the default values in the *Category* field, you can ignore it.

In the **Due Date** field, enter either the date on which you anticipate the risk may occur, or the date on which the *Assigned To* person must complete the risk management plan. Because this field is optional, your organization may use it any way you deem appropriate.

In the **Probability** field, enter a percentage value between 0 and 100, representing the likelihood that the risk may actually occur. In the **Impact** field, enter a number between 1 and 10 to describe the magnitude of the consequences should the risk actually occur. Remember that SharePoint multiplies these two numbers to calculate the *Exposure* value for the risk. In the **Cost** field, enter any additional cost incurred to the project if the risk occurs.

Information: Entering a value of *0%* in the *Probability* field makes no sense. If there is no chance of the risk occurring, there is no risk. Entering a value of *100%* in the *Probability* field makes no sense either. If the risk is absolutely certain, it is not a risk. Instead, it is an issue.

Click in the **Description** field to activate the text editing buttons at the top of the field. Enter additional information in this field about the risk, such as causes of the risk and consequences to the project, should the risk occur. You can optionally format the text using the formatting buttons at the top of the field.

Click in the **Mitigation Plan** field and enter your plan for reducing the likelihood of the risk occurring, or to reduce the impact of the risk should it occur. Click in the **Contingency Plan** field and enter your backup plan of action should the risk actually occur.

Click in the **Trigger Description** field and enter the description of the conditions that determine whether the risk is about to occur, is currently occurring, or has already occurred. If you want to enter additional trigger information, click the **Trigger** pick list button and select an item from the list. Alternately, you may also select the **Specify your own value** option and enter text in the accompanying field.

When you finish creating your new risk, click the **Save** button at the bottom of the page. SharePoint saves the new risk, then returns to the main *Risks* page and displays your new risk, such as shown in Figure 12 - 8.

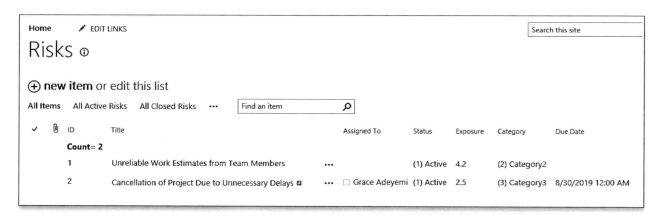

Figure 12 - 8: New risk created

Hands On Exercise

Exercise 12 - 2

Create a new risk in the Project Site for your class project.

1. Click the **Risks** link in the *Quick Launch* menu to display the *Risks* page.
2. At the top of the *Risks* page, click the **+ new item** link.
3. On the *Risks* page for a new risk, enter a realistic name for a risk in the **Title** field.
4. Enter or select your own name in both the **Owner** and **Assigned To** fields.
5. Leave the **Status** field value set to **Active** and leave the default value selected for the **Category** field.
6. Set a date in the **Due Date** field that is 10 working days from today.
7. Enter **50%** in the **Probability** field and enter **6** in the **Impact** field.
8. Enter **$25,000** in the **Cost** field.
9. Enter a brief amount of descriptive information in the **Description**, **Mitigation Plan**, **Contingency Plan**, and **Trigger Description** fields.
10. Click the **Save** button to save your new risk.

Working with Existing Risks

While working with existing risks, SharePoint allows you to do the following:

- Apply a view to the *Risks* page
- Sort and AutoFilter risks by the data in any column
- View a risk
- Edit a risk
- Delete a risk
- Subscribe to e-mail alerts about changes to a risk

I cover each of these topics individually.

Working with Risk List Views

To apply a view to the *Risks* page, click the name of one of the three built-in views at the top of the page. The three built-in views include the *All Items* view (the default view for the *Risks* page), the *All Active Risks* view, and the *All Closed Risks* view. You can also click the **More Views** button (the **...** button) to the right of these three views and select a view from the pick list, such as shown in Figure 12 - 9. Additional views on the pick list include *All Postponed Risks*, *All Risks Assigned to Me*, *All Risks Opened by Me*, and *All Risks Owned by Me*.

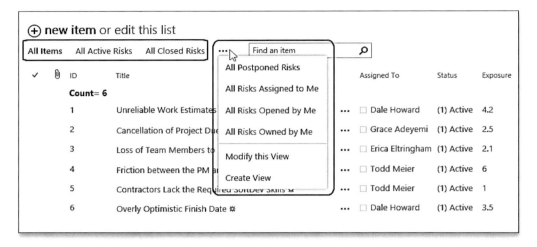

Figure 12 - 9: Apply a View to the Risks page

When you apply any view, SharePoint applies bold formatting to the name of the view, such as shown in Figure 12 - 10. Notice that I applied the *All Risks Assigned to Me* view, which displays only two risks associated with the project.

Figure 12 - 10: All Risks Assigned to Me view applied

Sorting and AutoFiltering Risks

SharePoint applies default sorting on the *ID* column, sorting the risks in the order users created them. To sort the risks another way, float your mouse pointer over the column header of the column you want to sort, and then click the pick list arrow button in the column header. For example, Figure 12 - 9 shows the pick list menu for the *Exposure* column, which allows me to sort the values in either *Ascending* or *Descending* order, and to apply AutoFilter as well.

Working with the Project Site

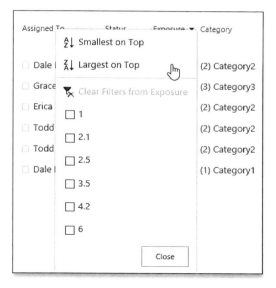

Figure 12 - 11: Sort on the Exposure column

When you apply sorting on the data in any column, SharePoint displays an up-arrow or a down-arrow in the column header to indicate you applied sorting to that column. The up-arrow means you applied an *Ascending* sort, while the down-arrow indicates you applied a *Descending* sort. If you apply an AutoFilter to the data in any column, SharePoint displays a funnel indicator to the right of the column name in the column header.

Viewing and Editing Existing Risks

To view a risk, simply click the name of the risk. You can also click the **More Options** icon (the **...** icon) for the risk and select **View Item** on the menu, as shown in Figure 12 - 12.

Figure 12 - 12: Pick list button and pick list for an existing risk

Figure 12 - 13 shows *Overly Optimistic Finish Date* risk open in view-only mode. When you finish viewing the risk data, click the **Close** button at the bottom of the page.

385

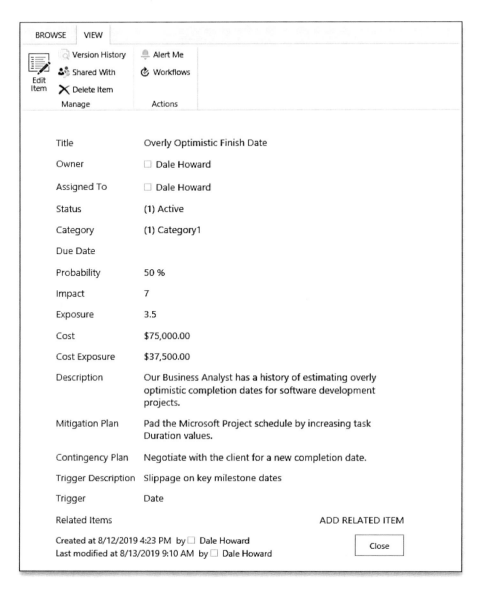

Figure 12 - 13: View a risk

To edit a risk that you have open in view-only mode, click the **Edit Item** button in the *View* ribbon. Otherwise, you can click the **More Options** icon (the ... icon) for the risk and select **Edit Item** on the menu. When you finish editing the risk, click either the **Save** button at the bottom of the page or the **Save** button in the *Edit* ribbon to save your changes.

Deleting a Risk

To delete a risk that you have open in view-only mode, click the **Delete Item** button in the *View* ribbon. Otherwise, you can click the **More Options** icon (the ... icon) for the risk and select **Delete Item** on the menu. SharePoint displays the confirmation dialog shown in Figure 12 - 14.

Working with the Project Site

Figure 12 - 14: Confirmation dialog

In the confirmation dialog, click the **OK** button to confirm the deletion. SharePoint sends the deleted risk to the Recycle Bin for the SharePoint site.

 Warning: If you delete a risk by accident, you can be relieved that it is not gone forever. Rather, it is in the Recycle Bin, where it will remain for up to 30 days before the system deletes it forever. To recover an accidentally deleted risk, ask your application administrator to restore it from the Recycle Bin to the Project Site for your project.

Subscribe to E-Mail Alerts about Changes to a Risk

To subscribe to e-mail alerts about changes to a risk, click the **More Options** icon (the ... icon) for the risk, select the **Advanced** item on the menu, and then click the **Alert me** item on the flyout menu, as shown in Figure 12 - 15.

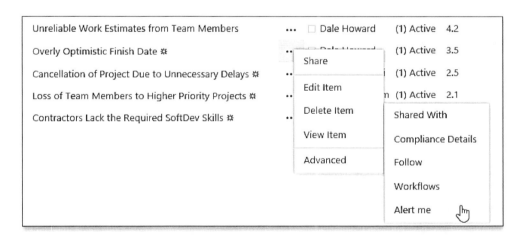

Figure 12 - 15: Select the Alert me item

SharePoint displays the *New Alert* dialog shown in Figure 12 - 16. In the *New Alert* dialog, specify your options in each section of the dialog, and then click the **OK** button. Notice that the *New Alert* dialog allows you to determine what type of change triggers SharePoint to send an e-mail message to you. You can determine the change when the system sends the e-mail message to you.

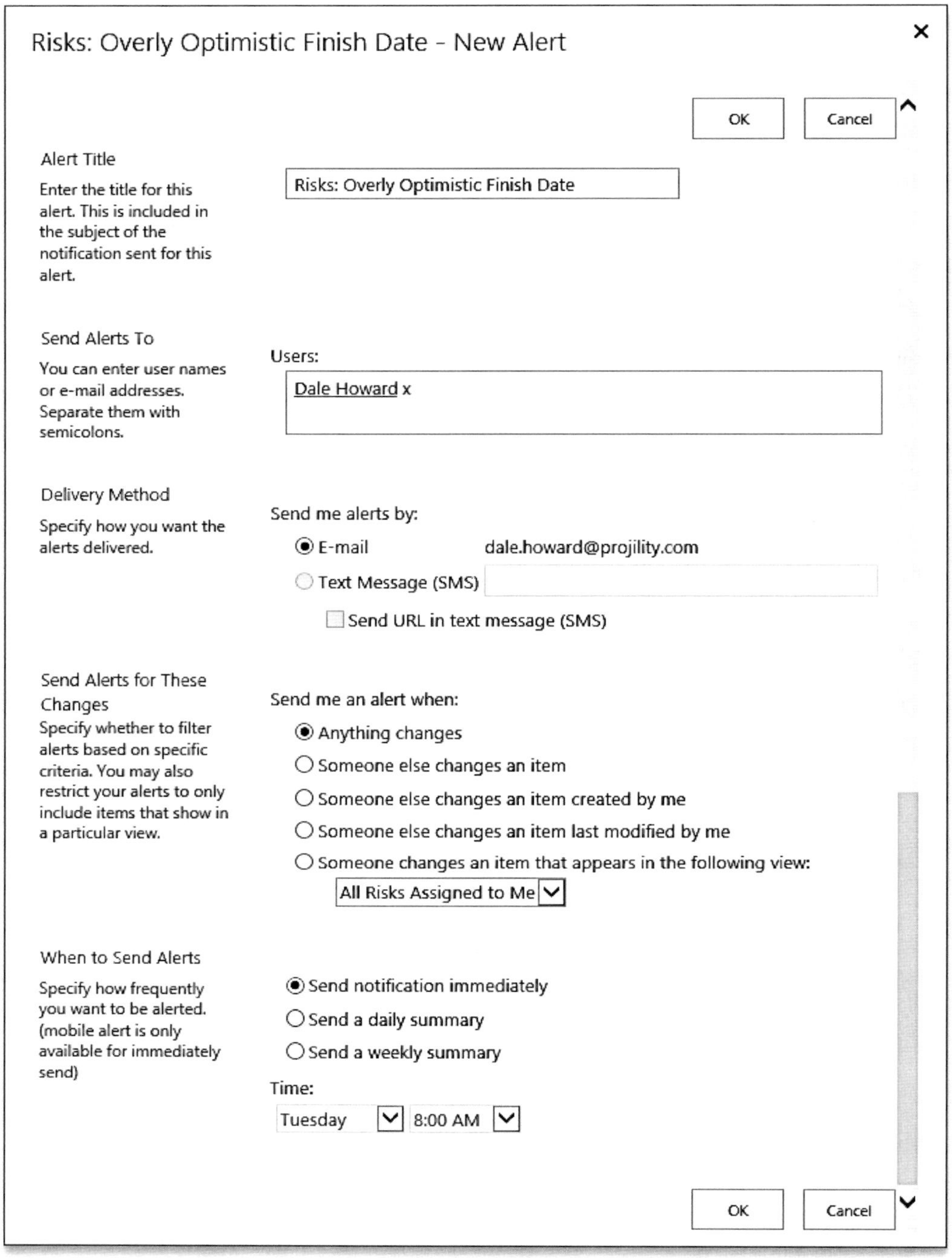

Module 12

| Risks: Overly Optimistic Finish Date - New Alert |

Alert Title
Enter the title for this alert. This is included in the subject of the notification sent for this alert.

Risks: Overly Optimistic Finish Date

Send Alerts To
You can enter user names or e-mail addresses. Separate them with semicolons.

Users:
Dale Howard x

Delivery Method
Specify how you want the alerts delivered.

Send me alerts by:
- ⦿ E-mail   dale.howard@projility.com
- ○ Text Message (SMS)
- ☐ Send URL in text message (SMS)

Send Alerts for These Changes
Specify whether to filter alerts based on specific criteria. You may also restrict your alerts to only include items that show in a particular view.

Send me an alert when:
- ⦿ Anything changes
- ○ Someone else changes an item
- ○ Someone else changes an item created by me
- ○ Someone else changes an item last modified by me
- ○ Someone changes an item that appears in the following view:
  All Risks Assigned to Me

When to Send Alerts
Specify how frequently you want to be alerted. (mobile alert is only available for immediately send)

- ⦿ Send notification immediately
- ○ Send a daily summary
- ○ Send a weekly summary

Time:
Tuesday   8:00 AM

OK   Cancel

Figure 12 - 16: New Alert dialog

At some point you may want to discontinue receiving e-mail alerts about changes to risks in the Project Site for your project. To manage your e-mail alerts, click the **List** tab at the top of the *Risks* page to expand the *List* ribbon. In the *Share & Track* section of the *List* ribbon, click the **Alert Me** pick list button and select the **Manage My Alerts** item, as shown in Figure 12 - 17.

Working with the Project Site

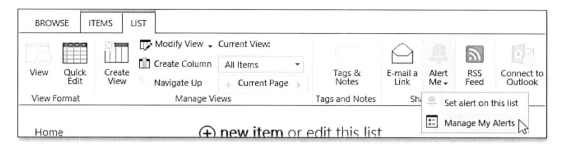

Figure 12 - 17: Select the Manage My Alerts item

SharePoint displays the *My Alerts on this Site* page, such as the one shown in Figure 12 - 18. To delete an existing alert, select the checkbox for the e-mail alert and then click the **Delete Selected Alerts** link at the top of the page. The system deletes the alert immediately and does not warn you of the impending deletion with a confirmation dialog.

Figure 12 - 18: My Alerts on this Site page

To edit an existing e-mail alert, click the name of the alert. The system displays the *Edit Alert* page in which you can modify the alert, as needed. Click the **OK** button when you finish modifying the alert.

Managing Project Issues

An issue is any type of problem or concern you might experience and need to manage during the life of the project. Another way to think of an issue is to consider it a realized risk. Whether or not you predicted their occurrence through proactive risk management, issues are events that cause problems that require management. The issues management features in Project Online allow you to identify, track, and manage issues in collaboration with your project team and stakeholders. Examples of project issues include a shortage of resources or an unanticipated hardware upgrade requirement.

Creating a New Issue

To create a new issue in the Project Site, click the **Issues** link on the *Quick Launch* menu. SharePoint displays the *Issues* page for the project. You can use the *Issues* page to view all current issues for the project, such as the two issues displayed on the *Issues* page shown in Figure 12 - 19, and to create new issues as well.

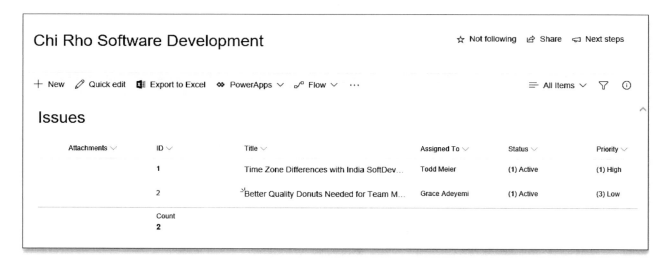

Figure 12 - 19: Issues page in a Project Site

 Information: SharePoint displays the *Issues* page with a newer, more modern look. If you prefer using the *Issues* page in *Classic* mode instead, click the **Return to Classic SharePoint** link at the bottom of the *Quick Launch* menu.

To create a new issue, click the **+ New** link button in the toolbar at the top of the *Issues* page. SharePoint displays the *Issues* page, ready for you to create a new issue, such as shown in Figure 12 - 20. The process for creating a new issue is nearly identical to the process for creating a new risk. The *Issues* page for creating a new issue contains fields into which you can enter relevant information about the issue. SharePoint requires you to enter information into only one field: the *Title* field.

In the **Title** field, enter a brief description of the issue. In the **Owner** field, enter the name of the individual who owns the new issue. Start typing the name or email address and SharePoint automatically starts searching for names, such as shown previously in Figure 12 - 7. Use your mouse pointer to select the name of the owner from the list of suggested names.

In the **Assigned To** field, enter the name of the person responsible for managing the issue. This is the person assigned to resolving the issue.

Click the **Status** pick list and select the status of the issue. Options on the pick list include *Active*, *Postponed*, and *Closed*. The *Active* status is the default for all new issues unless you select an alternate value.

 Warning: Do not change the default values on the *Status* pick list and replace them with your own values, as doing so can break some of the integration between your Project Site and Project Online.

Working with the Project Site

The **Category** field is a field your organization should modify to display data relevant to your organization's issue management methodologies. The default values of *Category 1*, *Category 2*, and *Category 3* do not provide any useful data, so unless you or your application administrator modifies the default values in the *Category* field, you can ignore it.

In the **Priority** field, select the priority for the issue, whether *High*, *Medium*, or *Low*. In the **Due Date** field, enter the date by which you want the issue resolved.

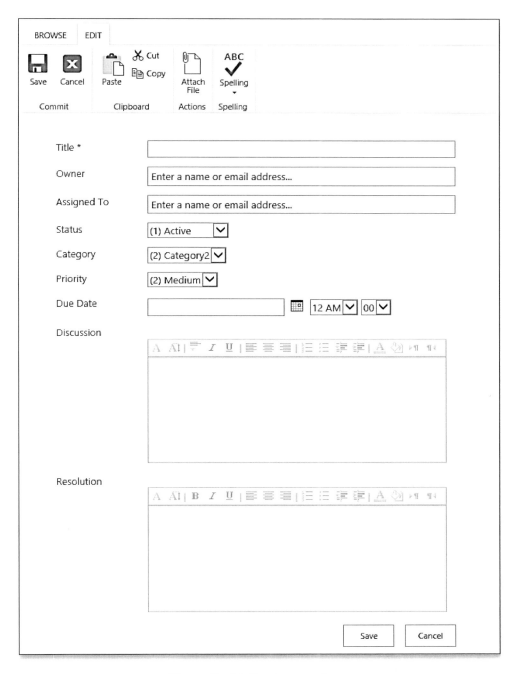

Figure 12 - 20: Create a new issue

391

Click in the **Discussion** field to activate the text editing buttons at the top of the field. Enter additional information in this field about the issue, such as causes of the issue and possible resolutions for the issue. You can optionally format the text using the formatting buttons at the top of the field.

When creating the issue initially, do not enter any text in the **Resolution** field. When the issue is resolved at a later time, the *Assigned To* person should change the **Status** value to **Closed**, and enter a description of the issue resolution in the **Resolution** field.

When you finish creating your issue risk, click the **Save** button at the bottom of the page. SharePoint saves the new issue, then returns to the main *Issues* page and displays your new risk.

Viewing and Editing Existing Issues

Because risks and issues are both List items in SharePoint, the process of viewing and editing is identical for both risks and issues. Refer back to the *Viewing and Editing Existing Risks* topical section for the detailed steps you can follow to view and edit issues.

Viewing Your Assigned Issues and Risks

Project Web App includes the *Issues and Risks* page in which users can see all of the risks and issues to which they are assigned. If your application administrator configured the system to display this link in the *Quick Launch* menu, click the **Issues and Risks** link to display the *Issues and Risks* page, such as the one shown in Figure 12 - 21.

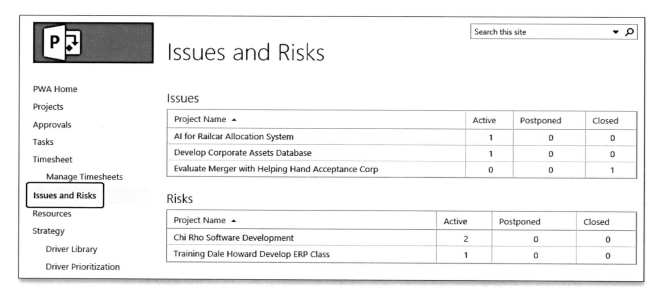

Figure 12 - 21: Issues and Risks page in PWA

Warning: The default configuration of the *Quick Launch* menu does not display the *Issues and Risks* link. In order for users to access this page, your application administrator must configure the *Quick Launch* menu to display the *Issues and Risks* link.

The *Issues and Risks* page displays the name of each project with risks or issues assigned to the user, along with the number of *Active, Postponed,* and *Closed* risks and issues by project. To access either the *Risks* page or the *Issues* page for one of the listed projects, click the project name in either the *Issues* or *Risks* section of the page. The system displays the *Risks* or *Issues* page for the selected project, and applies either the *All Risks Assigned to Me* view or the *All Issues Assigned to Me* view, as needed.

Hands On Exercise

Exercise 12 - 3

Create a new issue in the Project Site for your class project.

1. Click the **Issues** link in the *Quick Launch* menu to display the *Issues* page.
2. At the top of the *Issues* page, click the **+ New** link.
3. On the *Issues* page for a new issue, enter a realistic name for an issue in the **Title** field.
4. Enter or select your own name in both the **Owner** and **Assigned To** fields.
5. Leave the **Status** field value set to **Active**.
6. Leave the default value selected for the **Category** field.
7. Set the **Priority** value to **High**.
8. Set a date in the **Due Date** field that is 15 working days from today.
9. Enter a brief amount of descriptive information about the issue in the **Discussion** field.
10. Click the **Save** button to save your new issue.

Managing Project Documents

During the typical project life cycle, you and your team create multiple documents associated with the project. During project definition, you might create a project charter and a statement of work document. During the execution stage of the project, you might create change control documents and expense reports. At project closure, you might create a lessons learned document to capture the knowledge gained during the project. Regardless of which type of project documents you create, each document is a part of your project's "electronic paper trail" and you can manage these documents in the Project Site associated with each enterprise project.

Accessing the Documents Library

Each Project Site includes a single document library, appropriately named *Documents*. To access this document library, click the **Documents** link in the *Quick Launch* menu of the Project Site. SharePoint displays the newer, modern look of the *Documents* library, such as the one shown in Figure 12 - 22. If you prefer using it in Classic mode instead, click the **Return to Classic SharePoint** link at the bottom of the *Quick Launch* menu.

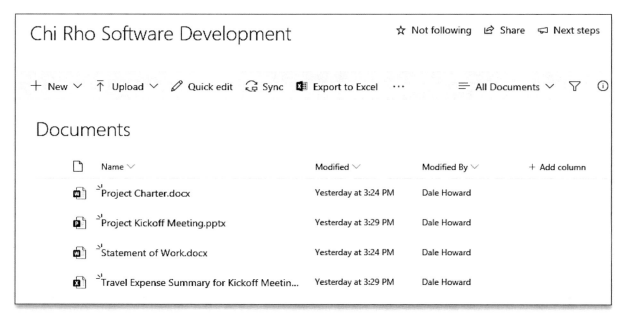

Figure 12 - 22: Documents page in a Project Site

Creating a New Document in a Document Library

Although uploading an existing document is the method many people prefer for adding documents to the *Documents* library, SharePoint does allow you to create new documents directly in this library. For example, suppose that you want to create a new Microsoft Word document named *Risk Management Plan* and you want to create it directly in the *Documents* library. You can create this document by clicking the **New** pick list button and selecting the **Word document** item on the menu, as shown in Figure 12 - 23.

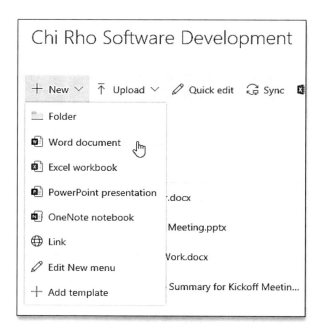

Figure 12 - 23: Create a new Word document

Working with the Project Site

SharePoint launches Word Online on a new web browser tab and creates a new blank Word document. Word Online immediately saves the document in the *Documents* library using the generic name *Document*, such as shown at the top of the application window in Figure 12 - 24.

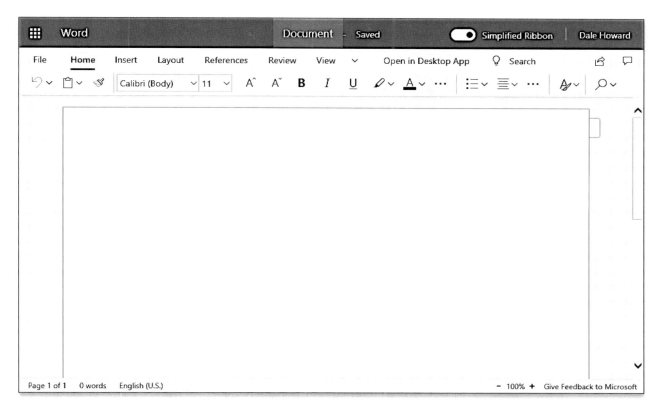

Figure 12 - 24: New blank document in Word Online

To rename the document with its correct name, such as *Risk Management Plan*, click the word **Document** at the top of the Word Online application window to place the file name into renaming mode. In the selected field, enter the new name for the document and then press the **Enter** key on your computer keyboard. Word Online renames the file to use the new name you supplied. As you edit your new document, Word Online automatically saves it. When finished, close the Word Online web browser tab to exit the editing session for the new document.

Creating a New Folder in a Documents Library

Although using content types and metadata are the methods many people prefer for organizing documents in the *Documents* library, SharePoint does allow you to create new folders in the *Documents* library. For example, suppose that your organization does not intend to use content types or metadata to organize project documents, and you need two new folders in the *Documents* library to organize meeting minutes from team meetings and expense reports related to the project. To create a new folder in the *Documents* library, click the **New** pick list button and select the **Folder** item on the menu, as shown in Figure 12 - 25.

395

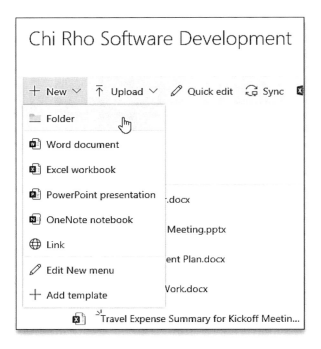

Figure 12 - 25: Create a new folder

SharePoint displays the *Create a Folder* dialog. Enter a name for the new folder in the *Create a Folder* dialog, such as shown in Figure 12 - 26, and then click the **Create** button.

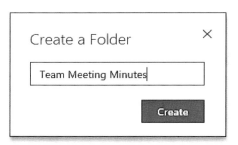

Figure 12 - 26: New Folder dialog

SharePoint displays any new folders you create at the top of the document list, sorted alphabetically in ascending order. You can create new documents or upload existing documents in a folder by clicking the name of the folder before you begin creating or uploading documents.

Uploading Templates and Documents to a Document Library

SharePoint allows you to upload both existing templates and existing documents into the *Documents* library of your Project Site. When you upload a template file, SharePoint adds the template to the list of items shown on the *New* pick list menu shown previously in Figure 12 - 23. This means that you can create new documents directly in the *Documents* library using your organization's preferred templates. To upload a template file, click the **Upload** pick list button and select the **Templates** item on the menu, such as shown in Figure 12 - 27.

Working with the Project Site

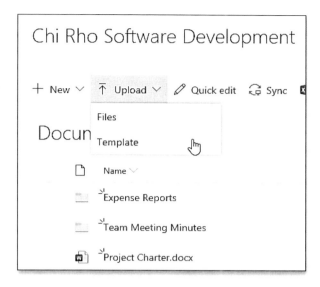

Figure 12 - 27: Upload a template

SharePoint displays the *Choose File to Upload* dialog shown in Figure 12 - 28. In this dialog, navigate to the folder containing the template you want to upload, select the template file, and then click the **Open** button. The system uploads the template into the Project Site and automatically adds it to the list of templates shown on the *New* menu in the *Documents* library.

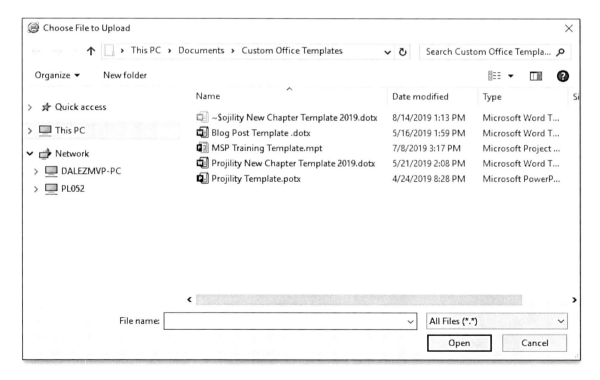

Figure 12 - 28: Choose File to Upload dialog

397

Module 12

Information: You can also upload an existing template to your *Documents* library by clicking the **New** pick list button and selecting the **+ Add Template** item on the menu.

Information: When you upload a template to your *Documents* library, SharePoint saves the template in the *Forms* folder of your *Documents* library. If you erroneously upload the wrong template and need to delete it, you can access the *Forms* subfolder by clicking the **All Documents** pick list button in the upper right corner of the *Documents* page and selecting the **View in File Explorer** item on the menu. Double-click the **Forms** folder to view all of the templates you uploaded to your *Documents* library.

You can also upload one or more existing document files to the *Documents* library simultaneously. To upload existing documents to the *Documents* library, click the **Upload** pick list button and select the **Files** item on the menu, such as shown in Figure 12 - 29.

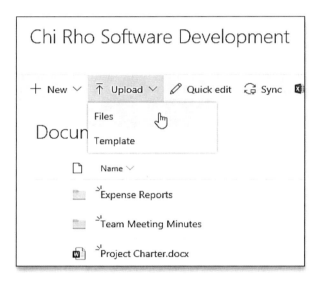

Figure 12 - 29: Begin uploading files

SharePoint displays the *Choose File to Upload* dialog shown previously in Figure 12 - 28. In this dialog, navigate to the folder containing the documents you want to upload. If you want to upload only one document, select the document and then click the **Open** button. If you want to upload multiple documents, use the **Shift** or **Control** keys on your computer keyboard to select multiple files, and then click the **Open** button. The system uploads the selected documents into your *Documents* library. For example, notice that Figure 12 - 30 show three Microsoft Excel expense reports uploaded to the *Expense Reports* folder.

Information: You can also upload documents to the *Documents* library by "dragging and dropping" them into the library using your File Explorer application.

Working with the Project Site

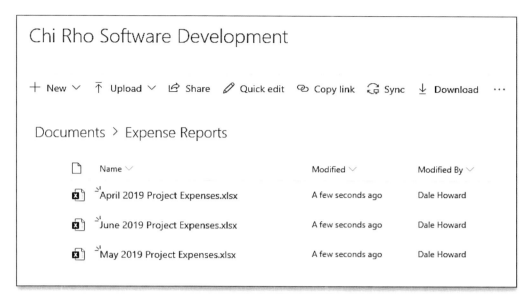

Figure 12 - 30: Expense reports uploaded to the Documents library

Renaming a Document

After uploading a document to the *Documents* library, at some point you may need to rename the document. To begin the process of renaming the document, select the checkbox to the left of the document name so that SharePoint changes the buttons on the toolbar at the top of the page. Click the **Rename** button in the toolbar, such as shown in Figure 12 - 31.

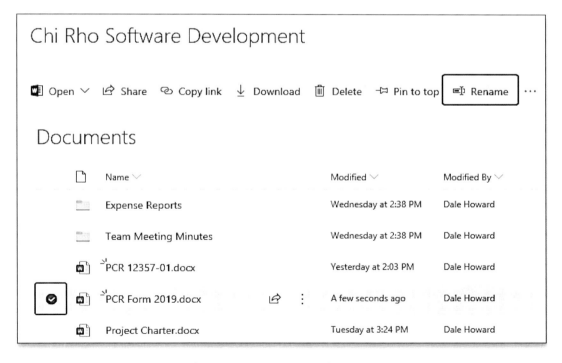

Figure 12 - 31: Rename a document

SharePoint displays the *Rename* dialog for the selected document. Enter the new name for the document in the *Rename* dialog, such as shown in Figure 12 - 32, and then click the **Save** button.

Figure 12 - 32: Rename dialog

Hands On Exercise

Exercise 12 - 4

Create a new document in the *Documents* library, upload an existing document to the *Documents* library, and then rename the uploaded document.

1. Click the **Home** link at the top of the *Quick Launch* menu to return to the *Home* page of the Project Site.

2. Click the **Documents** link the *Quick Launch* menu to navigate to the *Documents* library for your enterprise project.

3. Click the **+ New** pick list button and select the **Word document** item on the menu.

Notice that SharePoint opens a new web browser tab that displays Word Online with a new blank document open in editing mode.

4. At the top of the application window, click the **Document** item on the title bar.

5. Enter the name **Communication Plan** in the text field and then press the **Enter** key on your computer keyboard.

6. Close the web browser tab that displays Word Online to return to your *Documents* library.

Notice that your *Documents* library contains the new *Communication Plan* document.

7. Click the **Upload** pick list button and select the **Files** option on the menu.

8. In the *Choose File to Upload* dialog, navigate to the folder containing your student sample files.

9. In the *Choose File to Upload* dialog, select the **PCR Form 2019.docx** document and then click the **Open** button.

Working with the Project Site

Notice that your *Documents* library now contains the *PCR Form 2019.docx* document.

10. Select the checkbox to the left of the *PCR Form 2019.docx* document and then click the **Rename** button in the toolbar at the top of the page.

11. In the *Rename* dialog, enter the name **Change Request 01**, and then click the **Save** button.

Sharing a Document

You may occasionally need to share a document in a *Documents* library with someone who does not have permission to access the document. By default, the only people who have access to a Project Site are the project manager of the project, the project team members, the executives in your organization, and your Project Online application administrator. Suppose that you need to share a document with a fellow project manager to ask for input on the document, but your fellow project manager does not have permission access your Project Site.

To share a document with someone who does not have permission to access to the document, float your mouse pointer over the name of the document and then click the **Share** icon to the right of the document name, such as shown in Figure 12 - 33. By the way, when you float your mouse pointer over the document, SharePoint displays a dialog with information about the document, including suggested names of people with whom you might want to share the document.

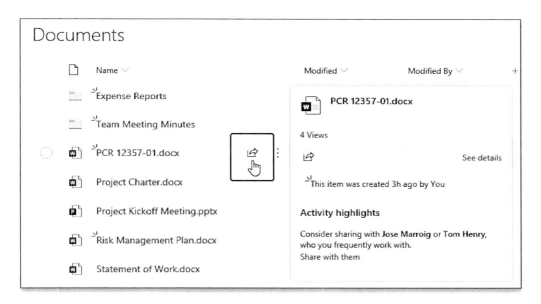

Figure 12 - 33: Click the Share icon

When you click the *Share* icon, SharePoint displays the *Send Link* dialog. In this dialog, enter the name or e-mail address of the person with whom you want to share the document, and then select the name of the person from the list of suggested names generated by SharePoint. Add an optional message if you wish. Figure 12 - 34 shows the completed *Send Link* dialog, ready to send. Click the **Send** button when finished. By the way, the person with whom you share the document receives special permission to edit the document, as indicated by the *People in _____ with the link can edit item* field at the top of the dialog.

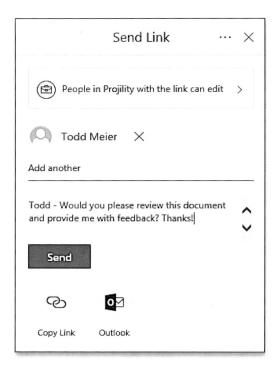

Figure 12 - 34: Send Link dialog

SharePoint uses your Microsoft Outlook application (or Outlook Online) to send the e-mail message that shares the document with the recipient. When the system successfully completes the send process, SharePoint displays the *Link sent* confirmation dialog shown in Figure 12 - 35. Close the dialog when finished.

Figure 12 - 35: Link send dialog

Checking Out a Document for Editing

Because Office Online allows multiple users to open and edit a document simultaneously, there may be times when you want to prevent others from editing the document while you are working with it. In situations such as this, you can check out the document for editing, which prevents others from editing the document while you have it open. To check out a document for editing, first select the checkbox to the left of the document name. In the toolbar at the top of the page, click the **Show Actions** button (the **...** button) and select the **Check Out** item on the menu, such as shown in Figure 12 - 36.

Working with the Project Site

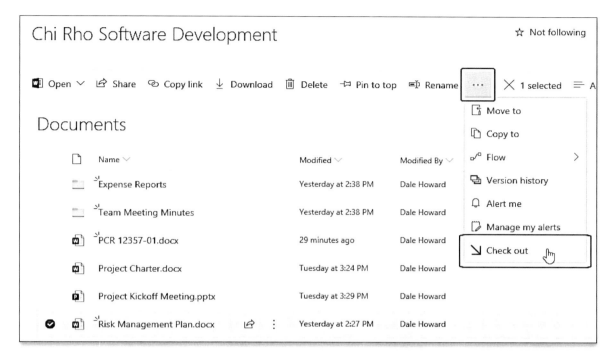

Figure 12 - 36: Select the Check Out item on the menu

To indicate that the document is checked out for editing, SharePoint displays "checked out" indicator (solid red circle indicator) to the right of the document name, such as shown in Figure 12 - 37. This indicator appears for the document to anyone who accesses the *Documents* library. Users can float their mouse pointer over the name of the checked out document and view an information dialog that shows you currently have the document checked out for editing.

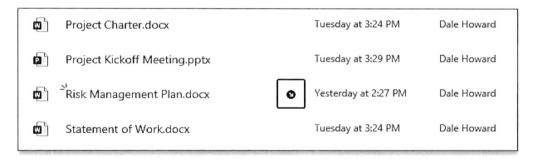

Figure 12 - 37: Checked out document indicator

Editing a Document

To edit a document, regardless of whether you checked out the document or not, click the **Show Actions** indicator to the right of the document name. At the top of the menu, select the **Open** item, and then select either the **Open in browser** item or the **Open in app** item in the flyout menu, such as shown in Figure 12 - 38.

If you select the *Open in browser* item, SharePoint opens the document for editing in the appropriate Office Online application that matches the document type. For example, it launches Excel Online if you attempt to open a workbook document. If you select the *Open in app* item, SharePoint opens the document for editing in the appropriate

403

Microsoft Office application that matches the document type. For example, it launches Microsoft Excel if you attempt to open a workbook document.

Warning: If you attempt to open a file not created in a Microsoft Office application, SharePoint hides the *Open* item on the menu and displays the *Preview* item instead. If you select the *Preview* item on the menu and SharePoint can identify a file viewer for the selected file, the system allows you to preview the file. If you attempt to preview a file not created in a Microsoft Office application and SharePoint cannot identify a file viewer for the file, the preview fails with an error message.

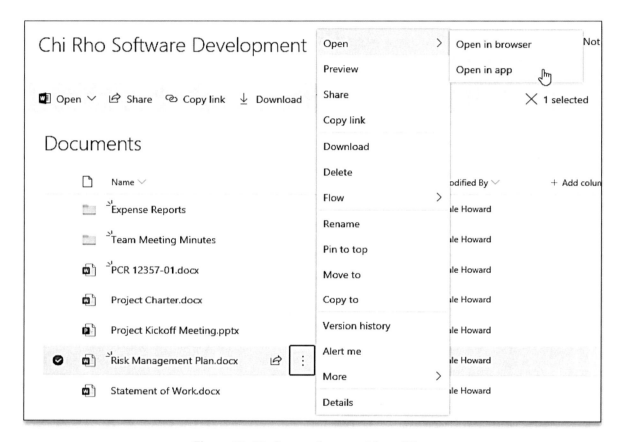

Figure 12 - 38: Open a document for editing

Information: Another way to open a document for editing is to select the checkbox to the left of the document. SharePoint changes the buttons on the toolbar at the top of the page. Click the **Open** pick list button on the toolbar and select either **Open in browser** or **Open in app** from the menu.

If you choose to edit the document using an Office Online application, the system opens the document on a new web browser tab, and automatically saves your document during the editing process. If you choose to edit the document using a Microsoft Office application, you may need to save the document occasionally by clicking the **File** tab and then clicking the **Save** tab in the *Backstage*.

Working with the Project Site

To close the document in the Offline Online application, simply close the web browser tab that displays the Office Online application. Keep in mind that when you close the web browser tab, SharePoint does not check in the document for you automatically, which means you must manually check in the document. To check in the document, float your mouse pointer over the name of the document so that SharePoint displays the information dialog about the document. In the information dialog, click the **Check in** hyperlink to check in the document, such as shown in Figure 12 - 39.

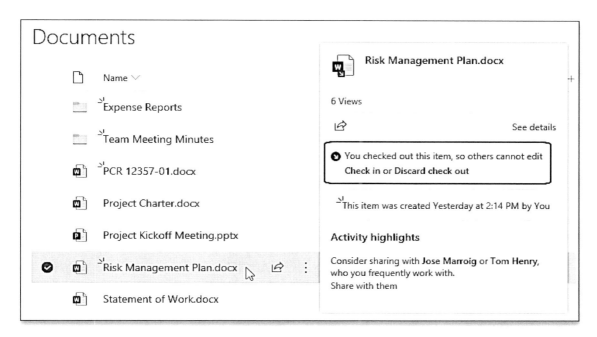

Figure 12 - 39: Click the Check in hyperlink in the dialog

SharePoint displays the *Check in* dialog shown in Figure 12 - 40. In the *Comments* field, add your comments about the editing session, and then click the **Check In** button. SharePoint checks in the document so that others can now edit the document.

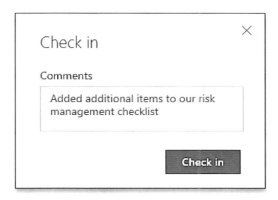

Figure 12 - 40: Check in dialog

To close the document in the Microsoft Office application, click the **File** tab and then click the **Close** tab in the *Backstage*. The system automatically displays the check in confirmation dialog shown in Figure 12 - 41.

405

Module 12

Figure 12 - 41: Check in confirmation dialog

In the confirmation dialog, click the **Yes** button. Microsoft project displays the *Check In* dialog shown in Figure 12 - 42. Enter your optional comments in the *Version Comments* field and then click the **OK** button.

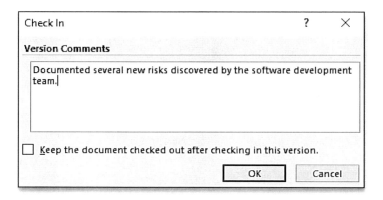

Figure 12 - 42: Check In dialog

 Hands On Exercise

Exercise 12 - 5

Check out a document, edit the document, and then check in the document when finished.

1. Select the checkbox to the left the **Change Request 01.docx** document.

2. In the toolbar at the top of the page, click the **Show Actions** button (the ... button) and select the **Check Out** item on the menu.

3. When the checkout process completes, leave the checkbox selected to the left of the *Change Request 01.docx* document.

4. In the toolbar at the top of the page, click the **Open** pick list button and then select the **Open in app** item on the menu to open the document in the Microsoft Word application.

Working with the Project Site

 Warning: Because the *Change Request 01.docx* document is actually a "fill-in form" created in the Microsoft Word application, SharePoint does not allow you to edit the document in Word Online. Instead, you ***must*** edit the document in the Microsoft Word application.

5. Enter the name of your class project in the first fill-in field, then press the **Tab** key on your computer keyboard to navigate to the next field.

6. Enter *your name* in the **Project Manager** field and then press the **Tab** key.

7. In the **Change Requested By** field, enter the name of an executive in your organization, and then press the **Tab** key.

8. In the **Change Request Date** field, click the calendar date picker arrow button and select the current date, then press the **Tab** key.

9. In the **Description of Change** field, enter the following text:

    Add resources to tasks to shorten their duration.

10. Click the **File** tab and then click the **Save** tab in the *Backstage*.
11. Click the **File** tab and then click the **Close** tab in the *Backstage*.
12. In the check in confirmation dialog, click the **Yes** button.
13. In the *Check In* dialog, enter the following text:

    Filled out the top of the form.

14. In the *Check In* dialog, click the **OK** button.

Working with the Version History of a Document

Microsoft enables the SharePoint versioning feature for all documents in the *Documents* library of every Project Site. Because of this, you can use the versioning feature to view the version history of revisions to any document. The version history of a document includes information such as:

- Date and time the document was first created or uploaded, along with who created it or uploaded it
- Date and time that the document was renamed, along with who renamed it
- Date and time of each editing session in the document, along with who made the edits
- Comments added to the document when the user checked in the document after editing it

To view the version history of a document, first select the checkbox to the left of the document name. In the toolbar at the top of the page, click the **Show Actions** button (the **...** button) and select the **Version history** item on the menu, such as shown in Figure 12 - 43.

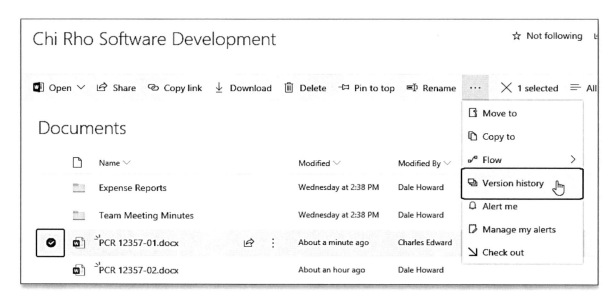

Figure 12 - 43: View the version history of a document

SharePoint displays the *Version history* dialog, such as the one shown in Figure 12 - 44. Notice that the dialog shows the running history of six versions of a document, sorted in descending order with the latest version at the top of the list. Notice that the version history shows the dates and times of each change to the document, along with the name of the user who made each of the changes.

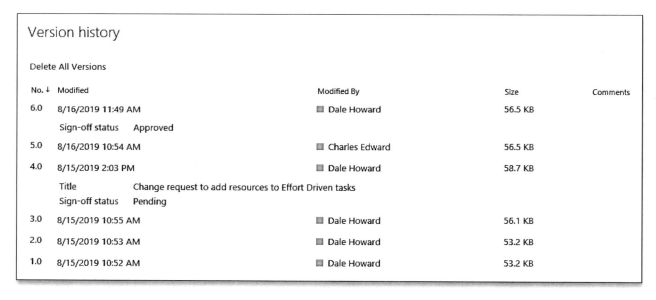

Figure 12 - 44: Version history dialog

 Information: SharePoint always opens the latest version of a document when you open it in either Office Online or a Microsoft Office application. Version 6.0, by the way, is the latest version of the document in the *Version history* dialog shown previously in Figure 12 - 44.

To manage the versions of the selected document, float your mouse pointer over the date and time of the version you wish to manage, then click the pick list button and select an item on the pick list, such as shown in Figure 12 - 45. Notice that the pick list offers you three options for managing a document version: *View, Restore,* or *Delete.*

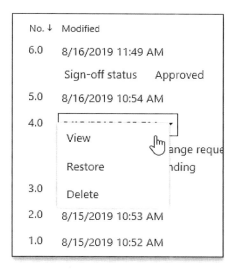

Figure 12 - 45: Manage a document version

If you select the **View** item from the pick list menu, SharePoint updates the *Version history* dialog with information about the selected version, such as shown in Figure 12 - 46. Notice that the *VIEW* ribbon in the top half of the dialog contains tools for working with the selected version. To return to the default *Version history* dialog, click the **Version History** button in the *VIEW* ribbon.

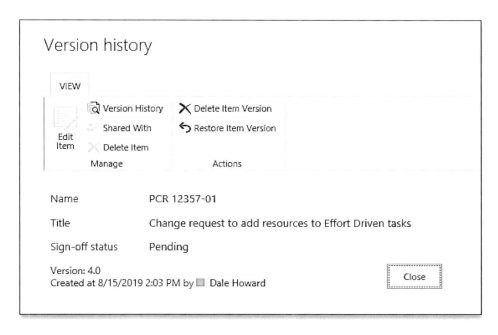

Figure 12 - 46: View Item page for a selected version of a document

SharePoint allows you to restore any older version of a document and make it the new current version of the document. To restore any previous version of a document, select the **Restore** item from the pick list menu or click the **Restore Item Version** button in the VIEW ribbon displayed in the *Version history* dialog. SharePoint displays the confirmation dialog shown in Figure 12 - 47.

Figure 12 - 47: Confirmation dialog for
restoring a previous version

In the confirmation dialog, click the **OK** button to complete the process of restoring a previous version and making it the current version of the selected document. The system creates a new current version as a copy the previous version. In the *Version history* dialog shown in Figure 12 - 48, version 7.0 is now the new current version of the selected document.

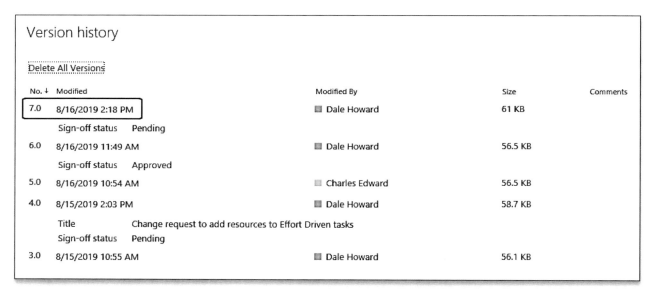

Figure 12 - 48: Restored previous version is now the current version of the document

Notice in the pick list shown previously in Figure 12 - 45 and in the *Version history* dialog shown previously in Figure 12 - 46 that you can also delete versions. The system allows you to delete any versions of the document except the current document version and to send them to the Recycle Bin. To delete versions of a document in a *Documents* library, click either the **Delete** item on the pick list menu or the **Delete Item Version** button in the VIEW ribbon displayed in the *Version history* dialog. SharePoint displays the confirmation dialog shown in Figure 12 - 49.

Working with the Project Site

Figure 12 - 49: Confirmation dialog to delete a version

Click the **OK** button in the confirmation dialog to delete the version and send it to the Recycle Bin.

 Warning: If you delete a document version by accident, it is not gone forever. Rather, it is in the Recycle Bin, where it will remain for up to 30 days before the system deletes it forever. To recover an accidentally deleted document version, ask your application administrator to restore it from the Recycle Bin to the Project Site for your project.

Subscribing to E-Mail Alerts about a Document

As with risks and issues, you can also subscribe to e-mail alerts about changes to a document in a *Documents* library. To subscribe to e-mail alerts about a document, first select the checkbox to the left of the document name. In the toolbar at the top of the page, click the **Show Actions** button (the ... button) and select the **Alert me** item on the menu, such as shown in Figure 12 - 50.

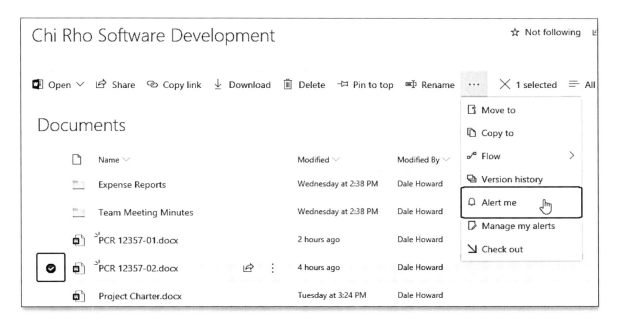

Figure 12 - 50: Select the Alert me item on the menu

SharePoint displays the *Alert me when items change* dialog for the document, such as the one shown in Figure 12 - 51. In the *Alert me when items change* dialog, specify your options in each section of the dialog, and then click the **OK**

411

button. Notice that the *New Alert* dialog allows you to determine what type of change triggers SharePoint to send an e-mail message to you. You can determine the change when the system sends the e-mail message to you.

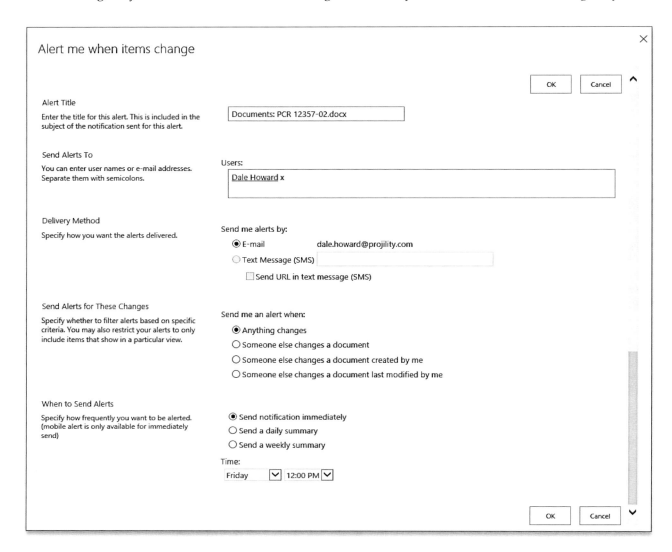

Figure 12 - 51: Alert me when items change dialog for a document

Attaching Risks, Issues, and Documents to Tasks

Perhaps the most challenging feature of the Project Site is attaching risks, issues, and documents to the tasks with which they are associated. This feature is challenging because the process is not very intuitive. For example, I have a risk named *Contractors Lack the Required SoftDev Skills* in the Project Site associated with the *Chi Rho Software Development* project, and I want to attach this risk to tasks that may be impacted if this risk actually occurs.

To attach a risk to a task, navigate to the *Risks* page and then click the name of the risk to open it in view-only mode. Scroll down to the bottom of the page and click the **ADD RELATED ITEM** link immediately above the *Close* button, such as shown in Figure 12 - 52.

Working with the Project Site

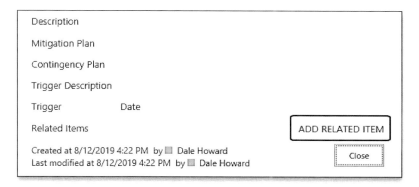

Figure 12 - 52: Click the ADD RELATED ITEM link

 Warning: SharePoint only displays the *ADD RELATED ITEM* link when you have the risk open in view-only mode. The link disappears when have the risk open in editing mode.

SharePoint displays the *Select an Asset* dialog. In the dialog, select the **Tasks** item, such as shown in Figure 12 - 53, and then click the ➔ button (the right-arrow button) in the lower-right corner of the dialog to continue.

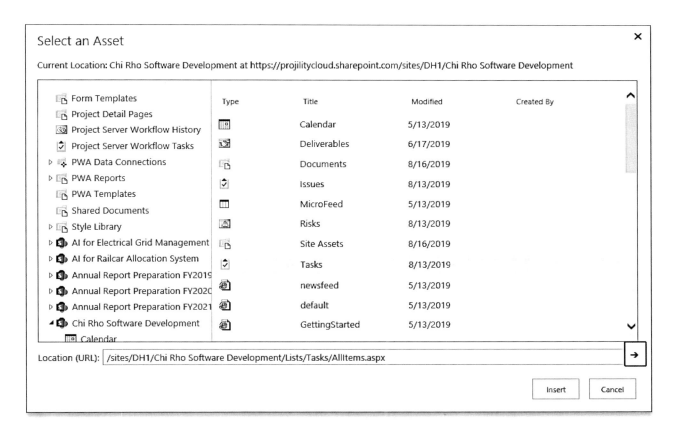

Figure 12 - 53: Select an Asset dialog

413

Module 12

 Warning: **Do not** click the *Insert* button in the *Select an Asset* dialog. You **must** click the ➔ button and **not** the *Insert* button to continue linking the risk to a task in your project schedule.

SharePoint updates the *Select an Asset* dialog by displaying all of the tasks in the project schedule. The system indicates completed tasks by displaying a checkmark in the checkbox to the left of the task name, and by applying the strikethrough font to the task name. You cannot attach a risk to a completed task, so if necessary, scroll down the list of tasks and select an unstarted task to which you want to attach the risk, such as shown in Figure 12 - 54. The system highlights the selected task using cell background formatting. Click the **Insert** button to attach the risk to the selected task.

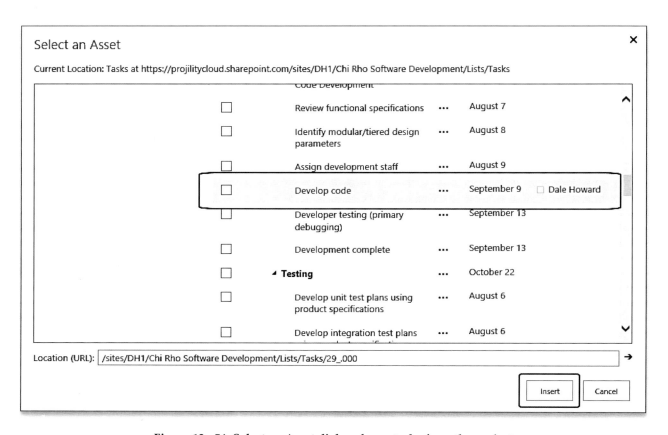

Figure 12 - 54: Select an Asset dialog shows tasks from the project

SharePoint returns to the page where you displayed the selected risk in view-only mode and adds the task in the *Related Items* section at the bottom of the page, such as shown in Figure 12 - 55. To attach the risk to additional tasks, you must repeat the preceding process for each additional task individually because the system ***does not*** allow you to select multiple tasks simultaneously.

414

Working with the Project Site

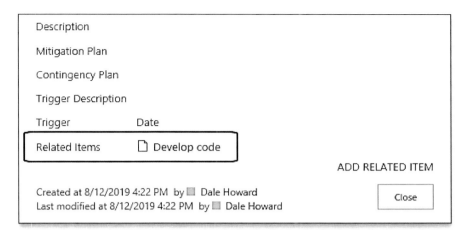

Figure 12 - 55: Risk attached to a task

The steps for attaching an issue to tasks are identical to the steps for attaching a risk to tasks, so I do not repeat the steps for issues. On the other hand, the process for linking documents to tasks is quite a bit different than for risks and issues. To begin the process of attaching a document to a related task, click the **Tasks** link in the *Quick Launch* menu. SharePoint displays the *Tasks* page, such as the one shown in Figure 12 - 56.

Figure 12 - 56: Tasks page in the Project Site

415

Module 12

Click the name of the task to which you want to attach the document. SharePoint displays the *Tasks* page for the selected task. At the bottom of the page, click the **SHOW MORE** link, such as shown in Figure 12 - 57.

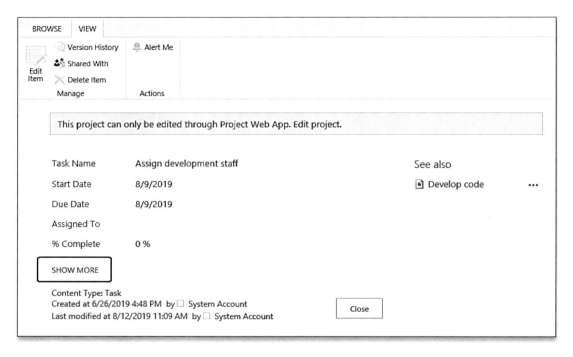

Figure 12 - 57: Click the SHOW MORE link

SharePoint expands the *Tasks* page to show additional fields for the selected task. In the *Related Items* section of the page, click the **ADD RELATED ITEM** link, such as shown in Figure 12 - 58.

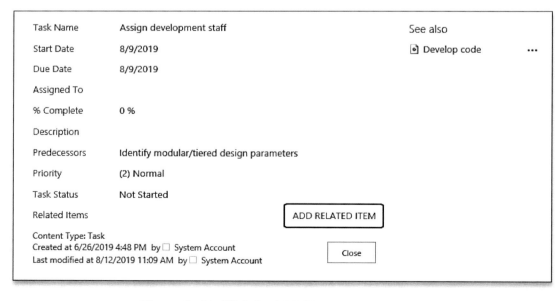

Figure 12 - 58: Click the ADD RELATED ITEM link

Working with the Project Site

The system displays the *Select an Asset* dialog for the selected project. In the *Select an Asset* dialog, select the **Documents** item, and then click the ➔ button (the right-arrow button) in the lower-right corner of the dialog to continue, such as shown in Figure 12 - 59.

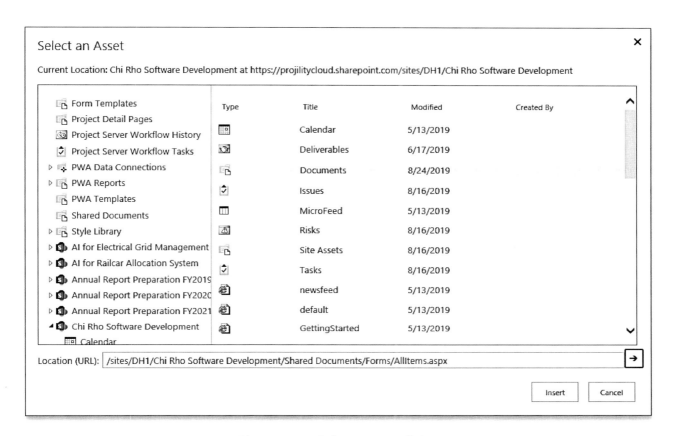

Figure 12 - 59: Select an Asset dialog

 Warning: **Do not** click the *Insert* button in the *Select an Asset* dialog. You **must** click the ➔ button and **not** the *Insert* button to continue linking the task to an associated document.

SharePoint updates the *Select an Asset* dialog to display the list of all documents currently in the *Documents* library. In the dialog, select a document and then click the **Insert** button, such as shown in Figure 12 - 60.

417

Module 12

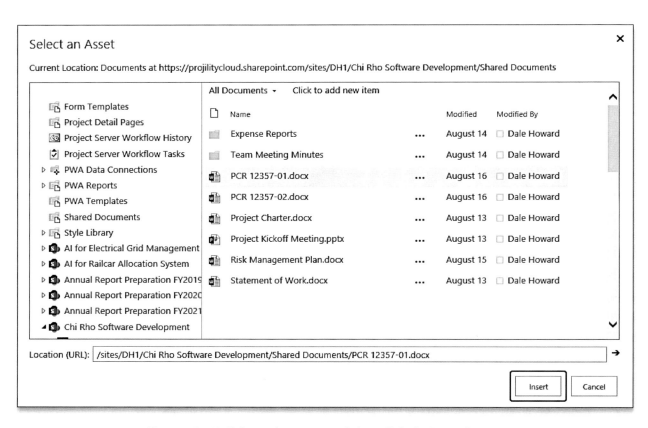

Figure 12 - 60: Select a document and then click the Insert button

SharePoint attaches the document to the task and then displays the *Tasks* page for the selected task with the attached document shown in the *Related Items* section of the page, such as shown in Figure 12 - 61. To attach the task to additional documents, repeat the preceding process for each additional document individually.

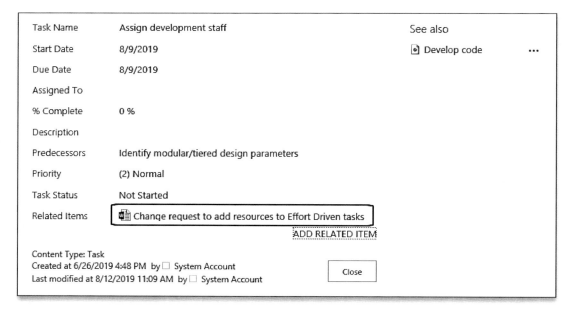

Figure 12 - 61: Document attached to the task

Working with the Project Site

 Warning: You can only attach risks, issues, and documents to tasks if your Project Online application administrator enabled the *Sync SharePoint Tasks Lists* option for the Enterprise Project Types (EPTs) that you use to create your enterprise projects. Once enabled, this option causes Project Online to automatically update the *Tasks* list in the Project Site every time you publish your enterprise project. If this option is disabled (the default setting), you cannot attach risks and issues to tasks in your project.

Using Other Features in the Project Site

Beyond risks, issues, documents, and deliverables, the Project Site does offer several additional features that you may find useful with Project Online. These features include the *OneNote* notebook page, the *Tasks* page, and the *Newsfeed* web part.

Using OneNote Notebooks

A new feature introduced in recent years to the Project Site allows you to use a shared OneNote notebook directly within the Project Site. To create a new OneNote notebook, click the **Notebook** link in the *Quick Launch* menu of the Project Site. SharePoint creates a new OneNote notebook in OneNote Online, saves the notebook, and then displays the shared *OneNote* notebook page, such as the one shown in Figure 12 - 62.

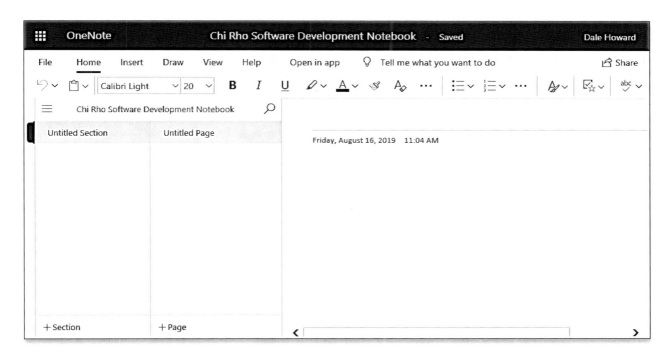

Figure 12 - 62: OneNote notebook page

The *OneNote* notebook page includes multiple ribbons at the top of the page that you can use for working with the shared notebook. To rename the notebook section, right-click on the **Untitled Section** tab and select the **Rename Section** item on the shortcut menu. OneNote displays the *Section Name* dialog. In the *Section Name* dialog, enter a new name for the section, such as shown in Figure 12 - 63, and then click the **OK** button.

419

Module 12

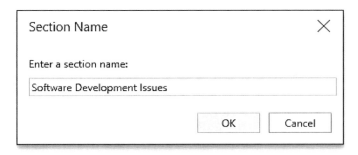

Figure 12 - 63: Section Name dialog

To enter a title for the unnamed page, simply type a title at the top of the shared document on the right side of the notebook. To create a new section, click the **+ Section** link in the lower left corner of the notebook. OneNote displays a *Section Name* dialog, such as the one shown previously in Figure 12 - 63. Enter a name for the new section in the *Section Name* dialog and then click the **OK** button.

To create a new page in the selected section, click the **+ Page** link in the lower left corner of the notebook. OneNote automatically creates a new unnamed page in the notebook. To enter a title for the unnamed page, simply type a title at the top of the shared document on the right side of the OneNote notebook.

Using the Tasks Page

To display the *Tasks* page in the Project Site, click the **Tasks** link in the *Quick Launch* menu. SharePoint displays the *Tasks* page, similar to the one shown in Figure 12 - 64. The *Tasks* page displays a list of every task in the project schedule, along with the *Due Date* of each task (the current *Finish* date of the task), and the names of the resources assigned to each task. The *Tasks* page indicates completed tasks by displaying a checkmark in the checkbox to the left of the task name, and by applying the strikethrough font to the task name.

BROWSE	TASKS	LIST			
✓		~~Incorporate feedback into functional specifications~~	...	July 29	
✓		~~Obtain approval to proceed~~	...	July 31	
✓		~~Design complete~~	...	July 31	
☐		▲ **Development**	...	September 6	
☐		Request Contractor Help for Code Development	...	August 6	☐ Dale Howard
☐		Review functional specifications	...	August 7	☐ Grace Adeyemi
☐		Identify modular/tiered design parameters	...	August 8	☐ Grace Adeyemi
☐		Assign development staff	...	August 9	☐ Grace Adeyemi
☐		Develop code	...	Friday	☐ Todd Meier
☐		Developer testing (primary debugging)	...	September 6	☐ Erica Eltringham

Figure 12 - 64: Tasks page in the Project Site

Working with the Project Site

Information: For the purpose of brevity, Figure 12 - 64 shows only a portion of the *Tasks* page for the sample project. At the top of the *Tasks* page, the figure does not show that SharePoint displays the *Timeline* view from the enterprise project if you created a *Timeline* view in Microsoft Project.

Warning: The *Tasks* page only displays project tasks if your Project Online application administrator enabled the *Sync SharePoint Tasks Lists* option for the Enterprise Project Types (EPTs) that you use to create your enterprise projects. Once enabled, this option causes Project Online to automatically update the *Tasks* list in the Project Site every time you publish your enterprise project. If this option is disabled (the default setting), the *Tasks* page is blank and displays no tasks from the associated enterprise project.

Unlike the *Tasks* page in Project Web App, in which team members can enter task progress, the primary purpose of the *Tasks* page in the Project Site is to provide the underlying functionality for attaching risks, issues, and documents to an associated task. Team members cannot submit task progress updates using the *Tasks* page in the Project Site. In fact, if team members attempt to enter task progress using the *Tasks* page in the Project Site, SharePoint warns them that the project can only be edited using Project Web App, a feature of Project Online only available to project managers.

The *Tasks* page includes seven built-in views that users can apply to the page. To see the default views available for the *Tasks* page, click the **List** tab at the top of the page to expand the *List* ribbon. In the *Manage Views* section of the *List* ribbon, click the **Current View** pick list, such as shown in Figure 12 - 65. Notice that the *All Tasks* view is the default view for the *Tasks* page. As the name implies, this view displays all unstarted, in-progress, and completed tasks in the project schedule.

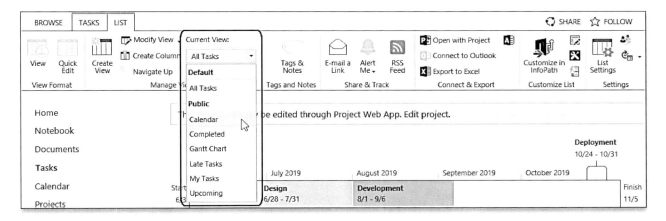

Figure 12 - 65: Current View pick list in the List ribbon

Select the **Calendar** item in the *Current View* pick list to display the *Calendar* view, such as shown in Figure 12 - 66. You can also display the *Calendar* view by clicking the **Calendar** link in the *Quick Launch* menu. To scroll from month to month, select a month in the small calendar displayed in the upper left corner of the *Tasks* page, or click the **Next Month** and **Previous Month** arrow buttons at the top of the *Tasks* page.

Module 12

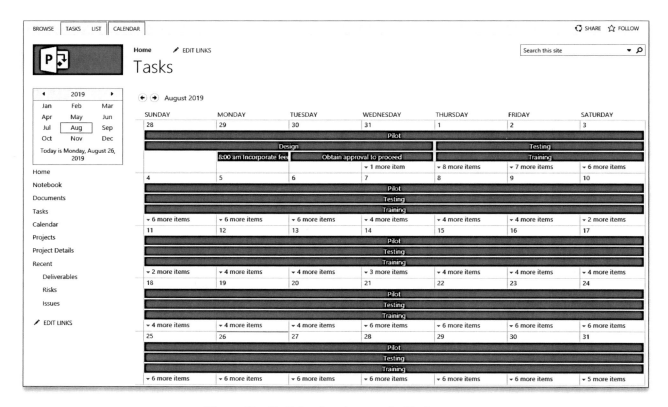

Figure 12 - 66: Calendar view on the Tasks page

Select the **Completed** item in the *Current View* pick list to display the *Completed* view. As the name implies, the *Completed* view displays only completed tasks.

Select the **Gantt Chart** item in the *Current View* pick list to display the *Gantt Chart* view. As you might expect, the *Gantt Chart* view displays all tasks in the project schedule, and includes a Gantt Chart pane on the right side of the view.

Select the **Late Tasks** item in the *Current View* pick list to display the *Late Tasks* view. As the name implies, the *Late Tasks* view displays only late tasks. In both Microsoft Project and Project Online, a late task is any task whose current *% Complete* value is less than the expected *% Complete* value, as defined by the *Status* date in the project.

Select the **My Tasks** item in the *Current View* pick list to display the *My Tasks* view. As the name implies, the *My Tasks* view displays only the tasks assigned to the current user logged into the Project Site.

Select the **Upcoming Tasks** item in the *Current View* pick list to display the *Upcoming Tasks* view. This view displays all unstarted and in-progress tasks with a *Due Date* value greater than or equal to the current date.

Using the Newsfeed Web Part

You find the *Newsfeed* web part at the bottom of the *Home* page of the Project Site. You can use this feature to create and participate in discussions with your project team members. Notice in the *Newsfeed* web part shown in Figure 12 - 67 that I initiated a conversation with my project team members about moving our weekly project status meetings from Mondays to Tuesdays instead. There are currently no replies to my discussion.

Working with the Project Site

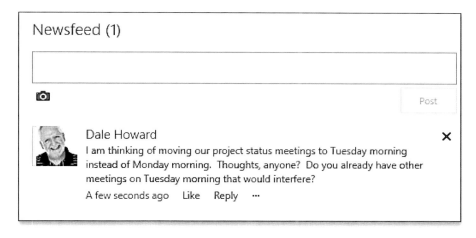

Figure 12 - 67

To create a new discussion, enter text in the **Newsfeed** field and then click the **Post** button. Others can reply to the discussion by clicking the **Reply** link or they can like a comment by clicking the **Like** link. To lock a closed discussion, click the **More Options** link (the ... link) and select the **Lock conversation** item on the menu. To completely delete a discussion, click the **Delete this conversation** button (the X button) in the upper right corner of the discussion you want to delete.

423

Module 12

Module 13

Working in the Project Center

Learning Objectives

After completing this module, you will be able to:

- Understand the Project Center page in Project Web App
- Apply Project Center views
- Open a project from the Project Center
- Apply detailed Project views for a project opened in the Project Center

Inside Module 13

Using the Project Center **427**
Using the Projects Ribbon in the Project Center *428*
Using Project Center Views *428*
Understanding Show/Hide Options *430*
Understanding Indicators *431*
Using the Open Menu Button *432*

Working With Projects in the Project Center **434**
Opening Projects in Microsoft Project from the Project Center *434*
Creating a Master Project from the Project Center *435*
Checking In a Project from the Project Center *436*
Setting Custom Permissions for a Project *438*
Working with Detailed Project Views *441*

Editing Projects in Project Web App **446**
Creating a New Project in Project Web App *446*
Understanding the PWA Task Planning Process *449*
Understanding the PWA Resource Planning Process *451*
Understanding the PWA Assignment Planning Process *452*
Finalizing Your PWA Project *453*
Closing and Reopening a PWA Project for Editing *454*

425

Using the Project Center

In Project Online, the **Project Center** is the central location for project and portfolio information, a launching point for new projects, and the gateway to editing projects on the web. To navigate to the *Project Center* page, click the **Projects** link in the *Quick Launch* menu or click the **Projects** tile in the *Track your work* carousel.

Figure 13 - 1 shows the *Project Center* page with the default *Summary* view applied. The *Project Center* page displays a timeline view at the top of the page, along with a data grid containing a project list on the left and a *Gantt Chart* view on the right. The project list displays a single line of information about each project, with multiple columns of information about each item. The *Gantt Chart* displays either one or two Gantt bars representing the life span of the project. When the system displays two Gantt bars, such as in the *Tracking* view (not shown), one Gantt bar represents the baseline schedule while the other represents the current schedule of the project.

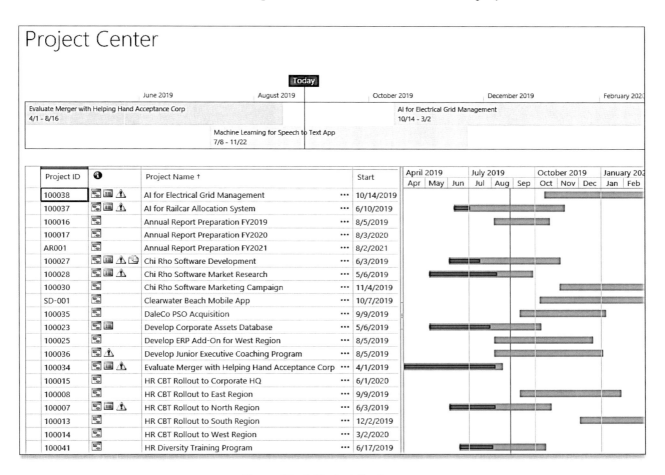

Figure 13 - 1: Project Center page

The *Project Center* page allows you to do each of the following:

- View the portfolio of active and proposed enterprise projects.

- Navigate to the Project Detail Pages (PDPs) or the Project Site for a project, or navigate directly to the *Risks*, *Issues*, or *Documents* page for the project.

- Open a single project or a group of projects in Microsoft Project.

- Check in a project stuck in a checked-out state.
- Set special permissions for a project.
- Drill down to a detailed *Project* view to view the project.
- Create a new project and manage it entirely within Project Web App.

Using the Projects Ribbon in the Project Center

In the upper left corner of the *Project Center* page, click the **Projects** tab to expand the *Projects* ribbon. Figure 13 - 2 shows the *Projects* ribbon for a user with project manager permissions.

Figure 13 - 2: Projects ribbon

The *Projects* ribbon contains eight ribbon sections: the *Project, Navigate, Zoom, Data, Timeline, Share, Show/Hide*, and *Project Type* sections. Use the commands in each ribbon section to do the following:

- In the **Project** section, you can create a new project from an Enterprise Project Type (EPT), open an existing project in Microsoft Project or in Project Web App, or create a new project from an existing *Tasks* list in a SharePoint site.

- In the **Navigate** section you can build a team in a project, set special security permissions for a project, and check in a project that is "stuck" in a Checked Out state.

- In the **Zoom** section, you can zoom the timescale of the *Gantt Chart* pane and scroll to a specific project in the *Gantt Chart* pane.

- In the **Data** section you can collapse outline levels in views that contain multi-level grouping, show rollup values when you apply grouping on specific fields, display a *Gantt Chart* pane on the right side of the view, select and apply views, and apply ad-hoc filters to the project list, and apply grouping to your views.

- In the **Timeline** section, you can add projects or individual project tasks to the *Timeline* view at the top of the page.

- In the **Share** section, you export the current view to Microsoft Excel or to send it to a printer.

- In the **Show/Hide** section, you choose whether to display subprojects along with their associated master projects, and you can specify whether *Project Center* views display the time along with the date in date fields.

- In the **Project Type** section, you can change the Enterprise Project Type (EPT) of the selected project.

Using Project Center Views

Using the default permissions in Project Web App, the *Project Center* page displays only the projects you have permission to see, based on your membership in the *Project Managers* security group. Unless your Project Online application administrator changes the default permissions for the *Project Managers* group, the only projects you have permission to see and edit are your own projects.

The first time you access the *Project Center* page, the system displays the default *Summary* view. The system provides five standard views that you can select from the *View* pick list in the *Data* section of the *Projects* ribbon. These default views include the *Summary, Tracking, Cost, Earned Value,* and *Work* views. The common columns included in each default view are the *ID* column, the *Indicators* column, and the *Task Name* column. The *Indicators* column displays indicators that are links to the risks, issues, or documents created in the Project Site for each project.

The **Summary** view, shown previously in Figure 13 - 1, displays the "vital statistics" for each project using the following fields: *Start* date, *Finish* date, *% Complete, Work, Duration, Owner,* and the *Last Published* date. The *Summary* view includes a *Gantt Chart* pane that displays a single Gantt bar for each project and a black stripe in each Gantt bar indicating project progress.

The **Tracking** view shown in Figure 13 - 3 displays variance information about each project using the following fields: *% Complete, Actual Cost, Actual Duration, Actual Finish, Actual Start, Actual Work, Baseline Finish, Baseline Start, Duration, Remaining Duration, Finish* date, and *Start* date. The *Tracking* view includes a *Tracking Gantt* pane with two Gantt bars for each project. The blue Gantt bar represents the current schedule for each project, while the gray Gantt bar represents the baseline schedule for each project. The dark blue section of each Gantt bar represents project progress.

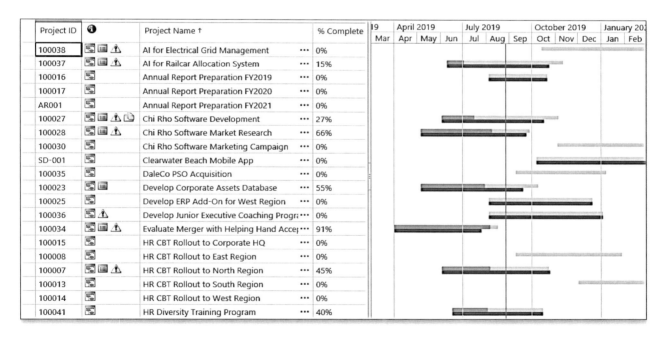

Figure 13 - 3: Tracking view

 Information: If you do not see a gray Gantt bar for a project in the *Tracking* view, this indicates that you did not save a baseline for the project before you published it.

The **Cost** view displays information about project costs using the following fields: *Finish* date, *Start* date, *Cost, Baseline Cost, Actual Cost, Fixed Cost, Cost Variance,* and *Remaining Cost*. The *Cost* view includes a *Tracking Gantt* pane identical to the one shown in the *Tracking* view.

The **Earned Value** view displays the calculated earned value data at the project level for each project. This view includes columns for *Finish* date, *Start* date, *Cost*, *Baseline Cost*, *BCWP*, *BCWS*, *SV*, *CV*, *ACWP*, and *VAC*. The *Earned Value* view includes the same *Tracking Gantt* pane shown in the *Tracking* and *Cost* views.

The **Work** view displays information about project work hours with columns for *% Work Complete*, *Finish* date, *Remaining Work*, *Start* date, *Work*, *Baseline Work*, *Actual Work*, and *Work Variance*. The *Work* view includes the same *Tracking Gantt* pane shown in the *Tracking*, *Cost*, and *Earned Value* views.

In addition to the five standard views, the *View* pick list may also include custom views created by your Project Online application administrator. The *Project Center* is a great forum for the use of graphical indicators in custom views. Figure 13 - 4 shows the data grid in the *Projects by Department* custom view. Notice that four custom fields display the overall status of the project using graphical indicators.

		Project Name ↑		Project Health	Cost Health	Duration Health	Work Health	Start	Finish
		▲Department: Finance							
		Annual Report Preparation FY2019	...	○	○	○	○	8/5/2019	10/21/2019
		Annual Report Preparation FY2020	...	○	?	?	?	8/3/2020	10/26/2020
		Annual Report Preparation FY2021	...		?	?	?	8/2/2021	9/30/2021
		▲Department: HR							
	⚠	Develop Junior Executive Coaching Program	...	○	○	○	○	8/5/2019	1/3/2020
		HR CBT Rollout to Corporate HQ	...	○	?	?	?	6/1/2020	10/22/2020
		HR CBT Rollout to East Region	...	○	?	?	?	9/9/2019	1/28/2020
	⚠	HR CBT Rollout to North Region	...	○	○	☺	○	6/3/2019	10/23/2019
		HR CBT Rollout to South Region	...	○	?	?	?	12/2/2019	4/23/2020
		HR CBT Rollout to West Region	...	○	?	?	?	3/2/2020	7/23/2020
		HR Diversity Training Program	...	○	○	○	○	6/17/2019	10/15/2019
		▲Department: IT							
	⚠	AI for Electrical Grid Management	...	○	?	?	?	10/14/2019	3/2/2020
	⚠	AI for Railcar Allocation System	...	○	○	○	○	6/10/2019	11/11/2019
	⚠	Chi Rho Software Development	...	○	○	○	○	6/3/2019	11/5/2019
		Clearwater Beach Mobile App	...	○	○	○	○	10/7/2019	3/9/2020
		Develop Corporate Assets Database	...	○	○	○	○	5/6/2019	10/9/2019
		Develop ERP Add-On for West Region	...	○	○	○	○	8/5/2019	12/20/2019
		Machine Learning for Speech to Text App	...	○	○	○	○	7/8/2019	11/22/2019

Figure 13 - 4: Projects by Department custom view

Understanding Show/Hide Options

The *Show/Hide* section of the *Projects* ribbon, shown in Figure 13 - 5, contains two checkbox options: *Subprojects* and *Time with Date*. The *Subprojects* option deals exclusively with master projects. If your organization allows project managers to create and publish master projects, the *Subprojects* option determines whether the *Project Center* page displays the subprojects of each published master project. When deselected, the default setting for this option, the *Project Center* displays each master project, but ***does not*** display any subprojects. When selected, the *Project Center* displays master projects and the subprojects of every master project as well.

When you select the checkbox for *Time with Date*, the system redisplays the *Project Center* page with both the date and time in every date field, such as the *Start* and *Finish* fields, for example. Select the options that best fit your needs for any work session.

Working in the Project Center

Figure 13 - 5: Show/Hide options

Understanding Indicators

To the left of the *Project Name* column, notice the indicators displayed in the *Indicators* column, as shown in Figure 13 - 6. These indicators reveal important information about each project, including the type of project and whether the project has risks, issues, or documents associated with it.

Figure 13 - 6: Project Center page shows indicators in the Indicators column

You can quickly see the meaning of each indicator by examining the tooltip displayed when you float your mouse over any icon in the *Indicators* column. Table 13 - 1 displays the indicators you may see in the *Indicators* column.

Indicator	Meaning
	Enterprise project
	Master Project
	Risks
	Issues
	Documents

Table 13 - 1: Project Center indicators

431

To access the Project Site for any project, click the **Risks**, **Issues**, or **Documents** indicator in the *Indicators* column. When you click the *Risks* icon, the system navigates you to the *Risks* page in the Project Site for the selected project, and when you click the *Issues* icon, the system navigates you to the *Issues* page. However, when you click the *Documents* icon, the system does not navigate you to the *Documents* library. Instead, the system navigates you to the *Home* page of the Project Site for the selected project.

Using the Open Menu Button

To the right of each project name, the *Project Center* page displays the **Open Menu** button (the **...** button). When you click the *Open Menu* button, Project Web App displays a dialog with high-level information about the project, such as for the project shown in Figure 13 - 7.

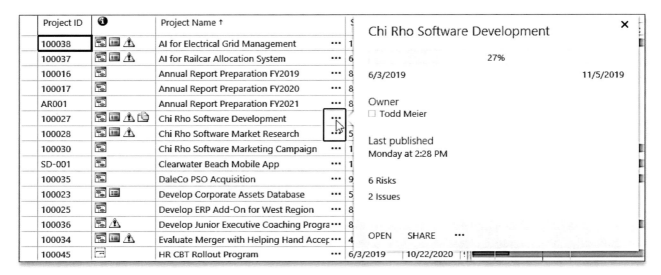

Figure 13 - 7: Click the Open Menu button

The dialog displays the current *Start* date and calculated *Finish* date of the project, the project's current *% Complete* value, the *Owner* of the project, the *Last Published* date for the project, and the current number of risks and issues associated with the project. If you click the link for either the number of risks or number of issues, the system navigates you to the appropriate page in the Project Site for the selected project.

The bottom of the dialog displays the *Open*, *Share*, and *More options* links. When you click the *Open* link, the system opens the project in Read-Only mode in Project Web App. When you click the *Share* link, the system navigates you to the *Project Permissions* page where you can set special permissions so that other users can access the project. When you click the *More options* link, PWA displays a menu with options to delete the project, build a team for the project, or to navigate to the Project Site for the project.

 Warning: Projility strongly recommends that you *never* delete an enterprise project, even though Project Web App may allow you to do so. Instead, if you erroneously create an enterprise project that you do not need, contact your Project Online application administrator, who can then delete the project for you. Because there is no "Undo" action available after deleting an enterprise project, it is better to allow the application administrator to handle project deletion.

Working in the Project Center

 Hands On Exercise

Exercise 13 - 1

Explore the *Project Center* page.

1. Return to the *Home* page of Project Web App from the Project Site for your class project, if necessary.
2. Click the **Projects** link in the *Quick Launch* menu to navigate to the *Project Center* page.
3. In the *Project Center* data grid, examine the *Gantt Chart* pane on the right side of the data grid.
4. In the data grid, drag the split bar to the right edge and examine the columns displayed in the default *Summary* view.
5. Drag the split bar back to the left and dock it on the right edge of the *Finish* column.
6. At the top of the *Project Center* page, click the **Projects** tab to expand the *Projects* ribbon.
7. Examine the groups in the ribbon and the commands in each ribbon group.
8. In the *Data* section of the *Projects* ribbon, click the **View** pick list button and select the **Tracking** view.
9. Examine the *Tracking Gantt* pane on the right side of the data grid.
10. In the data grid, drag the split bar to the right edge and examine the columns displayed in the *Tracking* view.
11. Drag the split bar back to the left and dock it on the right edge of the *% Complete* column.
12. In the *Data* section of the *Projects* ribbon, click the **View** pick list and reselect the **Summary** view.
13. In the *Data* section of the *Projects* ribbon, click the **Group By** pick list and select the **Owner** item.

Notice how Project Web App groups the projects in the *Project Center* by the *Owner* (project manager) of each project.

14. In the *Data* section of the *Projects* ribbon, click the **Group By** pick list and select the **No Group** item.
15. In the *Indicators* column, float your mouse pointer over the *Risks, Issues,* and *Documents* indicators for your class project and view the tooltips the system displays.
16. Click the **Open Menu** button (the **...** button) to the right of the name of your class project and then click the **1 Risk** link in the dialog.

Notice how Project Web App navigated you to the **Risks** page of the Project Site for your class project.

17. In the *Quick Launch* menu of your Project Site, click the **Projects** link to return to the *Project Center* page.

433

Module 13

Working With Projects in the Project Center

The *Project Center* page offers you a number of options for working with projects. These options include opening projects in either Microsoft Project or Project Web App, creating a master project in Microsoft Project from a selected group of projects, checking in a project "stuck" in a Checked Out state, setting special permissions for a project so that other users can access it, and applying detailed *Project* views for a project.

Opening Projects in Microsoft Project from the Project Center

The *Project Center* page offers you four methods for opening projects. You can open the project in Project Web App in either Read-Only or Read/Write mode, or you can open the project in Microsoft Project, again in either Read-Only or Read/Write mode.

To begin the process of opening a project from the *Project Center* page, click the header row (white box) at the left end of the row for the project you want to open. In the *Project* section of the *Projects* ribbon, click the **Open** pick list. The system displays the pick list show in Figure 13 - 8.

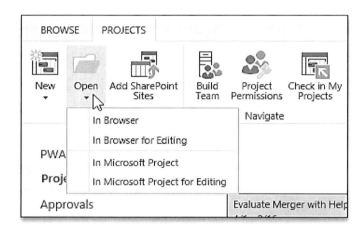

Figure 13 - 8: Open pick list

Notice that the *Open* pick list menu offers you the four options I mentioned previously. If you select the *In Browser* item in the menu, the system opens the project Read-Only in Project Web App. If you select the *In Browser for Editing* item, the system opens the project in Read/Write mode in PWA.

If you select the *In Microsoft Project* item, the system opens the project in Read-Only mode in Microsoft Project. If you select the *In Microsoft Project for Editing* item, the system opens the project in Read/Write mode in Microsoft Project. If you select either of the Microsoft Project items on the *Open* menu and the software is not currently open, the system launches Microsoft Project and connects to Project Online automatically.

 Information: You can quickly open an enterprise project in Microsoft Project in Read/Write mode by clicking the **enterprise project** icon displayed in the *Indicators* column to the left of the name of the project you want to open. If Microsoft Project is not currently open, the system launches Microsoft Project and connects to Project Web App automatically, then it opens and checks out the project for editing.

Creating a Master Project from the Project Center

One of the most useful features of the *Project Center* page gives you the ability to quickly create a **master project** in Microsoft Project. A master project, by the way, is a single project consisting of other projects known as **subprojects**. Master projects are useful for portfolio and program analysis, and you can use a master project to set cross-project dependencies by linking tasks in one subproject to tasks in another subproject.

To begin the process of creating a master project from the *Project Center* page, use either the **Control** key or **Shift** key on your computer keyboard to select the row headers (white boxes) for two or more projects displayed in the data grid. For example, notice in Figure 13 - 9 that I selected the row headers for the five projects that constitute the *HR CBT Rollout* program.

Project ID	ⓘ	Project Name ↑	Start
100036		Develop Junior Executive Coaching Progra ···	8/5/2019
100034		Evaluate Merger with Helping Hand Accep ···	4/1/2019
100015		HR CBT Rollout to Corporate HQ ···	6/1/2020
100008		HR CBT Rollout to East Region ···	9/9/2019
100007		HR CBT Rollout to North Region ···	6/3/2019
100013		HR CBT Rollout to South Region ···	12/2/2019
100014		HR CBT Rollout to West Region ···	3/2/2020
100041		HR Diversity Training Program ···	6/17/2019
100024		Machine Learning for Speech to Text App ···	7/8/2019
100022		Network Upgrade for Corporate HQ ···	10/5/2020

Figure 13 - 9: Five projects selected

In the *Project* section of the *Projects* ribbon, click the **Open** pick list. To insert the subprojects in Read-Only mode, select the In **Microsoft Project** item on the pick list. To insert the subprojects in Read/Write mode, click the **In Microsoft Project for Editing** item on the menu. Figure 13 - 10 shows a master project consisting of five subprojects inserted in a master project that represents the *HR CBT Rollout* program.

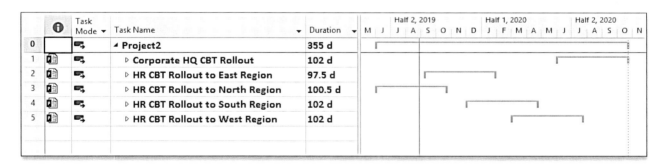

Figure 13 - 10: Master project includes five subprojects

After you create the master project, expand each of the subprojects by clicking the **Expand/Collapse** indicator (the ▶ symbol) to the left of each project name. To limit the amount of scrolling needed to expand each subproject, expand the last subproject first, and then work your way from the bottom to the top of the project list. Edit each subproject as per your requirements, including setting cross-project dependencies between the various subprojects.

435

When you close the master project, Microsoft Project displays a confirmation dialog that gives you the option of saving it. If you elect to save the project, you must enter a name for the project, and then both save and publish the master project. In you inserted the subprojects in Read/Write mode, the system prompts you to save the subprojects as well.

If you set cross-project dependencies between the subprojects in the master project, and your organization does not allow project managers to save master projects in Project Online, then simply close the master project *without saving it*. Microsoft Project displays a confirmation dialog about saving the subprojects in the master project, such as shown in Figure 13 - 11. In the confirmation dialog, click the **Yes to All** button to save the changes in each of the subprojects.

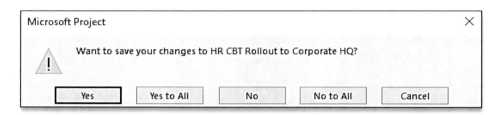

Figure 13 - 11: Confirmation dialog to save the subprojects

 Information: Remember that when you specify cross-project dependencies, Microsoft Project sets the dependencies *in each subproject* and not in the master project itself. This is why you do not need to save the master project, even when you set cross-project dependencies using the master project.

 Warning: You can only set cross-project dependencies in those projects for which you have Read/Write access. You normally have Read/Write access to your own projects, but not to projects owned by other project managers. Depending on your security permissions in Project Web App, you may have Read-Only access to projects owned by other project managers, or no access at all to their projects.

If you save and publish a master project in Project Online, the master project appears in the *Project Center* page with an indicator that shows it is a master project. Remember to select the *Subprojects* checkbox in the *Show/Hide* section of the *Projects* ribbon so that you can also see the subprojects as well as the master project in the *Project Center* page.

Checking In a Project from the Project Center

Occasionally an enterprise project can become "stuck" in a checked out state in the Project Online database. This can occur because of a glitch during the check in process or because of a workstation crash. Regardless of the cause, the project remains in a checked out state, even though it should be in a checked in state. The system does not warn you directly about this situation, however, but you can infer that a project is "stuck" in a checked out state the next time you attempt to open the project in Microsoft Project. If you attempt to open and check out a project that is "stuck" in a checked out state, Microsoft Project displays a warning dialog, such as the one shown in Figure 13 - 12.

Working in the Project Center

Figure 13 - 12: Warning dialog – project is checked out

If you see a warning dialog such as shown previously in Figure 13 - 12, this means that you need to perform a force check-in procedure from the *Project Center* page to manually check in the enterprise project. To force check-in an enterprise project that is "stuck" in a checked out state, click the **Project** tab to expand the *Projects* ribbon at the top of the *Project Center* page. In the *Navigate* section of the *Projects* ribbon, click the **Check In My Projects** button, such as shown in Figure 13 - 13.

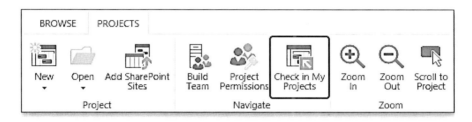

Figure 13 - 13: Check In My Projects button

Project Web App displays the *Force Check-in Enterprise Projects* page shown in Figure 13 - 14. This page displays all projects currently checked out to you, regardless of whether you checked them out intentionally or they are "stuck" in a checked out state.

Figure 13 - 14: Force Check-in Enterprise Projects page

To perform a force check-in of the enterprise project, select the checkbox to the left of the project name, and then click the **Check In** button. Project Web App displays the confirmation dialog shown in Figure 13 - 15. In the confirmation dialog, click the **OK** button to check in the enterprise project. Project Web App refreshes the *Force Check-in Enterprise Projects* page and removes the enterprise project from the data grid.

437

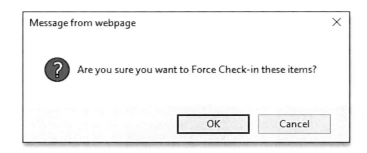

Figure 13 - 15: Confirmation dialog

Setting Custom Permissions for a Project

Project Web App allows project owners to set custom permissions for an enterprise project so that other users can access it. If your organization uses the default security permissions in Project Online, remember that the system does not allow other project managers to access the projects owned by you. This is because the default permissions allow project managers to access only the projects they own, but do not allow them to access projects owned by any other project manager.

Suppose you are leaving on two weeks of PTO and you need a fellow project manager to manage a project for you in your absence. The default permissions in Project Online do not allow your fellow project manager to access your project. In a situation such as this, you can use the *Project Center* page to grant special permissions for your enterprise project so that your fellow project manager can access, open, edit, save, and publish your enterprise project while you are on PTO. When you return from PTO, you can revoke these special permissions, if necessary.

To grant special permissions for an enterprise project, click the row header (white box) to the left of the project name in the *Project Center* data grid. In the *Navigate* section of the *Projects* ribbon, click the **Project Permissions** button shown in Figure 13 - 16.

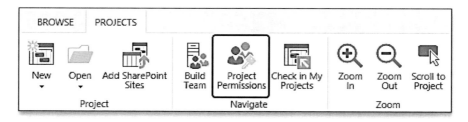

Figure 13 - 16: Project Permissions button

Project Web App displays the *Project Permissions* page for the selected project, such as shown in Figure 13 - 17.

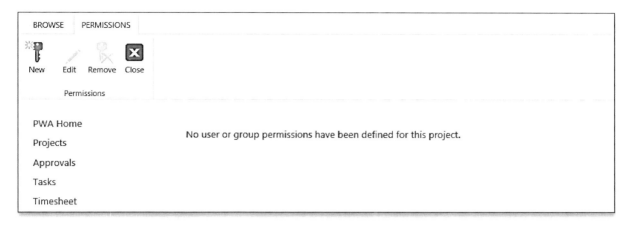

Figure 13 - 17: Project Permissions page

To begin the process of setting special permissions for the selected project, click the **New** button in the *Permissions* section of the *Permissions* ribbon. PWA displays the *Edit Project Permissions* page shown in Figure 13 - 18.

In the *Users and Groups* section of the page, you can specify permissions for entire security groups in as well as for individual users. Select an individual user or a security group from the *Available Users and Groups* list on the left and click the **Add** button to add them to the *Users and Groups with Permissions* list on the right.

 Information: If you add the *Project Managers* group to the *Users and Groups with Permissions* list, you can grant special permissions for your project in a single action to every project manager who is a member of the *Project Managers* group.

Next, select the permissions that you want to grant to the selected user(s) by selecting one or more checkbox options in the *Permissions* section of the page. Notice that you can grant the following special permissions:

- **Open the project within Project Professional or Project Web App** – This permission allows the user to open the project in Read-Only mode.

- **Edit and Save the project within Project Professional or Project Web App** – This permission allows the user to check out the project in Read/Write mode and save changes to it.

- **Edit Project Summary Fields within Project Professional or Project Web App** – This permission allows the user to specify values in the default and custom *Project* fields for the project.

- **Publish the project within Project Professional or Project Web App** – This permission allows the user to publish changes made to the project.

- **View the Project Summary in the Project Center** – This permission makes the project visible to the user in the *Project Center* page.

- **View the Project Schedule Details in Project Web App** – This permission allows the user to click the name of the project in the *Project Center* page and apply detailed *Project* views to the project.

- **View the Project Site** – This permission allows the user to access the Project Site for the project.

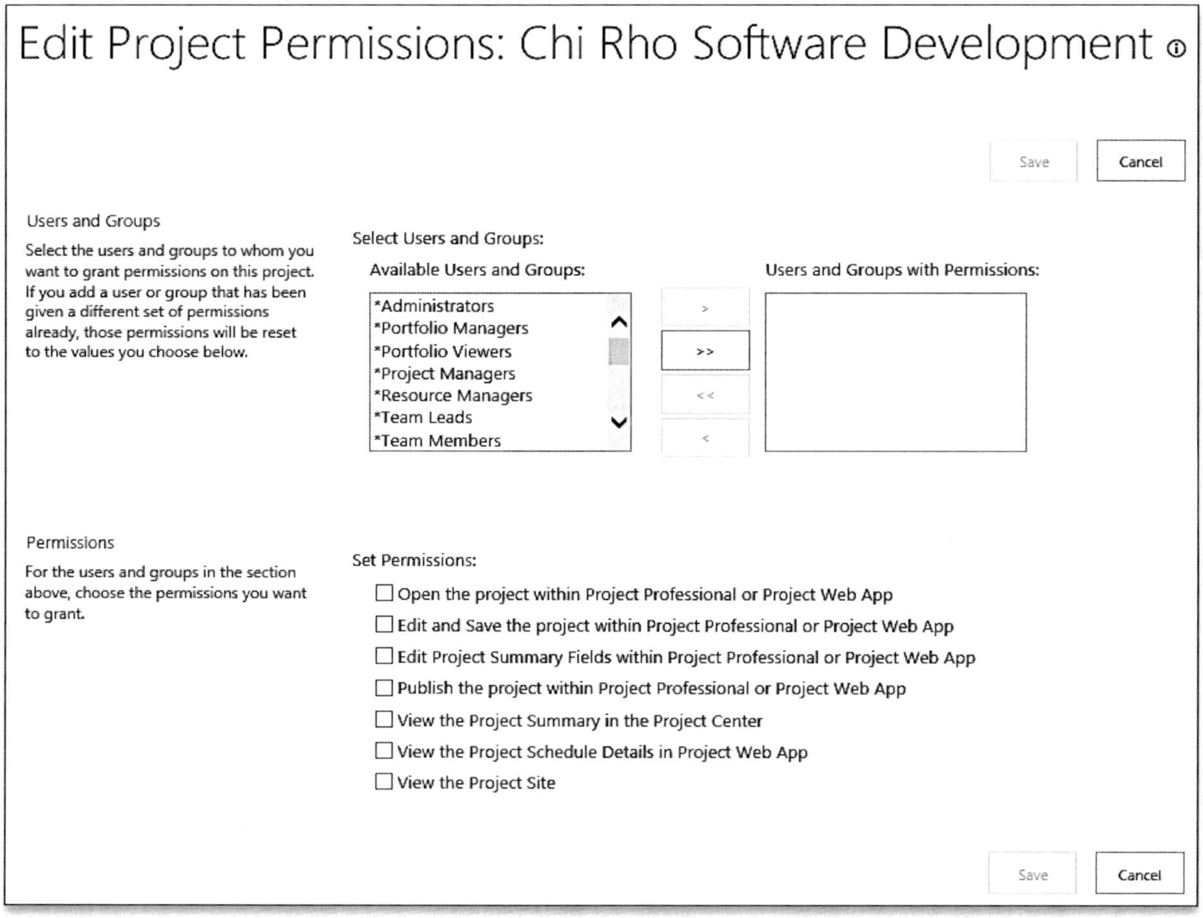

Figure 13 - 18: Edit Project Permissions page for the selected project

Select the checkboxes for the special permissions you want to grant to the selected user(s). In my earlier example, I would need to select at least the following permissions for the project manager who will manage my project while I am on PTO:

- Open the project within Project Professional or Project Web App

- Edit and Save the project within Project Professional or Project Web App

- Publish the project within Project Professional or Project Web App

When finished, click the **Save** button to activate the special permissions you set for your project. The system updates the *Project Permissions* page with the special permissions you specified, such as shown in Figure 13 - 19.

Working in the Project Center

Figure 13 - 19: Special permissions granted to the user

Notice that I gave Grace Adeyemi the three permissions that she needs in order to manage this project while I am away on PTO. To edit a set of special permissions, select the checkbox to the left of the permissions you want to edit and click the **Edit** button. To remove a set of special permission completely, select the checkbox to the left of the permissions you want to remove and click the **Remove** button. Click the **Close** button to return to your previous screen.

 Information: You can also grant special permissions for a project from Microsoft Project. Open the project for which you want to grant special permissions, then click the **File** tab to display the *Backstage*. In the *Permissions* section of the *Info* page, click the **Manage Permissions** button. The system launches your web browser, logs you into Project Web App, and then displays the *Project Permissions* page shown previously in Figure 13 - 17.

Working with Detailed Project Views

Project Web App allows you to display detailed *Project* views from the *Project Center* page for published projects displayed in the data grid. However, before you can apply detailed *Project* views for any project, you need to display the *Schedule* PDP (Project Detail Page) for the project. To display the *Schedule* PDP, click the name of a project in the *Project Center* data grid.

The system displays your last-used Project Detail Page. Notice in Figure 13 - 20 that I used the *Project Details* PDP the last time I accessed this project from the *Project Center* page. Notice also that the *Quick Launch* menu contains a special "drill down" section at the top of the menu which includes links to the Project Detail Pages associated with the Enterprise Project Type (EPT) used to create the project. Depending on how your application administrator configured your Project Web App, you may see a variation of the links shown in the "drill down" section of your organization's *Quick Launch* menu.

 Information: If you do not see a *Schedule* link in the "drill down" section at the top of the *Quick Launch* menu, this means that the *Schedule* PDP is not currently available for you to access during the current stage of the SharePoint workflow process. This also means that you cannot display detailed *Project* views at this time.

441

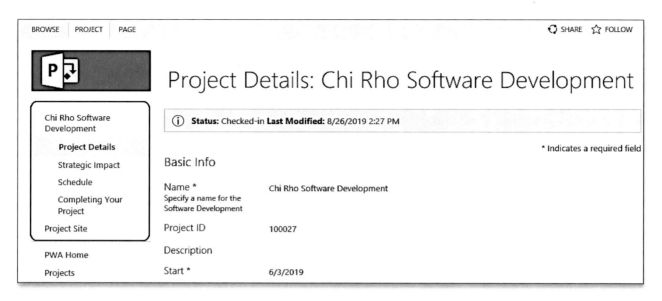

Figure 13 - 20: Project Detail Pages in the Quick Launch menu

In the "drill down" section at the top of the *Quick Launch* menu, click the **Schedule** link. Project Web App displays the *Schedule* Project Detail Page (PDP) for the project, such as shown in Figure 13 - 21. The *Schedule* PDP includes the *Timeline* view at the top of the page and the project schedule at the bottom of the page.

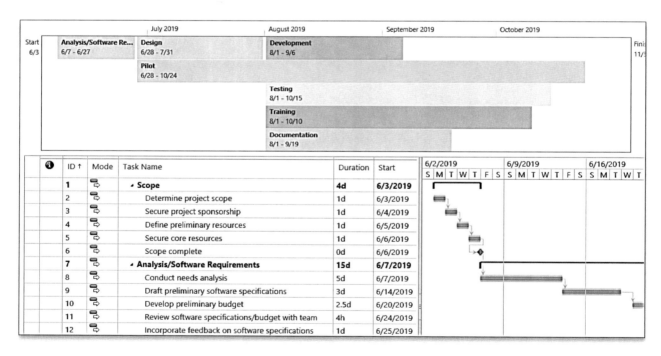

Figure 13 - 21: Schedule PDP

At the top of the *Schedule* PDP, the system displays four ribbon tabs: *Project*, *Page*, *Task*, and *Options*. Click the **Task** tab to expand the *Task* ribbon shown in Figure 13 - 22.

Working in the Project Center

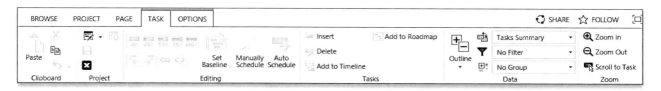

Figure 13 - 22: Task ribbon expanded

In the *Data* section of the *Task* ribbon, click the **View** pick list, such as shown in Figure 13 - 23. The *View* pick list offers 21 default *Project* views, along with any custom *Project* views created by your application administrator. By default, Project Web App initially applies the *Tasks Summary* view the first time you display the *Schedule* PDP.

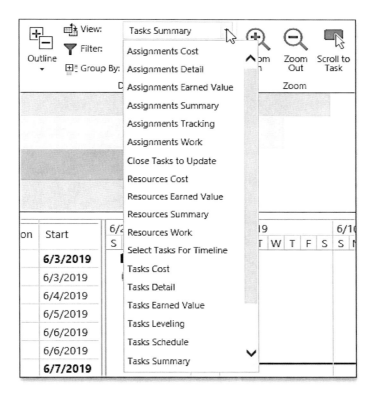

Figure 13 - 23: View pick list

Project Web App offers three types of detailed *Project* views: *Task* views, *Resource* views, and *Assignment* views. As you might surmise, *Task* views display task data and *Resource* views display resource data. *Assignment* views, on the other hand, combine both task and resource data in the same view. Both *Task* views and *Assignment* views include a *Gantt Chart* pane on the right side of the view, while *Resource* views do not include a *Gantt Chart* pane.

 Information: Project Web App remembers which detailed *Project* view you selected the last time you displayed the *Schedule* PDP, and applies that view when you select another project from the *Project Center* page.

Module 13

In the *View* pick list, select the view you want to apply. Table 13 - 2 lists the available views of each type. For the sake of comparison, notice that I list similar types of views on each row of the table. As the names imply, *Task* views display *Task* data, while *Resource* views display *Resource* data. *Assignment* views display a combination of both *Task* data and *Resource* data.

Task Views	Assignment Views	Resource Views
Tasks Cost	Assignments Cost	Resources Cost
Tasks Detail	Assignments Detail	
Tasks Earned Value	Assignments Earned Value	Resources Earned Value
Tasks Leveling		
Tasks Schedule		
Tasks Summary	Assignments Summary	Resources Summary
Tasks Top-Level		
Tasks Tracking	Assignments Tracking	
Tasks Work	Assignments Work	Resources Work
Close Tasks to Update		
Select Tasks For Timeline		

Table 13 - 2: Available detailed Project views

Figure 13 - 24 shows the *Resources Work* view. This view is very similar to the *Resource Sheet* view in Microsoft Project with the *Work* table applied.

Unique ID	Resource Name ↑	% Work Complete	Work	Overtime Work	Baseline Work	Work Variance	Actual Work	Remaining Work	Start
27	Charles Edward	100%	24h	0h	0h	24h	24h	0h	6/24/2019
24	Chuck Kirkpatrick	100%	40h	0h	164h	-124h	40h	0h	6/7/2019
26	Dale Howard	70%	292h	0h	0h	292h	204h	88h	6/14/2019
19	Dan Morton	100%	16h	0h	248h	-232h	16h	0h	8/1/2019
18	David Erickson	0%	0h	0h	144h	-144h	0h	0h	
28	Erica Eltringham	0%	176h	0h		176h	0h	176h	8/5/2019
21	George Stewart	75%	64h	0h	96h	-32h	48h	16h	6/5/2019
29	Grace Adeyemi	0%	96h	0h		96h	0h	96h	8/7/2019
25	Greg Owens	100%	8h	0h	120h	-112h	8h	0h	6/25/2019
23	Larry McKain	0%	80h	0h	96h	-16h	0h	80h	10/16/2019
17	Mike Andrews	8%	208h	0h	256h	-48h	16h	192h	8/1/2019
22	Richard Sanders	100%	60h	0h	52h	8h	60h	0h	6/3/2019
20	Ruth Andrews	7%	168h	0h	168h	0h	12h	156h	8/1/2019
30	Thomas Hopper	0%	16h	0h		16h	0h	16h	8/5/2019
31	Todd Meier	0%	216h	0h		216h	0h	216h	8/12/2019

Figure 13 - 24: Resources Work view

Working in the Project Center

Figure 13 - 25 shows the *Assignments Summary* view. Notice that this view includes both task data and resource data in the table on the left side of the view, plus a *Gantt Chart* pane on the right side. One big disadvantage of *Assignment* views is that the system sorts the list of tasks by *Task Name* in alphabetical order. This means that the system does not display the task list in the same order as in the Microsoft Project schedule.

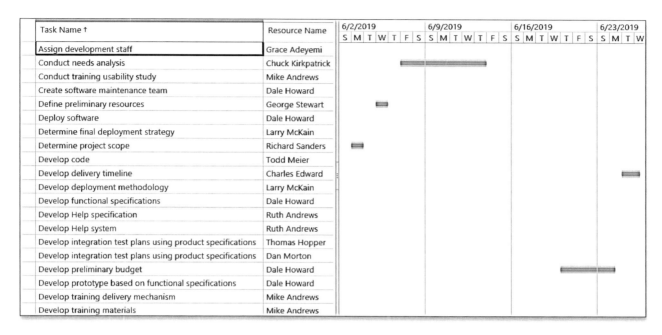

Figure 13 - 25: Assignments Summary view

Like *Project Center* views, you can apply grouping and filtering to any detailed *Project* view. For both *Task* and *Assignment* views, use the *Zoom In* and *Zoom Out* buttons to change the Gantt chart timescale, and use the *Scroll to Task* button to scroll the *Gantt Chart* pane to the *Start* date of the selected task.

 # Hands On Exercise

Exercise 13 - 2

Explore detailed *Project* views.

1. Click the **Projects** link in the *Quick Launch* menu to display the *Project Center* page, if necessary.

2. In the *Project Center* data grid, click the name of the enterprise project you created during this class.

3. In the "drill down" section at the top of the *Quick Launch* menu, click the **Schedule** link to display the *Schedule* PDP (Project Detail Page).

4. Click the **Task** tab to expand the *Task* ribbon.

5. In the *Data* section of the *Task* ribbon, click the **View** pick list and select the **Tasks Work** view.

445

Module 13

6. Examine the columns available in this view and then scroll through the *Gantt Chart* pane to view the schedule of each task.

7. In the *Data* section of the *Task* ribbon, click the **View** pick list and select the **Tasks Tracking** view.

8. Examine the columns available in this view and then scroll through the *Tracking Gantt* pane to view the schedule of each task compared with its original baseline schedule.

9. In the *Data* section of the *Task* ribbon, click the **View** pick list and select the **Resources Summary** view.

10. Examine the data shown in each column for your project team members.

11. In the *Data* section of the *Task* ribbon, click the **View** pick list and select the **Assignments Summary** view.

Notice how this view displays both task and resource information in the same view. Notice also how the task list is sorted in alphabetical order by the *Task Name* column.

12. Examine the columns available in this view and then scroll through the *Gantt Chart* pane to view the schedule of each task.

Notice how the Gantt bars for each task do not display the familiar "waterfall" pattern because the task list is sorted in alphabetical order and not by task *ID* numbers.

13. In the *Data* section of the *Task* ribbon, click the **View** pick list and reapply the **Tasks Summary** view.

14. Click the **Projects** link in the *Quick Launch* menu to return to the *Project Center* page.

Editing Projects in Project Web App

Not only can you view information in detailed *Project* views, you can also edit projects using these views because Project Online offers the capabilities of creating and editing projects using the Project Web App user interface. Keep in mind, however, that the project editing tools available in the Project Web App interface are a small subset of the features and functionality available in Microsoft Project. For example, you cannot change the *Status Manager* value for tasks using the Project Web App project editing interface.

A reasonable question might be to ask, "Why use the project editing features at all in Project Web App?" Microsoft designed this functionality for project managers who primarily manage small and simple projects, and who do not need the full feature set found in Microsoft Project. If you are a project manager who manages any kind of project other than small and simple projects, then you definitely need to use Microsoft Project to plan and manage your projects. If you only manage small and simple projects, you may find that the project editing tools in Project Web App are sufficient for your project management needs.

Creating a New Project in Project Web App

The process for creating a new project in Project Web App is nearly identical to the process I presented previously in Module 04. Navigate to the *Project Center* page in Project Web App and then click the **Projects** tab to expand the *Projects* ribbon. In the *Project* section of the *Projects* ribbon, click the **New** pick list button and select the **Enterprise**

Working in the Project Center

Project item to create a new blank project, such as shown in Figure 13 - 26. If your organization provides templates for you to use for creating small and simple projects, simply select the correct project type in the *New* pick list.

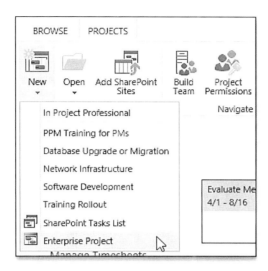

Figure 13 - 26: Create a new blank project

Project Web App displays the *Create a new project* page, such as the one show in Figure 13 - 27.

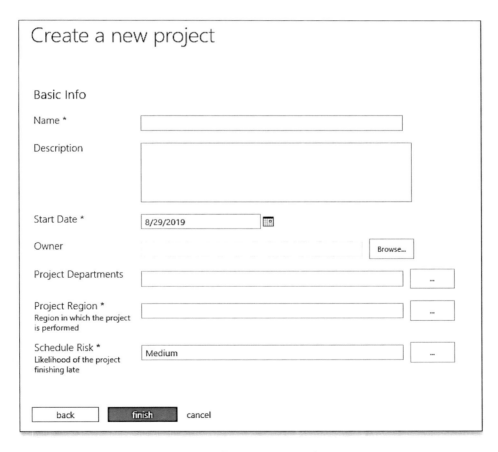

Figure 13 - 27: Create a New Project page

447

Module 13

The *Create a new project* page includes several default *Project* fields, plus custom fields created by your application administrator. Fields marked with a red asterisk indicate required fields in which you must supply a value before you can finish the process of creating your new project. Supply a value in every required and optional field and then click the **finish** button at the bottom of the page. The system creates a new blank project, saves the project using the name you provided, publishes the project initially, and then displays the *Schedule* PDP for the project, such as shown in Figure 13 - 28. Notice that the *Schedule* PDP includes a blank *Timeline* view at the top of the page and a blank project schedule in the bottom half of the page.

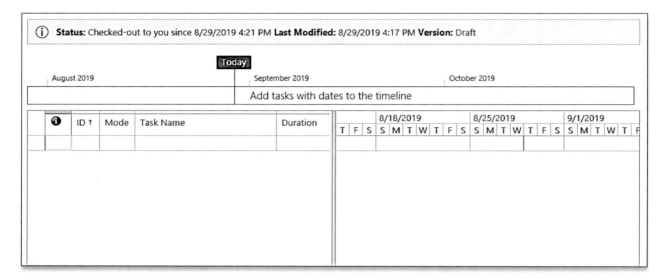

Figure 13 - 28: Schedule PDP for a new blank project

To display the Project Summary Task in your new blank project, click the **Options** tab to display the *Options* ribbon. In the *Show/Hide* section of the *Options* ribbon, select the **Project Summary Task** checkbox, such as shown in Figure 13 - 29.

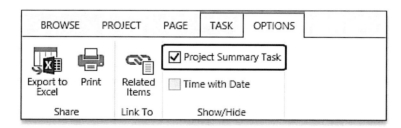

Figure 13 - 29: Options ribbon – display the
Project Summary Task

Project Web App displays the Project Summary Task (aka Row 0 or Task 0) in the blank project schedule, using the name of the project as the name of the Project Summary Task. For example, notice in Figure 13 - 30 that I created a new project named *Tech Ed Team Office Reconfiguration*, indicated by the name shown in the Project Summary Task.

448

Working in the Project Center

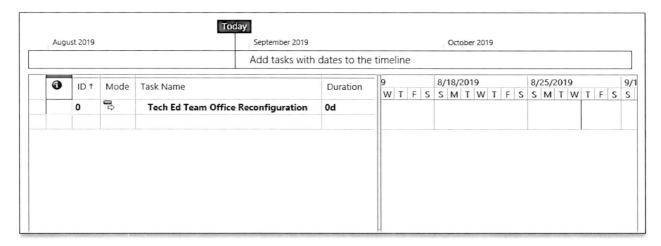

Figure 13 - 30: Project Summary Task displayed in the project

Click the **Project** tab to expand the *Project* ribbon. In the *Project* section of the *Project* ribbon, click the **Save** button shown in Figure 13 - 31 to save the changes to your new blank PWA project.

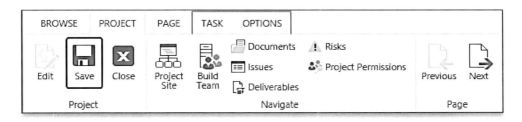

Figure 13 - 31: Save your new project

Understanding the PWA Task Planning Process

 Warning: Before you begin adding tasks to the new project, click the **Task** tab to expand the *Task* ribbon. In the *Data* section of the *Task* ribbon, click the **View** pick list and confirm that you have the **Tasks Summary** view applied. If not, apply this view before you begin the task planning process.

Before you begin the task planning process with your enterprise project, click the **Task** tab to display the *Task* ribbon. The commands on the *Task* ribbon in Project Web App are similar to the commands found on the *Task* ribbon in Microsoft Project. Figure 13 - 32 shows a partial *Task* ribbon that focuses on the task editing commands.

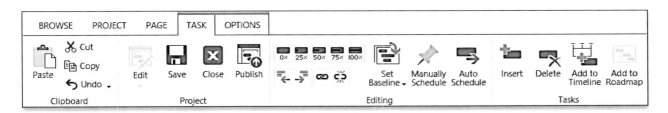

Figure 13 - 32: Task ribbon – task editing commands

449

Notice that the *Editing* section of the *Task* ribbon includes commands for setting a *% Complete* value for a task, indenting and outdenting tasks, linking and unlinking tasks, setting a baseline for the project, as well as setting the *Task Mode* for a task to either *Manually Scheduled* or *Auto Scheduled*. Notice that the *Tasks* section of the *Task* ribbon includes commands for inserting a new task, deleting an existing task, adding a task to the *Timeline* view, and adding a task to a *Roadmap*. I do not discuss the *Roadmap* feature in this book, however, as it is not specifically a feature of Project Online.

To complete the task planning process, I recommend you follow these steps:

1. Enter the name of each task.
2. Indent or outdent tasks to create summary tasks, if needed.
3. Create milestone tasks by setting their *Duration* value to *0 days*.
4. Link the detailed tasks in the schedule with dependencies, but do not link summary tasks to anything.
5. Specify a *Duration* estimate for each task.
6. Save your PWA project.

Figure 13 - 33 shows the *Tasks Summary* view for a very simple project consisting of only 10 tasks. Because of its simplicity, notice that I did not create any summary tasks.

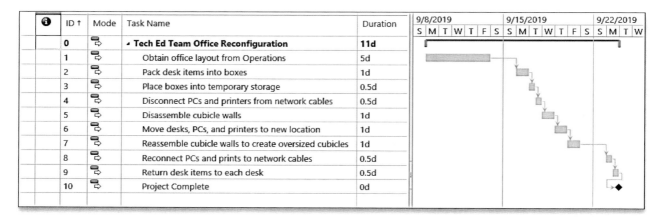

Figure 13 - 33: Simple project schedule

There are many task planning features available in Microsoft Project that are not available in Project Web App. Some of the features not available in PWA include the following:

- You cannot apply task calendars to tasks.
- You cannot set a task to *Inactive* status.
- You cannot apply cell background formatting to a task.
- You cannot add notes to tasks.
- You cannot add *Deliverables* to tasks.
- You cannot apply constraints to tasks unless the applied view includes the *Constraint Type* and *Constraint Date* columns.

Working in the Project Center

- You cannot apply deadlines unless the applied view includes the *Deadline* column.

- You cannot apply a *Task Type* setting to tasks unless the applied view includes the *Type* column.

- You cannot specify the *Effort Driven* status of each unless the applied view includes the *Effort Driven* column. The default value in the *Effort Driven* column for every task is *No*, which means that every task is *non-Effort Driven* by default.

Information: In several of the bulleted items above, I mention that you cannot perform a particular task planning action unless the applied view includes a column that relates to that action. If you organization intends for some of your project managers to create and manage projects using the Project Web App user interface, I recommend that your application administrator create a custom *Task* view used for the task planning process. Columns in this custom task planning view should include *Constraint Type*, *Constraint Date*, *Deadline*, *Type*, and *Effort Driven*.

Understanding the PWA Resource Planning Process

After completing the task planning process, you are ready to begin the resource planning process, which consists of building your project team. Click the **Project** tab to display the *Project* ribbon. In the *Navigate* section of the *Project* ribbon, click the **Build Team** button shown in Figure 13 - 34.

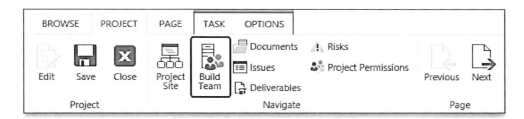

Figure 13 - 34: Project ribbon – Build Team button

Project Web App displays the *Build Team* page with the *All Resources* view applied, such as shown in Figure 13 - 35. The *Build Team* page includes a data grid on the left side with the list of all resources in the Enterprise Resource Pool, grouped into *Cost*, *Material*, and *Work* groups. The right side of the view includes existing project team members. The list of existing project team members should be blank if you started with a blank project, by the way.

To add resources to your project team, select the checkbox to the left of one or more resources, and then click the **Add** button in the middle of the page. If your application administrator enabled skill matching in your Project Web App system, you can add a Generic resource to the project team, select the checkbox for the Generic resource, and then click the **Match** button. The system updates the list of resources on the left side to show the resources who have the same skill or role in the organization as the Generic resource. You can replace the Generic resource with a designated human resource by selecting the checkbox for the human resource, selecting the checkbox for the Generic resource, and then clicking the **Replace** button. Set the **Booking** value to **Committed** for each human resource to make sure they can see their assigned tasks on either the *Timesheet* page or the *Tasks* page in PWA when you publish the project schedule.

To remove the skill matching filter and display the list of all resources again, click the **Clear Match** button. To remove an existing resource from the project team, select the checkbox for the resource you want to remove, and then click the **Remove** button.

Module 13

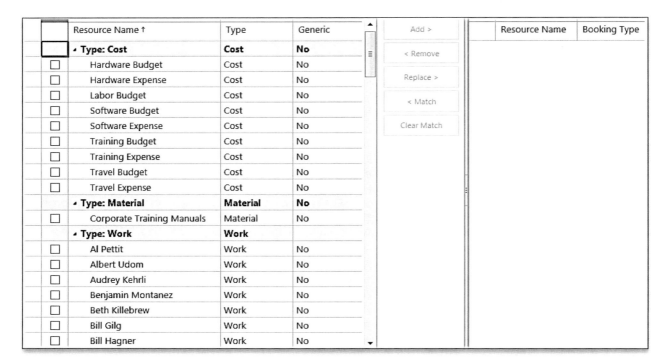

Figure 13 - 35: Build Team page with the All Resources view applied

When you finish building your project team, click the **Save & Close** button in the *Team* section of the *Team* ribbon shown in Figure 13 - 36. When the system finishes the save process, it returns to the *Schedule* PDP.

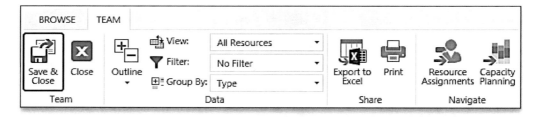

Figure 13 - 36: Team ribbon – Save & Close button

Understanding the PWA Assignment Planning Process

After completing the process of building your project team, you are ready to assign resources to each detailed task in the project schedule. The process for assignment planning with the Project Web App user interface is very simple. In your project schedule, drag the split bar to the right edge of the *Resource Names* column. For each task individually, click the pick list arrow button in the **Resource Names** cell, and then select the name of each resource you want to assign to the task, such as shown in Figure 13 - 37. The system automatically selects the checkbox for each resource name you select. When you finish selecting resources for the task, press the **Enter** key on your computer keyboard to complete the assignments. To remove a resource from the task, simply deselect the checkbox for the resource you want to remove. Be sure to save your project schedule when you finish the assignment planning process.

452

Working in the Project Center

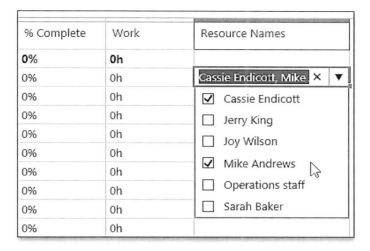

Figure 13 - 37: Assign resources to a task

 Warning: Project Web App provides you with no direct method for setting a *Units* value for each resource you assign to a task using the default *Project* views. If you need to specify a *Units* value for the resources you assign to tasks, ask your application administrator to create a custom *Assignments* view that includes the *Units* column. Keep in mind that when you change the *Units* value for a resource assignment, you may need to also change the *Work* value as well.

Finalizing Your PWA Project

As you finalize the planning process in your PWA project, you may want to view the Critical Path in the schedule. To display the Critical Path, click the **Task** tab to display the *Task* ribbon. In the *Data* section of the *Task* ribbon, click the **View** pick list show previously in Figure 13 - 23 and select either the **Tasks Detail** view or the **Tasks Schedule** view. As you might expect, Project Web App displays the schedule of Critical tasks using red Gantt bars and non-Critical tasks using blue Gantt bars.

Before you publish your project to "go live" with the project, you should save a baseline in the project. To save a baseline in your project, click the **Task** tab to display the *Task* ribbon, if needed. In the *Editing* section of the *Task* ribbon, click the **Baseline** pick list button, select the **Set Baseline** item on the pick list, and then select the **Baseline** item on the flyout menu, such as shown in Figure 13 - 38. Remember that you should always save your operating baseline for the life of the project in the *Baseline* set of fields. This is because Project Web App uses the data in the *Baseline* set to fields to calculate all variance in the project, and uses the *Baseline* schedule to draw the gray Gantt bars shown in the *Tasks Tracking* view.

After saving your operating baseline in the *Baseline* set of fields, you should also back up your original baseline in the Baseline 1 set of fields. To do this, click the **Baseline** pick list button again, select the **Set Baseline** item on the pick list, and then select the **Baseline 1** item on the flyout menu.

 Information: The Project Web App user interface does not provide any method for baselining selected tasks. This means that during the *Execution* stage, if you add new tasks to your project schedule through a change control procedure, you may need to rebaseline the entire project in the *Baseline* set of fields, losing all historical variance up to that point in the process.

453

Module 13

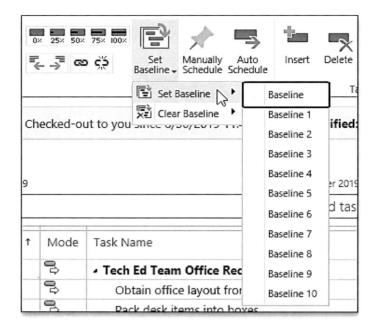

Figure 13 - 38: Save a baseline in the project

After baselining your project, you are ready to "go live" with your PWA project by publishing it. When you publish your project, Project Web App saves all data about the project in the Project Online database, displays project and resource data in views and reports, and displays the tasks assigned to your project team members on the *Timesheet* page and the *Tasks* page for each team member. To publish a project, click the **Publish** button in the *Project* section at the left end of the *Task* ribbon, such as shown in Figure 13 - 39.

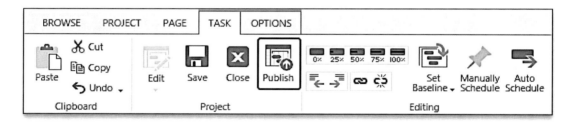

Figure 13 - 39: Publish the project

Closing and Reopening a PWA Project for Editing

At the end of each session of planning in your PWA project, you need to close and check in the project, similar to how you close and check in an enterprise project in Microsoft Project. To begin the process of closing and checking in a project in Project Web App, click the **Project** tab to expand the *Project* ribbon. In the *Project* section of the *Project* ribbon, click the **Save** button to save the latest changes to your project, and then click the **Close** button shown in Figure 13 - 40. You can also find the *Save* and *Close* buttons in the *Task* ribbon, by the way.

454

Working in the Project Center

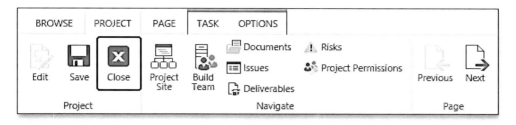

Figure 13 - 40: Project ribbon – Close button

Project Web App displays the *Close* dialog shown in Figure 13 - 41. In the *Close* dialog, leave the **Check it in** option selected and then click the **OK** button. When the check in process completes, the system displays the *Project Center* page in PWA.

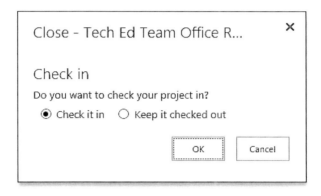

Figure 13 - 41: Close dialog

When you are ready to begin another session of working with your PWA project, navigate to the *Project Center* page and click the name of your project. The system opens the project in Read-Only mode. In the "drill down" section at the top of the Quick Launch menu, click the **Schedule** link to display the *Schedule* PDP, if necessary. Click the **Project** tab to display the *Project* ribbon. In the *Project* section of the *Project* ribbon, click the **Edit** button shown in Figure 13 - 42 to check out the project for editing in Read/Write mode. You can also find an *Edit* pick list button in the *Task* ribbon, by the way. In the *Project* section of the *Task* ribbon, click the **Edit** pick list button and select the **In Browser** item to check out the project for editing in Project Web App.

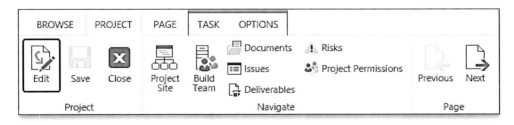

Figure 13 - 42: Project ribbon – Edit button

455

Module 14

Working in the Resource Center

Learning Objectives

After completing this module, you will be able to:

- View project resources in the Resource Center page
- Apply Resource Center views
- View resource availability across all projects in the portfolio
- View resource assignments in all projects in the portfolio
- View and respond to Resource Engagement requests

Inside Module 14

Using the Resource Center.. **459**
    Applying Resource Center Views.. *460*
    Selecting and Deselecting Resources.. *461*
Viewing Resource Availability.. **463**
Viewing Resource Assignments.. **471**
Working with Resource Engagements **473**
    Viewing Pending Resource Engagements *473*
    Viewing Resource Availability *476*
    Responding to Pending Resource Engagements *476*

Using the Resource Center

When you click the **Resources** link in the *Quick Launch* menu, Project Online displays the *Resource Center* page, such as the one shown in Figure 14 - 1. The *Resource Center* page contains a data grid that displays all of the resources in the Enterprise Resource Pool that your security permissions allow you to see. By default, the system applies the *All Resources* view when you display the *Resource Center* page the first time. This view displays all resources, grouped by resource type, including *Cost*, *Material*, and *Work* resources.

Resource Name ↑	ID	Checked Out	Email Address	Generic	Timesheet Manager	Type	Active
▲ Type: Cost		No		No		Cost	Yes
Hardware Budget	7,110	No		No		Cost	Yes
Hardware Expense	7,115	No		No		Cost	Yes
Labor Budget	7,112	No		No		Cost	Yes
Software Budget	7,111	No		No		Cost	Yes
Software Expense	7,116	No		No		Cost	Yes
Training Budget	7,114	No		No		Cost	Yes
Training Expense	7,117	No		No		Cost	Yes
Travel Budget	7,113	No		No		Cost	Yes
Travel Expense	7,118	No		No		Cost	Yes
▲ Type: Material		No		No		Material	Yes
Corporate Training Manuals	7,109	No		No		Material	Yes
▲ Type: Work		No				Work	Yes
Al Pettit	6,991	No		No	Al Pettit	Work	Yes
Albert Udom	6,992	No		No	Albert Udom	Work	Yes
Audrey Kehrli	6,993	No		No	Audrey Kehrli	Work	Yes
Benjamin Montanez	6,994	No		No	Benjamin Montanez	Work	Yes
Beth Killebrew	6,995	No		No	Beth Killebrew	Work	Yes
Bill Gilg	6,996	No		No	Bill Gilg	Work	Yes
Bill Hagner	6,997	No		No	Bill Hagner	Work	Yes

Figure 14 - 1: Resource Center page

The *Resource Center* page provides a central location from which you can view all aspects of the enterprise resources that you have permission to view. From this page, you can do each of the following:

- View detailed information for each resource.

- View resource assignments across all enterprise projects.

- View resource availability across all enterprise projects.

Before you begin working with resources in the *Resource Center* page, click the **Resources** tab to expand the *Resources* ribbon shown in Figure 14 - 2. Notice the commands available in the *Data* section of the ribbon. Use the commands in this section to quickly filter, group, or search through the resources shown in the *Resource Center* page.

Module 14

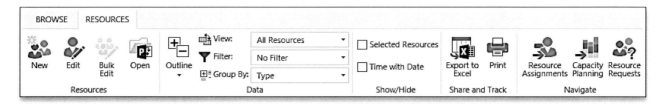

Figure 14 - 2: Resources ribbon

To filter for resources that meet certain criteria, click the **Filter** pick list in the *Data* section of the *Resources* ribbon and select the **Custom Filter** item. Project Web Ap displays the *Custom Filter* dialog shown in Figure 14 - 3. Build your custom filter by selecting or entering values in the **Field Name**, **Test**, **Value**, and **And/Or** fields. Click the **OK** button to apply the custom filter.

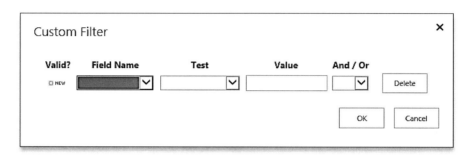

Figure 14 - 3: Custom Filter dialog

To organize the resources into groups with similar attributes, click the **Group By** pick list in the *Data* section of the *Resources* ribbon and select the field on which you want to group. You can group by any field displayed in the view, but you cannot group on fields not included in the view.

You can sort on any field by floating your mouse pointer over any column header, clicking the pick list in the column header, and then selecting either the **Ascending** or **Descending** item from the pick list. You can also use the **Outline** pick list in the *Data* section of the *Resources* ribbon to show any level of data when you have grouping applied. For example, select the *Outline Level 1* option in the *Outline* pick list to collapse the groups at the first level of grouping.

Applying Resource Center Views

To apply a view in the *Resource Center* page, click the **View** pick list in the *Data* section of the *Resources* ribbon as shown in Figure 14 - 4. The *View* pick list includes all standard and custom views that your security permissions allow you to see. Project Web App offers five default views, which include the *All Resources, Material Resources, Work Resources, Resources by Team,* and *Cost Resources* views.

The *All Resources* view displays all resources, grouped by resource type. The *Material Resources* view displays only material resources, while the *Work Resources* view displays only work resources. The *Resources by Team* view displays resources grouped by team. The *Cost Resources* view displays only *Budget Cost* and *Expense Cost* resources.

The last three views on the pick list shown in Figure 14 - 4 are custom views created by our application administrator. As you might surmise, the *_Generic Resources* custom view displays only Generic resources. The *_Work Resources by RBS* custom view displays only *Work* resources, grouped by their *RBS* values. The *_Work Resources by Role* custom view displays only *Work* resources, grouped by their *Role* values.

Working in the Resource Center

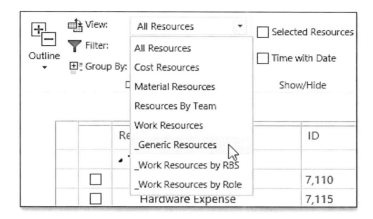

Figure 14 - 4: View pick list

 Information: In addition to the five default *Resource Center* views, you may see custom views on the *View* pick list that display and organize enterprise resources according to the needs of your organization. Because your application administrator controls the views available in the *Resource Center*, you should contact your application administrator for assistance if you need a custom *Resource Center* view.

Selecting and Deselecting Resources

After applying a view, you must select one or more resources to activate the other features available in the *Resource Center* page, such as viewing resource availability or viewing resource assignments. To select a resource, select the checkbox to the left of the resource's name. To quickly select all resources or to quickly clear your current resource selections, float your mouse pointer over the column header at the top of the checkboxes column to reveal the pick list arrow button. Select either the **Select All** or **Clear All** item on the pick list as needed, as shown in Figure 14 - 6.

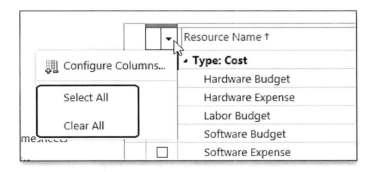

Figure 14 - 5: Select All or Clear All

After selecting resources, you can view availability or assignment information for the selected resources. If you have a large Enterprise Resource Pool with hundreds or thousands of resources, and you previously selected multiple resources in the *Resource Center*, it can require a lot of scrolling to determine which resources you selected previously. To simplify the process of determining which resources you selected, select the **Selected Resources** checkbox in the *Show/Hide* section of the *Resources* ribbon. The system displays the *Selected Resources* sidepane on the right side of the page, such as shown in Figure 14 - 6. Notice that the sidepane displays the list of all resources currently selected. Deselect the **Selected Resources** checkbox to hide the sidepane.

461

Module 14

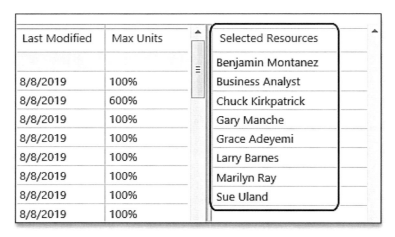

Figure 14 - 6: Resource Center with selected resources panel

 Hands On Exercise

Exercise 14 - 1

Explore the features of the *Resource Center* page in PWA.

1. Click the **Resources** link in the *Quick Launch* menu to display the *Resource Center* page.
2. Notice the grouping of resources in the default *All Resources* view.
3. Study the information shown in each of the columns displayed in the *All Resources* view.
4. In the *Data* section of the *Resources* ribbon, click the **View** pick list and individually apply each of the following views, then study the information shown in the columns in each view:

    - Cost Resources
    - Material Resources
    - Resources by Team
    - Work Resources

5. In the *Data* section of the *Resources* ribbon, click the **View** pick list and then reapply the **All Resources** view.
6. Select the checkboxes to the left of the names of any five resources.
7. In the *Show/Hide* section of the *Resources* ribbon, select the **Selected Resources** checkbox.

Notice how the system displays the *Selected Resources* sidepane on the right side of the page, which shows you the names of the resources currently selected.

Viewing Resource Availability

During the resource planning process, you may need to determine which resources are available to work in your project, along with their availability by time period. You can use the *Resource Center* to analyze availability for one or more resources across all projects in Project Web App. Begin this process by selecting the checkboxes for the resources whose availability you want to analyze. In the *Navigate* section of the *Resources* ribbon, click the **Capacity Planning** button. The system displays the *Capacity Planning* page, such as the one shown in Figure 14 - 7.

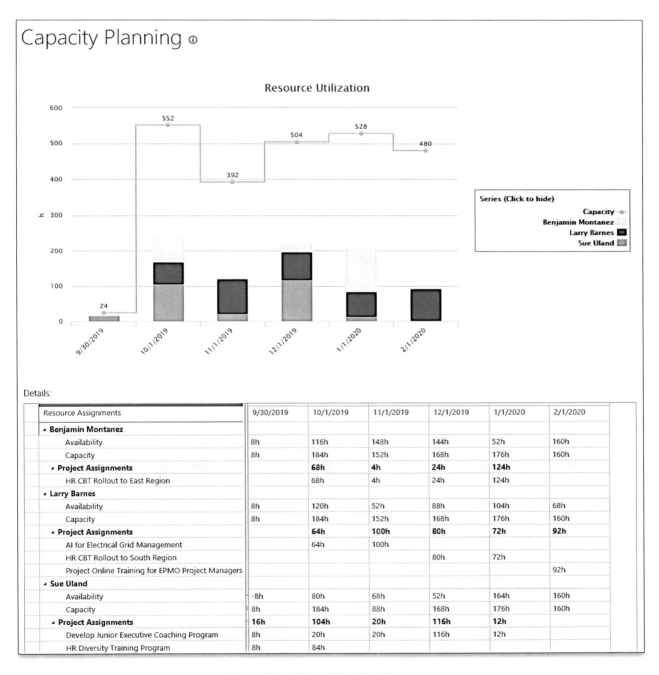

Figure 14 - 7: Capacity Planning page

The *Capacity Planning* page contains two sections: a chart at the top and a *Details* section with a timesheet grid at the bottom. For each selected resource, the timesheet grid displays the following information:

- Availability
- Capacity
- Assignment work for each project in which the resource is a team member

The chart displays a graphic representation of the data displayed in the timesheet grid. The *Capacity Planning* page uses Project Online terminology that may be new to you. **Capacity** is the amount of time a resource is committed to project work in general, minus any company holidays and planned nonworking time such as vacation. **Availability** is the amount of time the resource is available to work in any specific project. The system calculates the availability for each resource in each time period using the formula **Capacity - Work**.

> **Understanding Capacity vs. Availability Calculations**
>
> In the Enterprise Resource Pool, Mickey Cobb has a *Max. Units* value of *100%* and her resource calendar shows a normal working schedule of 8 hours/day. This means her typical *Capacity* for project work is 8 hours/day, 40 hours/week, and approximately 168 hours/month in an average month. During the month of July 2020, her organization recognizes Independence Day as a company holiday, and Mickey Cobb also has 5 days of planned vacation. July 2020 has 23 possible working days during the month or 184 possible working hours. The Independence Day holiday reduces the possible working hours to 176 (184 – 8). Her week of planned vacation reduces the *Capacity* for Mickey Cobb to 136 hours (176 – 40) during the month of July 2020.
>
> During the month of July 2020, Mickey Cobb is assigned to work in three projects, totaling 120 hours. Project Online calculates the *Availability* for Mickey Cobb as 16 hours (136 – 120) during the month of July 2020. If you need Mickey Cobb to work full-time in your project during the month of July 2020, the *Capacity Planning* page shows you that she has very limited availability to work in your project because of commitments to other projects.

By default, the chart and the data grid on the *Capacity Planning* page display a 2-week date range with daily time periods beginning from the current date. To analyze resource availability, you may want to select a different date range, and show different time periods, such as weeks or months. To change the time periods, click the **Timescale** pick list in the *Filters* section of the *Availability* ribbon, as shown in Figure 14 - 8, and select the time periods you want to see, such as *Months* for example. To change the date range, click the Set **Date Range** button in the in the *Date Range* section of the *Availability* ribbon, such as shown in Figure 14 - 8.

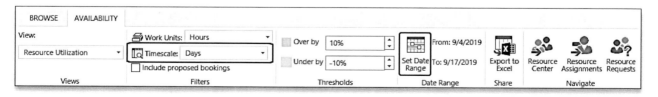

Figure 14 - 8: Availability ribbon – Change the date range and timescale

Project Web App displays the *Set Date Range* dialog shown in Figure 14 - 9. In the dialog, enter the dates of the date range in the **From** and **To** fields, and then click the **OK** button. The system updates the *Capacity Planning* page with the new date range and new time periods.

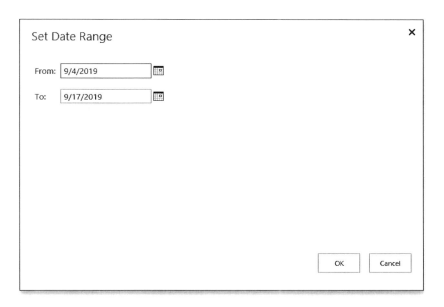

Figure 14 - 9: Set Date Range dialog

In the *Views* section of the *Availability* ribbon, the *View* pick list allows you to select one of five available views:

- Resource Utilization (the default view)
- Resource Utilization by Project
- Remaining Availability
- Work by Resource
- Capacity and Engagements Heatmap

Each of these five views controls the presentation of data shown in the chart. The **Resource Utilization** view shows a stacked column chart of the assignment work for each selected resource, along with a line chart of the total *Capacity* for all selected resources. If you float your mouse pointer over any of the columns, Project Web App displays a tooltip with the *Capacity* for all of the resources, along with the total *Work* assigned to each resource during that particular time period. In the *Series* section to the right of the chart, click and unclick the resource names to dynamically include or exclude the resource's information in the chart.

 Warning: When you click a resource name to exclude the resource data from the chart, Project Web App ***does not*** deduct the resource's information from the *Capacity* line. The *Capacity* line is static, and always shows the total *Capacity* for all selected resources, regardless of how many you include or exclude in the chart. Because of this limitation, it is difficult to visually determine the availability for any resource simply by examining the chart. The best way to determine resource availability is to examine the data for the resource shown in the *Details* data grid.

Module 14

The **Resource Utilization by Project** view shows a stacked column chart for all of the projects assigned to the selected resources, along with a line chart for the total *Capacity* for all selected resources. Figure 14 - 10 shows the *Capacity Planning* page with the *Resource Utilization by Project* view applied.

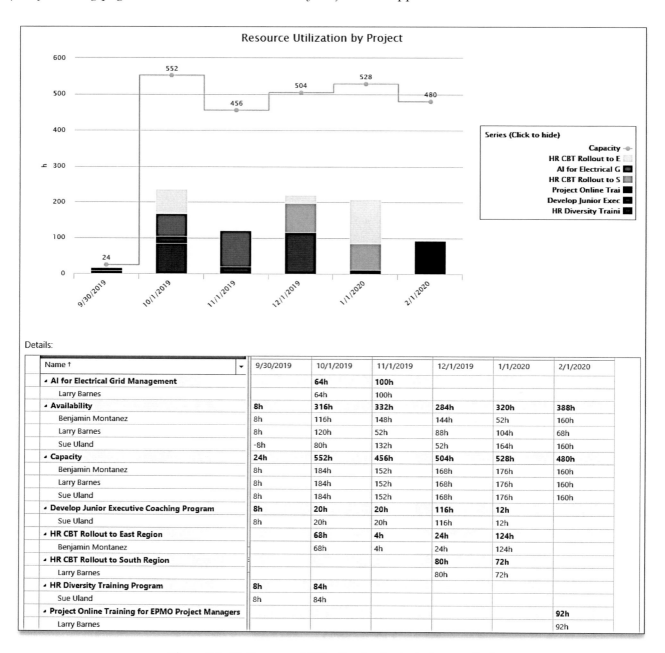

Figure 14 - 10: Resource Utilization by Project view applied

The **Remaining Availability** view displays a column chart that shows the availability for each selected resource by time period. With multiple resources selected, this view is difficult to use; therefore, you should select resources individually to see their availability by time period. Figure 14 - 11 shows the *Capacity Planning* page with the *Remaining Availability* view applied with only Larry Barnes selected in the *Series* section to the right of the chart.

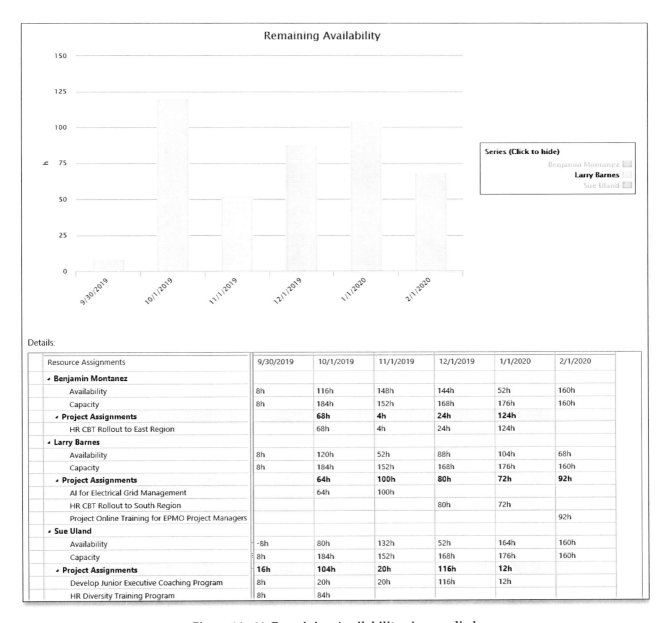

Figure 14 - 11: Remaining Availability view applied

The **Work by Resource** view displays a column chart showing the total assigned project work for each selected resource. With multiple resources selected, this view is also difficult to use; therefore, you should select resources individually to see their assigned work by time period. Figure 14 - 12 shows the *Capacity Planning* page with the *Work by Resource* view applied with only Larry Barnes selected in the *Series* section to the right of the chart.

Module 14

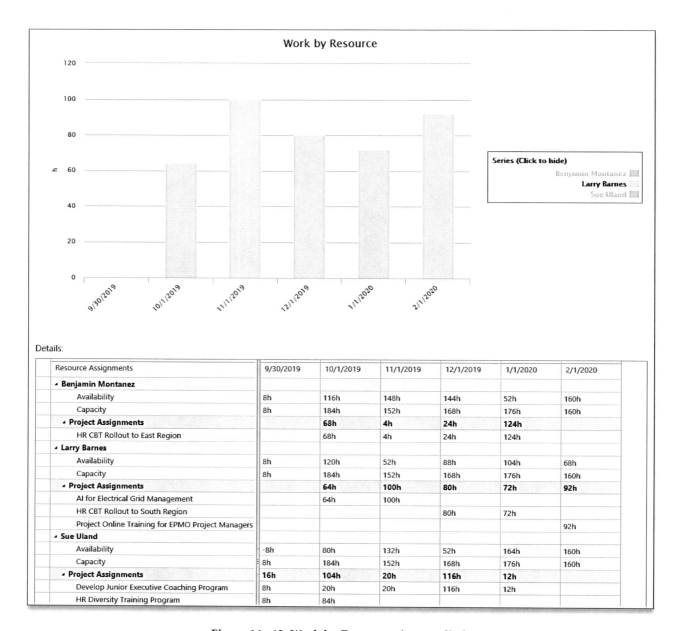

Figure 14 - 12: Work by Resource view applied

The **Capacity and Engagements Heatmap** view displays a data grid that compares approved Resource Engagements against the *Capacity* of each resource. This view is only useful to organizations that meet two requirements:

- The organization requires project managers to create Resource Engagements for some or all of the resources used in projects.
- The organization requires resource managers to approve or reject all pending Resource Engagements.

If your organization meets the two preceding requirements, you may be able to benefit from using the *Capacity and Engagements Heatmap* view shown in Figure 14 - 13. In the data grid, the system displays the number of hours requested for each resource in the Resource Engagements for each of their projects. If the Resource Engagement hours exceed the *Capacity* of the resource by more than 10%, indicating that the resource is overallocated, the system formats the time period with a red cell background color. If the Resource Engagement hours are more than 10%

468

under the *Capacity* for the resource, indicating that the resource is underutilized, the system formats the time period with the blue cell background color. All other cases, where the resource falls in the range between 10% overallocated and 10% underutilized, the system formats the time period with the green cell background color.

Capacity and Engagements Heatmap View the discrepancies between resource capacity and committed engagements for these resources.							
Resource Engagements		9/30/2019	10/1/2019	11/1/2019	12/1/2019	1/1/2020	2/1/2020
▲ Benjamin Montanez		8h	184h	152h	168h	0h	0h
	HR CBT Rollout to East Region	8h	184h	152h	168h	0h	0h
▲ Larry Barnes		0h	112h	152h	336h	336h	320h
	AI for Electrical Grid Management	0h	112h	152h	168h	0h	0h
	HR CBT Rollout to South Region	0h	0h	0h	168h	176h	160h
	Project Online Training for EPMO Project Managers	0h	0h	0h	0h	160h	160h
▲ Sue Uland		0h	112h	80h	0h	0h	0h
	Develop Junior Executive Coaching Program	0h	0h	80h	0h	0h	0h
	HR Diversity Training Program	0h	112h	0h	0h	0h	0h

Figure 14 - 13: Capacity and Engagements Heatmap view in Hours

Information: The system also formats with blue cell background color any time period that does not have a Resource Engagement specified.

To fully comprehend the *Capacity and Engagements Heatmap* view, I recommend that you view the data using Full Time Equivalents (FTEs) rather than hours. In the *Filters* section of the *Availability* ribbon, click the **Work Units** pick list and select the **Full-time Equivalent** item. Figure 14 - 14 displays *Capacity and Engagements Heatmap* view with FTE values. During the month of October 2019, notice that the Resource Requests for Larry Barnes utilize only 61% of his *Capacity*, meaning that he is underutilized. In the time periods from December 2019 through February 2020, notice that the Resource Requests for Larry Barnes are 200%, 191%, and 200% of his *Capacity*, meaning that he is overallocated during these three months. During the month of November 2019, notice that the Resource Engagements for Sue Uland utilize 91% of her *Capacity*, meaning that she is almost fully utilized.

Capacity and Engagements Heatmap View the discrepancies between resource capacity and committed engagements for these resources.							
Resource Engagements		9/30/2019	10/1/2019	11/1/2019	12/1/2019	1/1/2020	2/1/2020
▲ Benjamin Montanez		1	1	1	1	0	0
	HR CBT Rollout to East Region	1	1	1	1	0	0
▲ Larry Barnes		0	0.61	1	2	1.91	2
	AI for Electrical Grid Management	0	0.61	1	1	0	0
	HR CBT Rollout to South Region	0	0	0	1	1	1
	Project Online Training for EPMO Project Managers	0	0	0	0	0.91	1
▲ Sue Uland		0	0.61	0.91	0	0	0
	Develop Junior Executive Coaching Program	0	0	0.91	0	0	0
	HR Diversity Training Program	0	0.61	0	0	0	0

Figure 14 - 14: Capacity and Engagements Heatmap view in FTEs

Module 14

The next time you return to the *Capacity Planning* page, Project Web App maintains your selected date range settings and your applied view. To return to the *Resource Center* page, click the **Resource Center** button in the *Navigate* section of the *Availability* ribbon.

Hands On Exercise

Exercise 14 - 2

Explore the *Capacity Planning* page in PWA.

1. Leave the checkboxes selected for the five resources you chose in the previous exercise.
2. In the *Navigate* section of the *Resources* ribbon, click the **Capacity Planning** button.
3. Examine the resource data shown in the chart at the top of the page and the *Details* grid at the bottom of the page.
4. In the *Series* section to the right of the chart, click and unclick several names to see how the system updates the chart.
5. In the *Filter* section of the *Availability* ribbon, click the **Timescale** pick list and select the **Months** item.
6. In the *Date Range* section of the *Availability* ribbon, click the **Set Date Range** button.
7. In the *Set Date Range* dialog, select the first working day of *next month* in the **From** field.
8. In the **To** field, select the *last working day of the month* for the month that is *six months later* than the date in the *From* field, and then click the **OK** button.

Setting the dates in the *From* and *To* fields in the two previous steps should provide you with a 6-month "look ahead" for resource availability starting from next month.

9. Study the chart at the top of the page, then study the monthly *Capacity, Availability,* and *Project Assignments* information shown for each resource in the *Details* grid.
10. In the *Views* section of the *Availability* ribbon, click the **View** pick list, select each of the following views, and then study the information shown in each view:
    - Remaining Availability
    - Work by Resource
    - Capacity and Engagements Heatmap – Apply this view only if your organization uses Resource Engagements and each engagement must be approved by a resource manager.
11. In the *Navigate* section of the *Availability* ribbon, click the **Resource Center** button to return to the *Resource Center* page.

Viewing Resource Assignments

After selecting one or more resources on the *Resource Center* page, in the *Navigate* section of the *Resources* ribbon, click the **Resource Assignments** button to view all project work currently assigned to the selected resources. Project Web App displays the *Resource Assignments* page shown in Figure 14 - 15. Using the *Resource Assignments* page, you can determine the total amount of task work assigned to each selected resource and determine the specific times during which each resource has scheduled work.

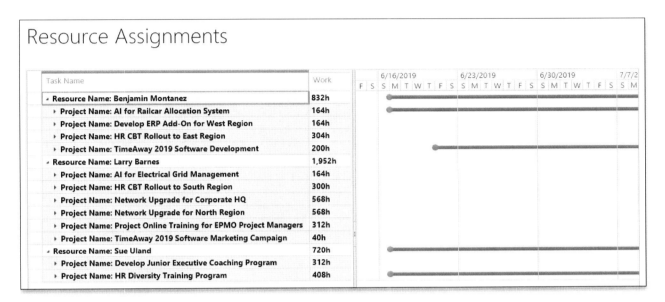

Figure 14 - 15: Resource Assignments page

The *Resource Assignments* page includes a table on the left that lists resource assignments, with either a *Gantt Chart* pane or a timephased grid on the right side that displays the time span for each resource assignment. In the *Display* section of the *Assignments* ribbon, click either the **Gantt Chart** button or the **Timephased Data** button to switch between these two views. In the *Date Range* section of the *Assignments* ribbon, click the **Set Date Range** button to set a date range for the timephased data.

By default, the system groups the resource assignment information by resource name and then by project name, with the projects collapsed by default. Expand any project to view the task assignments for that project.

In the *Show/Hide* section of the *Assignments* ribbon, select the **Summary Tasks** option to include summary tasks in the display. Select the **Time with Date** option to display dates with times.

 Warning: As of the writing of this book, the *Summary Tasks* checkbox is non-functional due to an unfixed bug in Project Online. This bug makes it impossible to view summary tasks in the *Resource Assignments* page.

When you apply the *Timephased Data* display, you can *deselect* the **Overtime Work** checkbox to hide the *Overtime* row in the timephased grid, or *deselect* the **Work** checkbox to hide the *Planned* row in the timephased grid, and show only the *Actual* row. When you apply the *Gantt Chart* display, you can use the **Zoom In**, **Zoom Out**, and **Scroll to Task** buttons in the *Zoom* section of the *Assignments* ribbon to navigate in the *Gantt Chart* pane.

In the *Data* section of the *Assignments* ribbon, the *Summary* view is the only available selection on the *View* pick list by default, unless your application administrator creates additional views for your organization. The table in the *Summary* view includes the following fields:

- Task Name
- Work
- Remaining Work
- Start
- Finish
- % Work Complete
- Comments
- Resource Name
- Project Name

To return to the *Resource Center* page, click the **Resource Center** button in the *Navigate* section of the *Assignments* ribbon.

 Hands On Exercise

Exercise 14 - 3

Explore the *Resource Assignments* page in PWA.

1. Leave the checkboxes selected for the five resources you chose in the first exercise.
2. In the *Navigate* section of the *Availability* ribbon, click the **Resource Assignments** button.
3. Study the assignment information shown in the *Summary* view with the *Gantt Chart* displayed.
4. In the *Display* section of the *Assignments* ribbon, click the **Timephased Data** button.
5. Study the assignment information shown in the *Summary* view with timephased data displayed.
6. Expand the task assignments for the projects assigned to several of the resources.
7. In the *Date Range* section of the *Assignments* ribbon, click the **Set Date Range** button.
8. In the *Set Date Range* dialog, set a date range that spans two months into the past and two months into the future, and then click the **OK** button.
9. In the *Show/Hide* section of the *Assignments* ribbon, *deselect* the **Overtime Work** checkbox.
10. In the *Navigate* section of the *Assignments* ribbon, click the **Resource Center** button to return to the *Resource Center* page.

Working with Resource Engagements

Warning: This topical section is **only** for project managers who also have resource management responsibilities, and assumes that your application administrator added you to the *Resource Managers* security group in Project Web App.

The Resource Engagements feature in Project Online serves two primary purposes:

- Resource Engagements with Generic resources allow a project managers to forecast the resource needs for a proposed project. For example, a project manager creates a Resource Engagement for a Network Engineer to work half-time during the months from June through August 2020. In response to this Resource Engagement request, a resource manager can "swap" the Network Engineer resource with a human resource named Calvin Baker who is a network engineer and is available to work half-time during the requested time period.

- Resource Engagements with human resources allow a project manager to request a specific resource for a specific time period. For example, a project manager adds Mickey Cobb to the project team and assigns her to tasks in the project. Then the project manager creates a Resource Engagement for Mickey Cobb spanning the time period she is assigned to tasks. A resource manager can respond to this Resource Engagement request by either approving or rejecting the use of Mickey Cobb, or by substituting another resource in place of Mickey Cobb.

As you can see from the above descriptions, the Resource Engagements feature formalizes the negotiation process between a project manager and resource managers for the use of resources in projects. Again, you will only find the content in this topical section relevant if you and your organization meet the following requirements:

- You are a project manager who also has resource management responsibilities, and your Project Online application administrator added you to the *Resource Managers* security group.

- Your organization uses Resource Engagements with Generic and/or human resources.

- Your organization requires resource managers to respond to Resource Engagement requests.

Warning: Users who are only a member of the *Project Managers* security group **do not** have permission to view Resource Engagements in Project Web App. If you are a project manager who also has resource management responsibilities, make sure that your application administrator adds you to the *Resource Managers* security group in PWA so that you have permissions to work with Resource Engagements.

Viewing Pending Resource Engagements

To view the pending Resource Engagement requests for the resources who report to you, navigate to the *Resource Center* page in Project Web App. Select the checkbox to the left of the name of every resource who reports to you, along with any Generic resources that describe the roles of your resources. For example, suppose that you manage a team of seven Business Analysts. In a situation such as this, you should select all of the human resources who have the Business Analyst role, along with the Generic resource named Business Analyst, such as shown in Figure 14 - 16.

Module 14

		Resource Name ↑	Checked Out	Booking Type	Active	Last Modified	Max Units
		▲ Corporate Role: Business Analyst	No	Committed	Yes		
✓		Benjamin Montanez	No	Committed	Yes	9/4/2019	100%
✓		Business Analyst	No	Committed	Yes	8/8/2019	600%
✓		Chuck Kirkpatrick	No	Committed	Yes	8/8/2019	100%
✓		Gary Manche	No	Committed	Yes	8/8/2019	100%
✓		Grace Adeyemi	No	Committed	Yes	8/8/2019	100%
✓		Larry Barnes	No	Committed	Yes	9/4/2019	100%
✓		Marilyn Ray	No	Committed	Yes	8/8/2019	100%
✓		Sue Uland	No	Committed	Yes	9/4/2019	100%
		▲ Corporate Role: Database Analyst	No	Committed	Yes		
☐		Charles Edward	No	Committed	Yes	8/8/2019	100%
☐		Chip Kelmell	No	Committed	Yes	8/8/2019	100%
☐		Dave Harbaugh	No	Committed	Yes	8/8/2019	100%

Figure 14 - 16: Select all resources with the Business Analyst role

After selecting the resources that report to you, along with relevant Generic resources, click the **Resource Requests** button in the *Navigate* section of the *Resources* ribbon. Project Web App displays the *Resource Requests* page, such as the one shown in Figure 14 - 17.

Resource Requests

	❶	Resource Name	Description	Project	Requester	State ↑
		▲ Resource Name: Benjamin Montanez				Proposed
		▲ Project: HR CBT Rollout to East Region				Proposed
☐		☐ Benjamin Montanez	BA work	HR CBT Rollout to East Region	Dale Howard	Proposed
		▲ Resource Name: Business Analyst				Proposed
		▲ Project: AI for Electrical Grid Management				Proposed
☐		☐ Business Analyst	Communicate with the client	AI for Electrical Grid Management	Dale Howard	Proposed
		▲ Project: Chi Rho Software Marketing Campaign				Proposed
☐		☐ Business Analyst	Business analysis support 2	Chi Rho Software Marketing Campaign	Dale Howard	Proposed
☐		☐ Business Analyst	Business analysis support 1	Chi Rho Software Marketing Campaign	Dale Howard	Proposed
		▲ Project: HR CBT Rollout to Corporate HQ				Proposed
☐		☐ Business Analyst	Business analysis support 1	HR CBT Rollout to Corporate HQ	Dale Howard	Proposed
☐		☐ Business Analyst	Business analysis support 2	HR CBT Rollout to Corporate HQ	Dale Howard	Proposed
		▲ Resource Name: Gary Manche				
		▲ Project: Clearwater Beach Mobile App				Committed
☐		☐ Gary Manche	Develop customer requirements	Clearwater Beach Mobile App	Dale Howard	Committed
		▲ Project: HR CBT Rollout to West Region				Proposed
☐		☐ Gary Manche	Post implementation closeout	HR CBT Rollout to West Region	Dale Howard	Proposed
☐		☐ Gary Manche	Determine customer requirements	HR CBT Rollout to West Region	Dale Howard	Proposed
		▲ Resource Name: Larry Barnes				Proposed
		▲ Project: HR CBT Rollout to South Region				Proposed

Figure 14 - 17: Resource Requests page

By default, the *Resource Requests* page displays the *Engagement Details* view. This view provides a detailed overview of all proposed and committed Resource Engagements for all selected resources across all projects. This view groups all engagements by the resource name and then by project name. By default, the *Engagement Details* view includes the *Resource Name, Description, Project, Requestor, State, Committed Units, Committed Start,* and *Committed Finish* columns.

Notice that the *State* column reveals that there is a mix of proposed and committed engagements for the human resources, and a number of proposed engagements for the Business Analyst generic resource. Notice also that the *Description* column contains a description of the type of work to be performed by the Business Analyst. Although

Working in the Resource Center

the *Description* value is optional for the project manager when creating a new Resource Engagement, I strongly recommend that every project manager supply a *Description* value when creating Resource Engagements. You can see the immediate benefit of this action when viewing the *Resource Requests* page in PWA.

The *Resource Requests* page includes two ribbons: the *Engagements* ribbon shown in Figure 14 - 18 and the *Options* ribbon shown in Figure 14 - 19. You can use the commands in the *Engagements* ribbon to accept or reject pending engagements, to work with engagements, to display either the *Sheet* or *Timephased* data display, and to change the view, apply a filter, or apply a group. When you click the **Timephased Data** button in the *Display* section of the *Engagements* ribbon, the system displays a timephased grid on the right side of the page. The timephased grid allows you to view the planned work hours by time period for each proposed Resource Engagement.

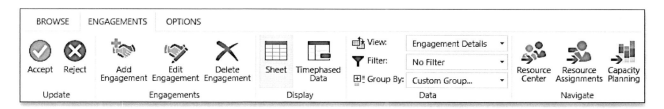

Figure 14 - 18: Engagements ribbon

With the *Timephased Data* display selected, the *Options* ribbon allows you to control the display of data in the timephased grid. Use the commands in this ribbon to specify the date range of the timephased grid, the time units (timescale) displayed in the timephased grid, and whether to display work in hours or in days. In the *Show/Hide* section of the *Options* ribbon, you can also **deselect** the **Proposed** checkbox to hide the *Proposed Work* row in the timephased grid and show only the *Committed Work* row.

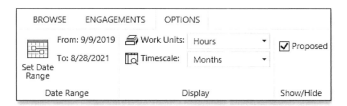

Figure 14 - 19: Options ribbon

To view a simpler arrangement of pending and approved Resource Engagements, you can apply the *Summary* view. In the *Data* section of the *Engagements* ribbon, click the **View** pick list and select the **Summary** view, then click the **Sheet** button in the *Display* section of the *Engagements* ribbon, if necessary. The *Summary* view, shown in Figure 14 - 20 displays high level information about all pending and approved resource engagements without using the two levels of grouping found in the *Engagement Details* view. By default, the *Summary* view includes the *Resource Name*, *Description*, *State*, *Committed Units*, *Committed Start*, and *Committed Finish* columns. Notice again the importance of the project manager supplying information in the *Description* field when creating a new Resource Engagement.

As with the *Engagement Details* view, you can display a timephased grid on the right side of the view by clicking the **Timephased Data** button in the *Display* section of the *Engagements* ribbon. With the timephased grid displayed, you can also use the commands in the *Options* ribbon to customize the display of the timephased data.

475

		Resource Name ↑	Description	State	Committed Units	Committed Start	Committed Finish
☐		Benjamin Montanez	BA work	Proposed	100%	9/9/2019	1/31/2020
☐		Business Analyst	Communicate with the client	Proposed	100%	10/18/2019	11/27/2019
☐		Business Analyst	Business analysis support 2	Proposed	100%	2/10/2020	7/10/2020
☐		Business Analyst	Business analysis support 1	Proposed	100%	6/1/2020	9/25/2020
☐		Business Analyst	Business analysis support 2	Proposed	100%	6/1/2020	9/25/2020
☐		Business Analyst	Business analysis support 1	Proposed	100%	2/10/2020	7/10/2020
☐		Gary Manche	Post implementation closeout	Proposed	100%	6/15/2020	7/24/2020
☐		Gary Manche	Develop customer requirements	Committed	100%	10/21/2019	3/6/2020
☐		Gary Manche	Determine customer requirements	Proposed	100%	3/2/2020	5/1/2020
☐		Larry Barnes	BA support work	Proposed	100%	8/17/2020	9/4/2020
☐		Larry Barnes	BA Work	Proposed	100%	12/2/2019	5/15/2020
☐		Larry Barnes	BA work	Proposed	100%	1/6/2020	10/30/2020
☐		Larry Barnes	Business analysis support	Proposed	100%	5/4/2020	8/27/2021
☐		Sue Uland	BA Work	Committed	100%	11/11/2019	11/22/2019
☐		Sue Uland	BA work	Committed	100%	10/1/2019	10/18/2019

Figure 14 - 20: Summary view with the Sheet display

Viewing Resource Availability

Simply by examining the Resource Engagement information shown on the *Resource Requests* page, it is nearly impossible to determine which engagements to approve as submitted, which ones to edit and approve, and which ones to reject. The missing information needed to make these decisions can be found on the *Capacity Planning* page in PWA. The easiest way to view these two pages is to display each page on its own web browser tab so that you can easily switch back and forth between browser tabs.

In your preferred web browser, leave the *Resource Requests* page displayed on the first browser tab, then add a new browser tab. In the Internet Explorer, you can right-click on the **Resource Requests** tab and then select the **New tab** item on the shortcut menu. In the Google Chrome browser, you can click the **New tab** button (the + button) at the top of the browser application window. On the new browser tab, navigate to the *Resource Center* page, which shows your batch of resources still selected. Click the **Resources** tab to display the *Resources* ribbon, then click the **Capacity Planning** button in the *Navigate* section of the ribbon.

On the *Capacity Planning* page, click the **View** pick list button and select the **Remaining Availability** view in the *Views* section of the *Availability* ribbon. Project Web App displays the *Remaining Availability* view shown previously in Figure 14 - 11. In the *Series* section to the right of the chart, click and unclick the names of resources individually to study their availability to help you make decisions about the approval of pending Resource Engagements for each resource.

Responding to Pending Resource Engagements

On the *Resource Requests* page, you can respond to each pending Resource Engagement in the following manner:

- You can approve the Resource Engagement as submitted.

- You can edit the details of the Resource Engagement and then approve it. For example, you can substitute a new resource for the requested resource, or you can edit other details of the engagement, such as *Committed Units*, *Committed Start*, and/or *Committed Finish*.

- You can reject the Resource Engagement as submitted.

Working in the Resource Center

To approve or reject a Resource Engagement request as submitted, click the **Resource Requests** browser tab and then select the checkboxes of the resources whose engagement requests you want to approve or reject. In the *Update* section of the *Engagements*, click the **Accept** button to approve the selected engagements or click the **Reject** button to reject the selected engagements. For example, notice in Figure 14 - 21 that I am preparing to approve all of the pending Resource Engagements for Benjamin Montanez and Gary Manche, as well as the first two engagements for Larry Barnes. I do not want to approve the other two engagements for Larry Barnes since the first two engagements require his full-time commitment to their projects.

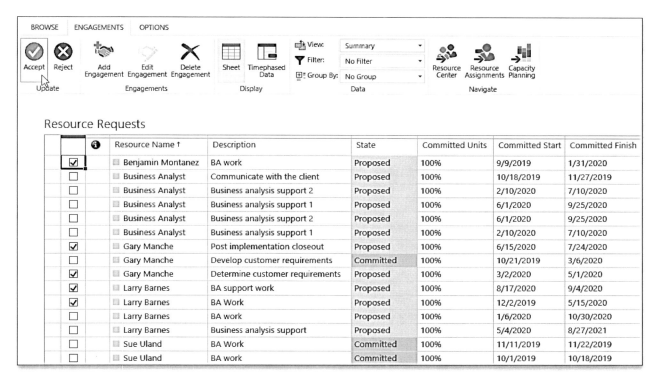

Figure 14 - 21: Prepared to approve multiple Resource Engagements

Project Web App displays the *Confirm Accept* dialog shown in Figure 14 - 22. In the dialog, add an optional comment in the *Comments* field and then click the **OK** button.

Figure 14 - 22: Confirm Accept dialog

477

Project Web App approves the pending Resource Engagements and sets the *State* value to *Committed*. When the project manager opens the enterprise project in Microsoft Project, the system changes the *Engagement Status* value from *Proposed* to *Committed* for the approved Resource Engagement. If you added a comment in the *Confirm Accept* dialog, the system displays a *Note* indicator in the *Indicators* column to the left of the engagement. The project manager can read your comment by double-clicking the engagement to display the *Engagement Information* dialog.

If a project manager creates a Resource Engagement for a Generic resource, your organization may require you to "swap" the Generic resource in an available human resource. To substitute a human resource for the Generic resource, click anywhere in the **Resource Name** cell of the Generic resource, click the pick list arrow button in the cell, and then select the name of the available human resource. Notice in Figure 14 - 23 that I intend to substitute Chuck Kirkpatrick for the Business Analyst generic resource on the *Communicate with the client* engagement.

		Resource Name ↑	Description	State	Committed Units	Committed Start	Committed Finish
☐		Benjamin Montanez	BA work	Committed	100%	9/9/2019	1/31/2020
☐	✳	Business Analyst × ▼	Communicate with the client	Proposed	100%	10/18/2019	11/27/2019
☐		☐ Benjamin Montanez	Business analysis support 2	Proposed	100%	2/10/2020	7/10/2020
☐		☐ Business Analyst	Business analysis support 1	Proposed	100%	6/1/2020	9/25/2020
☐		☐ Chuck Kirkpatrick	Business analysis support 2	Proposed	100%	6/1/2020	9/25/2020
☐		☐ Gary Manche	Business analysis support 1	Proposed	100%	2/10/2020	7/10/2020
☐		☐ Grace Adeyemi	Post implementation closeout	Committed	100%	6/15/2020	7/24/2020
☐		☐ Larry Barnes	Develop customer requirements	Committed	100%	10/21/2019	3/6/2020
☐		☐ Marilyn Ray	Determine customer requirements	Committed	100%	3/2/2020	5/1/2020
☐		☐ Sue Uland	BA support work	Committed	100%	8/17/2020	9/4/2020
☐			BA Work	Committed	100%	12/2/2019	5/15/2020
☐		Larry Barnes	BA work	Proposed	100%	1/6/2020	10/30/2020
☐		Larry Barnes	Business analysis support	Proposed	100%	5/4/2020	8/27/2021
☐		Sue Uland	BA Work	Committed	100%	11/11/2019	11/22/2019
☐		Sue Uland	BA work	Committed	100%	10/1/2019	10/18/2019

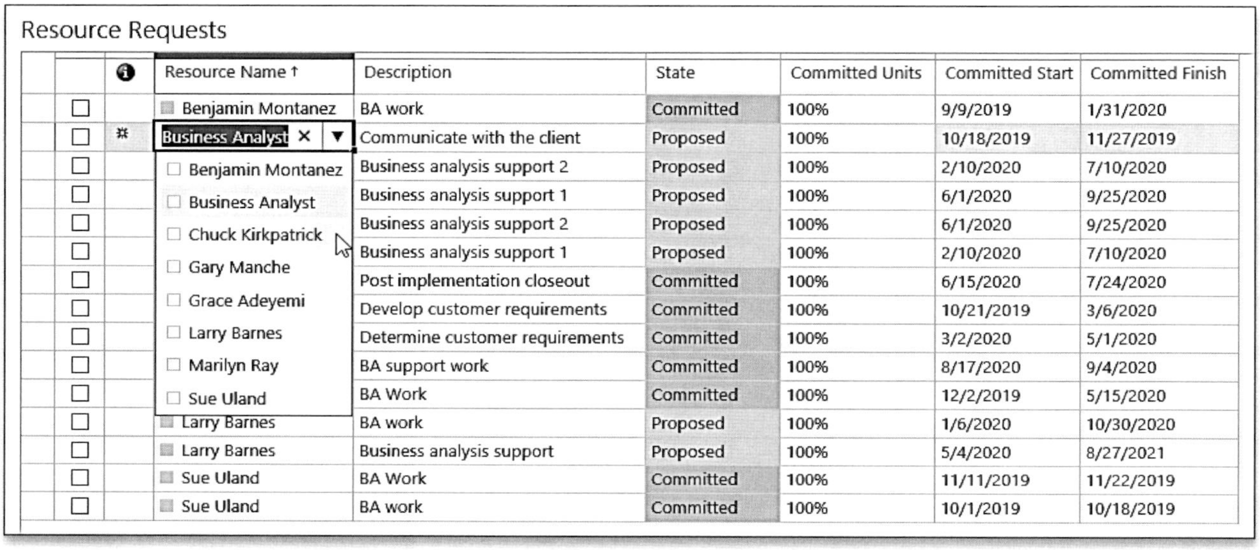

Figure 14 - 23: Swap a human resource for a Generic resource

After substituting the human resource for the Generic resource, Project Web App selects the checkbox for the impacted engagement. In the *Update* section of the *Engagements* ribbon, click the **Accept** button to approve the edited Resource Engagement. In the *Confirm Accept* dialog, add an optional comment in the *Comments* field and then click the **OK** button.

The next time the project manager opens the enterprise project in Microsoft Project, the system displays a yellow band at the top of project schedule, announcing the commitment of new resources in the project, such as shown in Figure 14 - 24. In the yellow band, click the **View Engagements** button to apply the *Resource Plan* view, if not already applied. The system leaves the original Generic resource in the project, but with no Resource Engagement assigned to it. Instead, the system adds the new human resource to the project team and then transfers the original Resource Engagement from the Generic resource to the human resource. In Figure 14 - 24, you can see that the system transferred the *Communicate with the client* engagement to Chuck Kirkpatrick, and changed the *Engagement Status* value from *Proposed* to *Committed*.

Working in the Resource Center

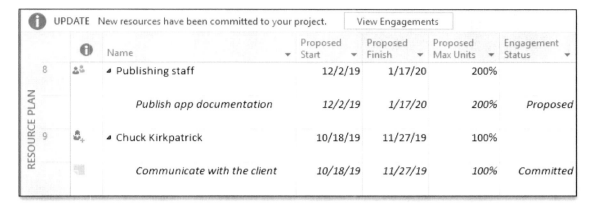

Figure 14 - 24: New resources in the project

In addition to substituting one resource for another on a Resource Engagement, Project Web App also allows you to change other details for the engagement, such as *Committed Units, Committed Start*, and/or *Committed Finish*, based on the availability of the resource in question. For example, notice in Figure 14 - 25 that I substituted Grace Adeyemi and Marilyn Ray for the last two Resource Engagements originally assigned to Larry Barnes, and changed the *Committed Units* value to *50%* for each of them.

		Resource Name ↑	Description	State	Committed Units	Committed Start	Committed Finish
☐		Benjamin Montanez	BA work	Committed	100%	9/9/2019	1/31/2020
☐		Business Analyst	Business analysis support 2	Proposed	100%	2/10/2020	7/10/2020
☐		Business Analyst	Business analysis support 1	Proposed	100%	6/1/2020	9/25/2020
☐		Business Analyst	Business analysis support 2	Proposed	100%	6/1/2020	9/25/2020
☐		Business Analyst	Business analysis support 1	Proposed	100%	2/10/2020	7/10/2020
☐		Chuck Kirkpatrick	Communicate with the client	Committed	100%	10/18/2019	11/27/2019
☐		Gary Manche	Post implementation closeout	Committed	100%	6/15/2020	7/24/2020
☐		Gary Manche	Develop customer requirements	Committed	100%	10/21/2019	3/6/2020
☐		Gary Manche	Determine customer requirements	Committed	100%	3/2/2020	5/1/2020
☐		Larry Barnes	BA support work	Committed	100%	8/17/2020	9/4/2020
☐		~~Larry Barnes~~	~~BA Work~~	~~Committed~~	~~100%~~	~~12/2/2019~~	~~5/15/2020~~
☑	※	Grace Adeyemi	BA work	Proposed	50%	1/6/2020	10/30/2020
☑	※	Marilyn Ray	Business analysis support	Proposed	50%	5/4/2020	8/27/2021
☐		~~Sue Uland~~	~~BA Work~~	~~Committed~~	~~100%~~	~~11/11/2019~~	~~11/22/2019~~
☐		Sue Uland	BA work	Committed	100%	10/1/2019	10/18/2019

Figure 14 - 25: Resource Engagement details edited

After editing Resource Engagement information, Project Web App selects the checkbox for the impacted engagement. In the *Update* section of the *Engagements* ribbon, click the **Accept** button to approve the edited Resource Engagement. In the *Confirm Accept* dialog, add an optional comment in the *Comments* field and then click the **OK** button.

Module 15

Managing Personal Settings

Learning Objectives

After completing this module, you will be able to:

- Set e-mail Alerts and Reminders for yourself
- Set e-mail Alerts and Reminders for your team members
- Manage queued jobs
- Create a delegation session
- Act as a delegate for another user

Inside Module 15

Personal Settings Overview 483
Managing Alerts and Reminders for Yourself 484
Managing Alerts and Reminders for Your Resources 487
Managing My Queued Jobs 490
Managing Delegates 491
Acting as Delegate 495

Personal Settings Overview

Depending on your permissions within the system, Project Web App provides you with a number of personal settings that you can modify to meet your needs. To use and modify these personal settings, click the **Settings** button (it looks like a gear wheel) in the upper right corner of Project Web App and select the **PWA Settings** item on the menu, such as shown in Figure 15 - 1.

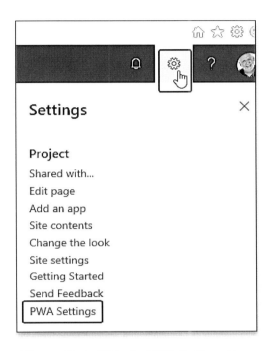

Figure 15 - 1: Click the PWA Settings item

Project Web App displays the *Personal Settings* group in the upper left-hand corner of the *PWA Settings* page as shown in Figure 15 - 2.

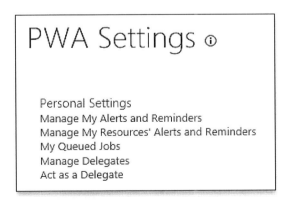

Figure 15 - 2: Personal Settings group

Depending on your permissions Project Web App and the configuration of your Project Online system, the *Personal Settings* group may offer you some or all the following options:

Module 15

- Manage My Alerts and Reminders
- Manage My Resources' Alerts and Reminders
- My Queued Jobs
- Manage Delegates
- Act as a Delegate

I discuss each of these options individually in the remainder of this module.

Warning: Your Project Online application administrator must enable the *Turn on notifications* feature so that the *PWA Settings* page displays the *Manage My Alerts and Reminders* option and the *Manage My Resource's Alerts and Reminders* option.

Managing Alerts and Reminders for Yourself

Project Web App allows you to set up a subscription to receive e-mail Alerts and Reminders from the system. An **Alert** is an e-mail message that the system sends immediately when an event occurs, such as when a project manager publishes a new project. A **Reminder** is an e-mail message that the system sends on a periodic cycle, usually once a day at midnight, to remind you of upcoming or overdue responsibilities, such as an overdue tasks.

To manage your e-mail subscriptions for Alerts and Reminders, click the **Manage My Alerts and Reminders** link in the *Personal Settings* section of the *PWA Settings* page. Project Web App displays the *Manage My Alerts and Reminders* page shown in Figure 15 - 3 and Figure 15 - 4. Because of the extreme length of this page, it was necessary to break up the page into two screenshots for this book.

Notice that the *Manage My Alerts and Reminders* page includes options in four sections. In the *Tasks* section of the page, the default options trigger Project Web App to send you an e-mail message immediately whenever the following occurs:

- A project manager publishes a new project containing one or more tasks assigned to you, or the project manager assigns you to a new task in an existing project and then publishes the project.
- The schedule changes for one or more of your task assignments in an existing project.

Between these two default e-mail subscriptions for Alerts on tasks, the first option is most valuable because you should always notify team members about new task assignments. The second e-mail subscription is problematic, however, because it can lead to a large number of e-mail messages sent to team members every time the project schedule changes. If team members receive too many e-mail messages from Project Web App, they may treat these messages as "spam" and create an Outlook rule to filter out all messages originating with Project Web App.

Best Practice: To reduce the number of e-mail messages that users receive from Project Web App, Projility recommends that *all* project managers and *all* team members **deselect** the *My project tasks are modified* option on the *Manage My Alerts and Reminders* page

The second set of options in the *Tasks* section allows you to subscribe to e-mail Reminders related to specific task criteria for project work. Each night the system tests your criteria and generates an e-mail message containing the task reminders for your subscriptions. You receive an e-mail only if you set reminder criteria and the system finds an appropriate match between your tasks in the system and your specified criteria. Think of these criteria as triggering conditions, which include each of the following:

- Before a task starts
- Before a task is due
- Until a task is complete or becomes overdue
- When you have an incomplete task
- When you have an overdue task
- Until an overdue task is complete

Notice in Figure 15 - 3 that the default settings do not select any of these options. If you select any of these options, however, you should also specify the frequency, as you do not have to receive these messages every day unless you prefer daily delivery.

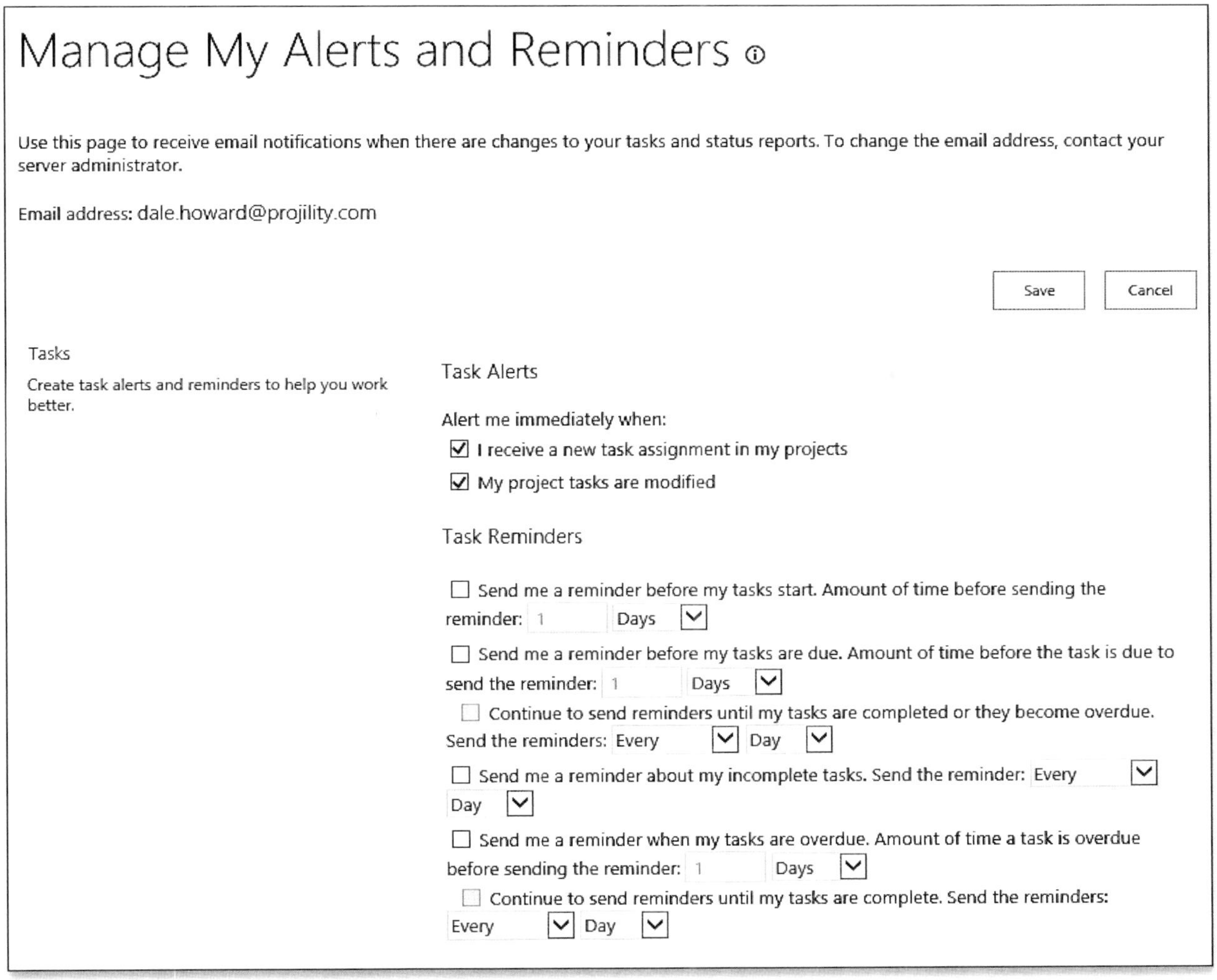

Figure 15 - 3: Manage My Alerts and Reminders – top of page

Module 15

The options in the *Status Reports* section shown in Figure 15 - 4 are similar to those in the *Tasks* section. The default permission for Alerts on status reports causes the system to send you an e-mail message immediately when a manager includes you in a new status report request. The second set of options allows you to subscribe to reminders for status reports.

Information: I do not discuss the Status Reports feature in this book as it is a feature most organizations do not use. In fact, since I began working with Microsoft's PPM tools beginning with Project Central in the year 2000, not one of my clients uses the Status Reports feature!

The *Queue Job Failures* section shown in Figure 15 - 4 includes only a single option. This option causes Project Web App to send you an e-mail message immediately if any job you send to the Queue fails in the queuing process. For example, if you submit a timesheet or a task update, each of these constitutes a job sent to the Queue for processing. If the job fails, the system immediately sends you an e-mail message.

The *Language Setting* section shown in Figure 15 - 4 contains a single option that allows you to set your language preference for e-mail messages sent to you by Project Web App. Select the language you want, if necessary, and then click the **Save** button to save the selections you specify.

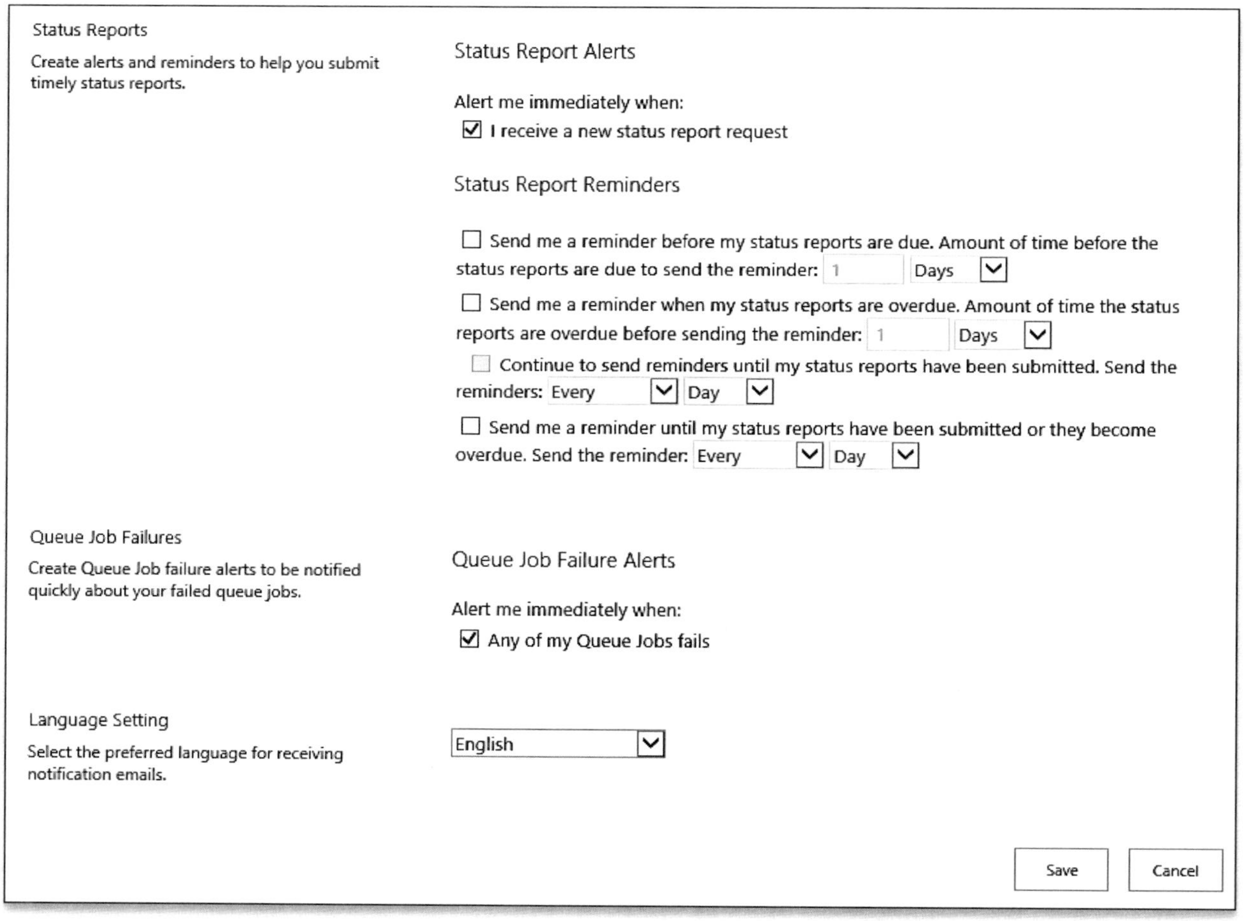

Figure 15 - 4: Manage My Alerts and Reminders – bottom of page

Managing Alerts and Reminders for Your Resources

In addition to managing e-mail subscriptions for Alerts and Reminders for yourself, Project Web App also allows you to manage e-mail subscriptions for your team members and your resources (if you are also a resource manager). The system defines "your team members" as those resources who are a team members in your projects and are assigned to at least one task. The system defines "your resources" as any resource included in a status report you created. To set e-mail subscriptions for your team members and resources, click the **Manage My Resource's Alerts and Reminders** link in the *Personal Settings* section of the *PWA Settings* page. Project Web App displays the *Manage My Resource's Alerts and Reminders* page shown in Figure 15 - 5 and Figure 15 - 6. Because of the extreme length of this page, it was necessary to break up the page into two screenshots for this book.

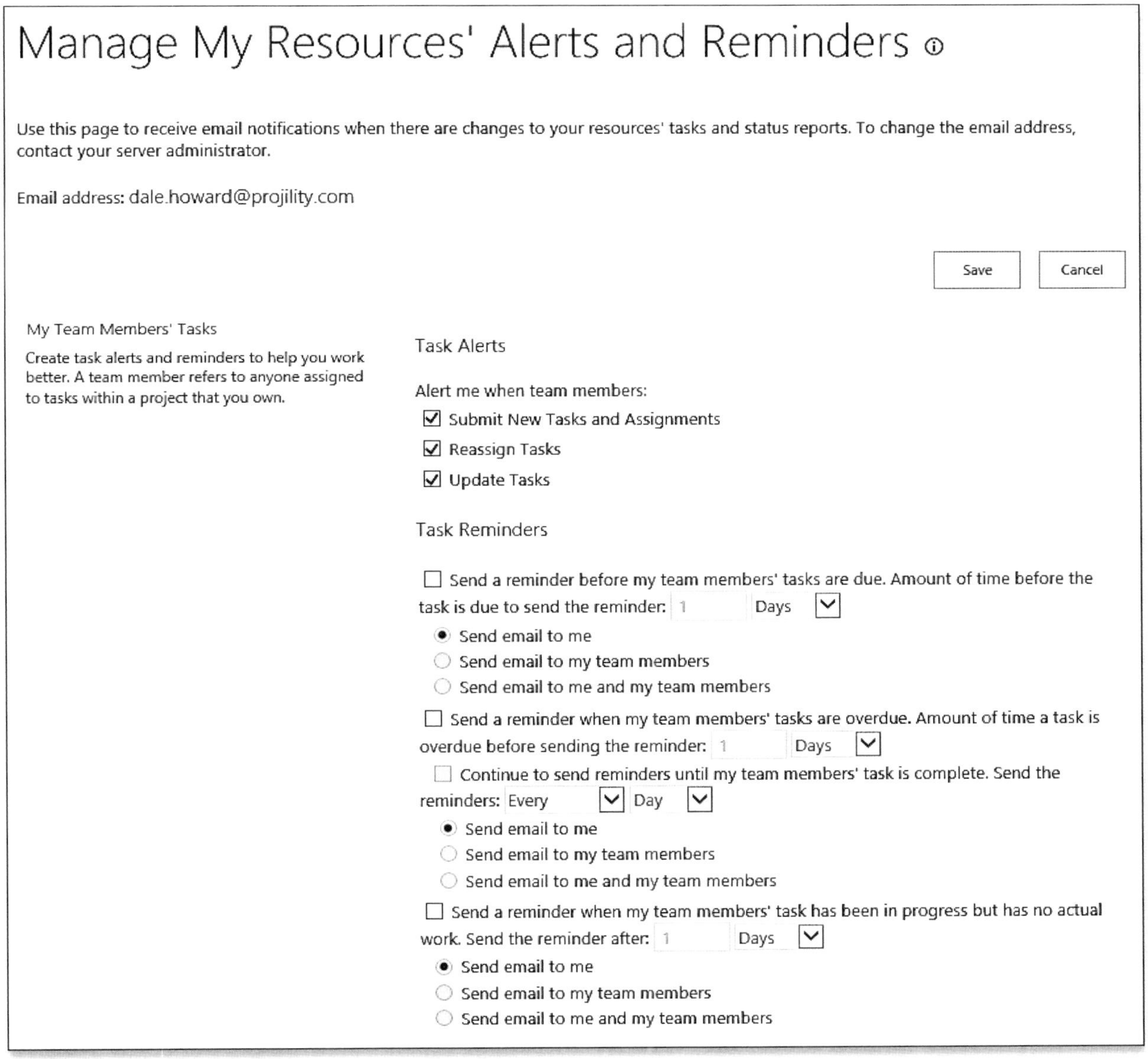

Figure 15 - 5: Manage My Resources' Alerts and Reminders – top of page

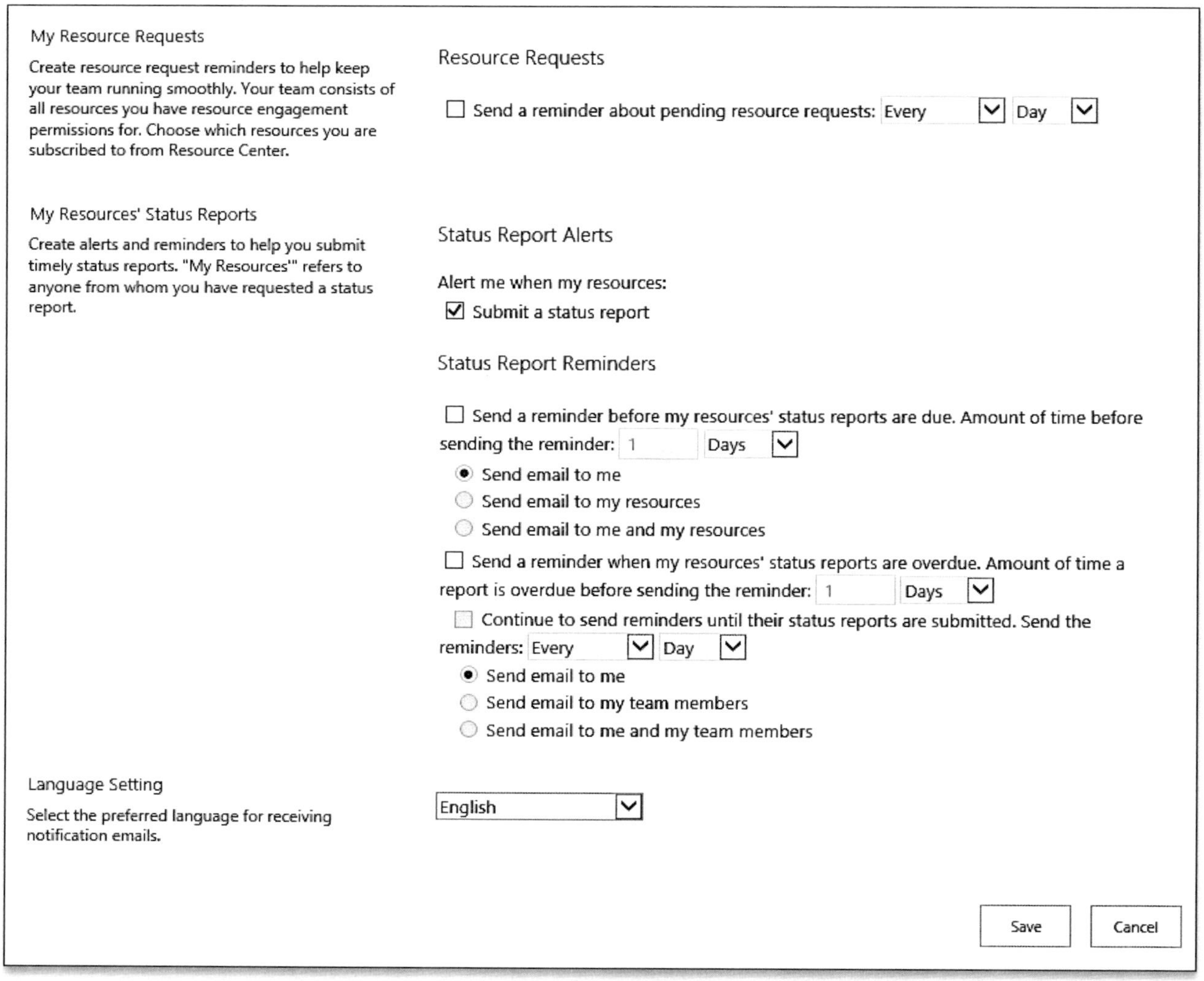

Figure 15 - 6: Manage My Resources' Alerts and Reminders – bottom of page

Notice in Figure 15 - 5 and Figure 15 - 6 shown previously that the *Manage My Resource's Alerts and Reminders* page layout is very similar to the *Manage My Alerts and Reminders* page. Notice that the default options in the *Task Alerts* section of the page cause Project Web App to send you an e-mail Alert immediately when one of the following triggering events occurs:

- A team member submits a New Task request or a New Assignment request to you.

- A team member reassigns a task to another team member.

- A team member submits task progress to you.

Because these three default options can lead to a high volume of e-mail messages sent to you by Project Web App, you may wish to deselect one or more of these options. Of the three, the *Update Tasks* option causes the system to send the most e-mail messages.

Managing Personal Settings

Best Practice: Projility recommends that you **deselect** the **Update Tasks** checkbox, and that you leave the other two checkboxes selected in the *Task Alerts* section of the page. Doing this reduces the volume of e-mails that you receive from Project Web App, alerting you only when an unusual event occurs, such as when a team member proposes a new task in one of your projects.

The *Task Reminders* section of the page allows you to set up subscriptions for e-mail Reminders for your team members about their project work. When you set up reminder subscriptions for your team members, you may choose to have the Reminders sent only to you, only to the team members, or to both you and your team members. Select the reminders for your team members and specify who receives the e-mail Reminders.

The single option in the *Resource Requests* section works in conjunction with the Resource Engagements feature in Microsoft Project. This option is relevant to you if both you and your organization meet the following criteria:

- You have resource management responsibilities, even though you are primarily a project manager.

- Project Managers in your organization create Resource Engagements for Generic or human resources in their enterprise projects.

- Your organization requires you to approve or reject pending Resource Engagements for the resources who report to you.

If you and your organization meet all three of the preceding criteria, then you probably want to receive Alerts about new Resource Engagements for the resources who report to you. If so, select the **Send a reminder about pending resource requests** checkbox and leave the frequency set to **Every Day**.

Information: If you have resource management responsibilities, then you **must** tell Project Web App which resources report to you. To do this, navigate to the *Resource Center* page and select the checkbox for every resource who reports to you. In the *Share and Track* section of the *Resources* ribbon, click the **Request Reminders** pick list and select the **Subscribe to selected resources** item on the pick list. Click the **OK** button in the confirmation dialog.

The *Status Report Alerts* section contains only a single option, selected by default. This option causes Project Web App to send you an e-mail Alert immediately when a resource submits a status report to you. Again, because this option can lead to a flurry of e-mail messages, you may want to deselect it.

The *Status Report Reminders* section allows you to set up subscriptions for e-mail Reminders for those resources assigned to Status Report requests you create. Again, you may choose to have the Reminders sent only to you, only to the resources, or to both you and your resources.

Warning: Be very wary of setting up subscriptions for e-mail Reminders for your team members. The Reminder options you select on the *Manage My Resource's Alerts and Reminders* page **override** the same options specified by your team members on their *Manage My Alerts and Reminders* page. This can cause Project Web App to send a flurry of e-mail messages to your team members, with no way for the team members to stop receiving the e-mails, resulting in very frustrated team members.

In the *Language Setting* section, specify your language preference. Click the **Save** button to save your settings.

Managing My Queued Jobs

Every time you stand in line at a fast food restaurant, you are waiting in a "queue." In Project Web App, the Queue is a waiting line that is necessary whenever the number of service requests in the system is greater than the system's optimum service capacity. Whenever you save and publish a project, or you submit a timesheet or a task update, the system places those jobs in the Queue for processing. The Queue processes jobs on a "first in/first out" basis, by the way.

Project Web App allows you to view your current jobs in the Queue by clicking the **My Queued Jobs** link in the *Personal Settings* section of the *PWA Settings* page. The system displays the *My Queued Jobs* page shown in Figure 15 - 7.

Figure 15 - 7: My Queued Jobs page

The *My Queued Jobs* page should normally appear blank, as shown previously in Figure 15 - 7. A blank page means, "no news is good news" and indicates that Project Web App is running without errors. If you see a queue job on the *My Queued Jobs* page, you can actually watch the job's progress in the system by clicking the **Refresh Status** button in the *View* section of the *Jobs* ribbon.

In addition to being able to view jobs currently processing in the Queue, Project Web App also allows you to view the history of completed jobs. In the *View* section of the *Jobs* ribbon, click the **View** pick list and select one of the following views:

- In Progress and Failed Jobs in the Past Week (the default view)
- All In Progress and Failed Jobs
- Successful Jobs in the Past Week
- All Successful Jobs
- All Jobs in the Past Week
- All Jobs

For example, Figure 15 - 8 shows the *My Queued Jobs* page with the *Successful Jobs in the Past Week* view applied. Notice that the page shows a number of different job types, such as *Project Publish*, *Reporting*, and *Project Checkin* jobs, for example

Managing Personal Settings

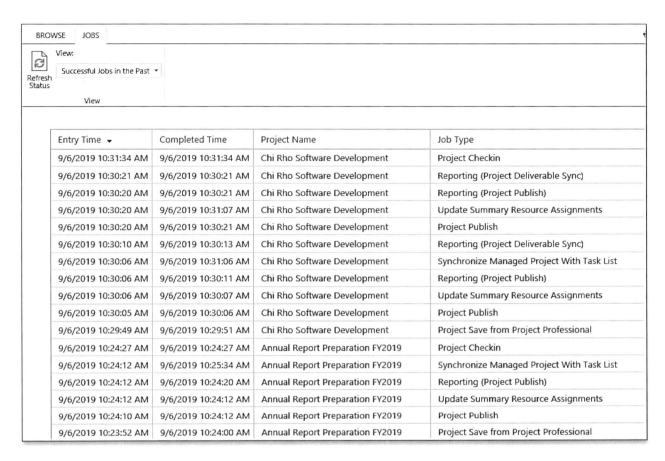

Figure 15 - 8: Successful Jobs in the Past Week view

If you notice a job that the system simply does not process, or a job that failed, you should contact your Project Online application administrator immediately for assistance. To help the application administrator, click the **Click to view the error details** link in the **Error** column. The *Queue Job Error Details* dialog contains valuable information that can help your application administrator to diagnose and resolve the Queue error. To share the error information with your application administrator, click the **Copy to Clipboard** button in the dialog and paste the error contents into an e-mail message.

Managing Delegates

Project Web App allows you to designate another user as a delegate who can act on your behalf in the system, such as when you know you will be away when timesheets are due. In this case, the **Delegation** feature in Project Web App would allow you to appoint a co-worker as your delegate so that the user can submit your timesheet for you. If you have expanded permissions in PWA set by your application administrator, you can also appoint one user to serve as a delegate for another user.

To begin the process of creating a delegation session, the **Manage Delegates** link in the *Personal Settings* section of the *PWA Settings* page. The system displays the *Manage Delegates* page shown in Figure 15 - 9.

491

Figure 15 - 9: Manage Delegates page

To create a new delegation session, click the **New** button in the *Delegate* section of the *Delegations* ribbon. The system displays the *Add Delegation* page shown in Figure 15 - 10.

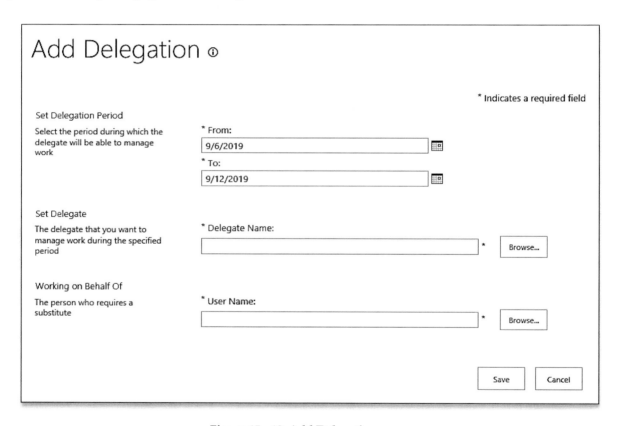

Figure 15 - 10: Add Delegation page

To add a new delegation session, begin the process by setting a date range during which the delegation is effective by selecting dates in the **From** and **To** fields. Specify the person whom you want to act on your behalf by clicking the **Browse** button for the **Delegate Name** field, selecting the user's name in the *Choose User* dialog, and then clicking the **OK** button. Click the **Browse** button for **User Name** field, select your own name in the *Choose User* dialog,

Managing Personal Settings

and then click the **OK** button. Click the **Save** button to save the delegation session. Project Web App displays the *Manage Delegates* page with the new delegation session listed, such as shown in Figure 15 - 11.

Figure 15 - 11: Manage Delegates page with new delegation

By default, the *Manage Delegates* page displays only delegations that occur during the current one-week time period beginning with today, but does not show any delegations planned two weeks or more in the future. To see all of your delegation sessions, click the **Filters** toggle button in the *Data* section of the *Delegations* ribbon. Project Web App displays the *Filters* pane shown in Figure 15 - 12.

Figure 15 - 12: Filters pane in the Manage Delegates page

In the *Filters* pane, you can show other delegation sessions by selecting the **Date range** option, and then setting the date range by selecting dates in the **From** and **To** fields. The system also gives you multiple options for filtering by users as well. After selecting your filtering options, click the **Apply** button.

493

Figure 15 - 13 shows the *Manage Delegations* page with a filter set to display delegation sessions through December of the current year. Notice that the *Manage Delegations* page now displays three delegation sessions. To hide the *Filters* pane, click the **Filters** toggle button again.

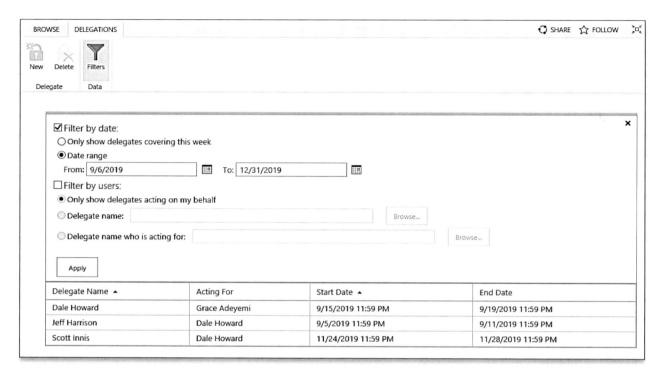

Figure 15 - 13: Manage Delegations page – filter through December 2019

To edit an existing delegation session, click the name of the user in the *Delegate Name* column. Project Web App displays the *Modify Delegation* page, which is identical to the *Add Delegation* page shown previously in Figure 15 - 10. Modify your delegation session as needed and then click the **Save** button when finished.

To delete an existing delegation session, click anywhere in the line for the delegation session you want to delete, but *do not* click the user name in the *Delegate Name* column. In the *Delegate* section of the *Delegations* ribbon, click the **Delete** button. Project Web App displays the confirmation dialog shown in Figure 15 - 14. Click the **OK** button in the confirmation dialog to delete the delegation session.

Figure 15 - 14: Confirmation dialog

Acting as Delegate

When another user selects you as a delegate by creating a delegation session, you can act on behalf of that person to perform most functions within Project Web App. To act as a delegate, click the **Act as a Delegate** link in the *Personal Settings* section of the *PWA Settings* page. The system displays the *Act as a Delegate* page, such as shown in Figure 15 - 15. Notice that Grace Adeyemi created a delegation session for me during the time she will be away on PTO. In this delegation session, I will work on behalf of Grace during the two-week period from September 5-19.

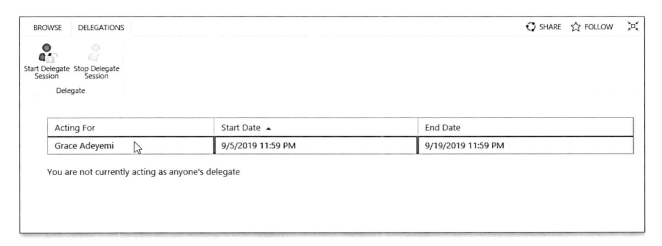

Figure 15 - 15: Act as a Delegate page

To start a delegation session, select the session you want to start by clicking anywhere in the row, such as shown previously in Figure 15 - 15, and then click the **Start Delegate Session** button in the *Delegate* section of the *Delegations* ribbon. The *Act as a Delegate* page displays a gold warning band at the top of the page with a message indicating that the delegation session is active, such as shown in Figure 15 - 16.

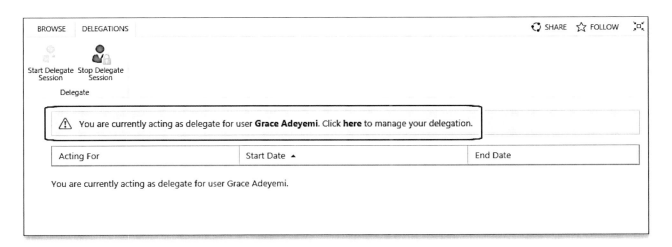

Figure 15 - 16: Act as Delegate page with delegation session in progress

Module 15

This warning band displays across the top of every page in Project Web App while you are acting as a delegate to remind you that you are now working in Project Web App on behalf of someone else. After you complete your work on behalf of the other user, you can return to the *Act as Delegate* page by clicking the **here** link in the gold warning band. In the *Delegate* section of the *Delegations* ribbon, click the Stop **Delegate Session** button to end the delegation session. The system exits the delegate session and the page displays the *You are not currently acting as anyone's delegate* message immediately below the data grid.

Warning: The Delegation feature ***only*** applies to the pages displayed in Project Web App, such as the *Tasks* page or the *Timesheet* page, for example. This feature ***does not*** apply to the Project Sites for any projects. If you navigate to a Project Site, the gold warning band disappears, indicating that you accessed the Project Site as yourself, not as the person named in the delegation session. This feature ***does not*** apply to Microsoft Project either. While a delegation session is active, if you launch Microsoft Project and attempt to connect to Project Web App, the system displays an error message and will not allow you to connect.

496

Module 16

Index

A

Act as a Delegate page 495–96

Adding Task Notes in Microsoft Project 124–25

Analyzing Variance

    analyzing Cost variance in Microsoft Project 338–39

    analyzing Date variance in Microsoft Project 335–37

    analyzing variance in Microsot Project 333–39

    analyzing Work variance in Microsoft Project 337–38

Approval Center page

    accessing 301–2

    approving pending task updates 313

    creating rules to auto approve task updates 314–20

    overview of the weekly approval process 306–8

    rejecting pending task updates 311–12

    reviewing pending task updates 309–11

    setting up for best use 301–5

    understanding 302–3

Approving Task Updates

    accessing the Approval Center page 301–2

    approving pending task updates 8, 313

    axioms for success with tracking task progress 328–29

    creating rules to auto approve task updates 314–20

    overview of the weekly approval process 306–8

    rejecting pending task updates 8, 311–12

    rescheduling incomplete work in the past 320–24

    reviewing pending task updates 309–11

    setting up the Approval Center page 301–5

    understanding the Approval Center page 302–3

    updating actuals for Expense Cost resources 326–28

Index

viewing pending task updates on the Home page of Project Web App 309–11

Assign Resources Dialog, assigning resources to tasks **174–76**

Assigning Resources to Tasks **169–95, 452–53**

Assignment Details page .... *See* **Tasks page**

Assignment Planning

    assigning Budget Cost resources 189–93

    assigning Cost resources 189–95

    assigning Expense Cost resources to tasks 193–95

    assigning Material resources to tasks 197–99

    assigning resources to tasks 169–95

    assigning resources to tasks in a PWA project 452–53

    understanding Effort Driven 183–87

    understanding Task Types 179–81

    understanding the Duration Equation 178–79

    using the Assign Resources dialog 174–76

    using the Task Entry view 169–72

Auto Scheduled Tasks **113–14**

Availability, defined **464**

B

Baselines

    backing up a saved baseline 226–27

    backing up a saved baseline in a PWA project 453–54

    backing up after updating the baseline 347–48

    baseline fields 224

    baselining only selected tasks 345–47

    clearing a saved baseline 227–28

    defined 224

    overview 224–25

    saving 225–26

    saving a baseline in a PWA project 453–54

    updating after a Change Control process 345–48

    updating after creating Deliverables 247

Budget Cost Resources, assigning to tasks **189–93**

Build Team Dialog

    adding Cost resources to the project team 189

adding enterprise resources to the project team	162
adding local resources to the project team	165–66
Booking Type values	163
building a project team	152–63
filtering enterprise resources	154–57
grouping enterprise resources	158
matching resources	162–63
Proposed vs Committed team members	163
skill matching	162–63
viewing enterprise resource information	158–59
viewing resource availability	159–61
Build Team page in PWA	**451–52**

C

Capacity Planning page	**159–61, 463–70**
Capacity, defined	**464**
Cell Background Formatting	**125–27**
Change Control	
adding new tasks to the project schedule	343
backing up a baseline after updating	347–48
baselining only selected tasks	345–47
overview	343
updating the baseline	345–48
Changing the Project Manager of a Project	**79–83**
Check In an Enterprise Project	**34–38, 436–38**
Check In, defined	**5**
Check Out an Enterprise Project	**37**
Check Out, defined	**5**
Close an Enterprise Project	**34–38, 454–55**
Combination View in Microsoft Project, defined	**17**
Constraints, setting in Microsoft Project	**137–40**
Creating a New Project	
from an Enterprise Project Type	70–77
from Project Web App	70–77, 446–49
using the Strategic Impact page	74–75

Index

    using the Workflow Status page 73–74

Creating a Project Web App Login Account 29–33

Critical Path

    defined 221

    viewing 221–23

    viewing in a PWA project 453

    viewing nearly Critical tasks 223

Critical tasks .... *See* **Critical Path**

D

Deadline dates, setting in Microsoft Project 141–42

Defining a New Project Using a 6-Step Method 85–98

Delegation, defined 491

Deleting a Task in Microsoft Project 116–17

Deliverables

    creating a Deliverable dependency 244–47

    creating a new Deliverable 235–39

    deleting a Deliverable 239–41

    editing a Deliverable 239–41

    overview 235

    updating the Baseline after creating 247

    viewing in the Project Site 242–44

Dependencies, setting in Microsoft Project 128–34

Documents .... *See* **Project Site**

Duration Equation 178–79

E

Effort Driven Tasks 183–87

Enterprise Global, defined 5

Enterprise Project

    assigning resources to tasks in a PWA project 452–53

    assigning resources to tasks in Microsoft Project 169–95

    building a project team 100–103, 152–63, 451–52

    checking in 5

    checking in with Microsoft Project 34–38

Index

checking in with Project Web App	77, 436–38
checking out	5
checking out in Microsoft Project	37
closing in Microsoft Project	34–38
closing in Project Web App	76
creating a Deliverable	235–39
creating a Deliverable dependency	244–47
creating a master project from the Project Center page	435–36
creating from an Enterprise Project Type	70–77
creating from Microsoft Project	83–84
creating from Project Web App	70–77, 446–49
creating Resource Engagements with Generic resources	103–8
defined	4
defining using a 6-step method	85–98
display the Project Summary Task	88–89
entering the project Properties	87–88
opening from the Project Center page	434–36
opening in Microsoft Project	34–38
publishing	6, 8, 229–30, 328, 453–54
publishing a PWA project	453–54
publishing overview	228–29
saving	6, 8, 97–98
saving a baseline	225–26
saving a baseline in a PWA project	453–54
setting a new Owner	79–81
setting custom permissions	230–34, 438–41
setting project Options	93–97
setting the Project calendar and Nonworking Time calendar	89–93
setting the project working schedule	89–93
setting the Resource Utilization value	100
setting the Start date	85–87
setting the Status Manager for every task	81–83
specifying enterprise field values	85–87
submitting a proposed project to a SharePoint workflow	75
submitting Resource Engagements	108–9
task planning process in a PWA project	449–51
task planning process, overview	113
transferring to a new project manager	79–81
using Resource Engagements to forecast resource utilization	99–109

501

Index

using the Strategic Impact page when creating a new project 74–75
using the Workflow Stage Status page when creating a new project 73–74

Enterprise Project Type (EPT), defined **69**

Enterprise Resource
defined 4–5

Enterprise Resource Pool
defined 5, 151–52
enterprise resource types 151–52
view in the Resource Center page 459

Expense Cost Resources
assigning to tasks 193–95
updating with actual cost data 326–28

F

Filters
using in Microsoft Project 21–25
using with Project Web App views 59–60

Fixed Duration tasks .... *See* **Task Types**

Fixed Units tasks .... *See* **Task Types**

Fixed Work tasks .... *See* **Task Types**

Force Check In an Enterprise Project 436–38

G

Groups
using in Microsoft Project 26–29
using with Project Web App views 60–61

I

Inserting Tasks in Microsoft Project 115–16

Issues .... *See* **Project Site**

L

Leveling .... *See* **Leveling Overallocated Resources**

Leveling Gantt view 208–10

Leveling Overallocated Resources
    clearing leveling results 210
    leveling an overallocated resource 207–8
    leveling defined 199
    setting leveling options 205–7
    setting task Priority numbers 210–11
    using a leveling methodology 204
    viewing leveling results 208–10

Local Cache
    adjusting cache settings 39–40
    cleaning up cache contents 42–44
    defined 39
    resolving cache corruption 44–45
    viewing cache contents 40–41

Local Project Cache ....*See* **Local Cache**

Local Resources, using in an enterprise project 165–66

M

Manage Delegates page 491–94

Manage My Alerts and Reminders page 484–86

Manage My Resource's Alerts and Reminders page 487–89

Manage Timesheets page, accessing 251–52

Manually Scheduled Tasks 113–14

Master Project, creating in Microsoft Project from the Project Center page 435–36

Material Resources, assigning to tasks 197–99

Microsoft Project Data Model 15

Milestones
    creating in Microsoft Project 121–22
    defined 121

Moving a Task in Microsoft Project 114–15

My Queued Jobs page 490–91

503

N

Newsfeed .... *See* **Project Site**

Non-Effort Driven Tasks 183–87

O

Offline Mode, when working with Microsoft Project 45–50

OneNote notebook .... *See* **Project Site**

Open an Enterprise Project 34–38

Overallocated Resources .... *See* **Resource Overallocations**

Overallocation .... *See* **Resource Overallocations**

Owner, changing to a new project manager 79–81

P

Planning Wizard Messages about Constraints in Microsoft Project 140–41

Portfolio, defined 3

Power BI

    creating natural language queries in Power BI 370–71

    filtering in Power BI reports 368–69

    reporting in Project Online 359–71

    using the Project Online Content Pack 359–67

PPM Terminology 3–4

Priority Numbers, tasks .... *See* **Leveling Overallocated Resources**

Program, defined 3

Project Center

    accessing 427

    assigning resources to tasks in a PWA project 452–53

    building a project team in a PWA project 451–52

    checking in a checked out project 436–38

    closing a PWA project 454–55

    creating a master project in Microsoft Project 435–36

    creating a new PWA project 446–49

    detailed Project views 441–45

    force check in a checked out project 436–38

indicators in the Indicators column, understanding 431–32
Open Menu button, using 432
opening a project in Microsoft Project 434–36
opening a PWA project 454–55, 454–55
Project views 441–45
Projects ribbon, using 428
publishing a PWA project 453–54
saving a baseline in a PWA project 453–54
setting custom permissions 438–41
Show/Hide options 430–31
task planning process in a PWA project 449–51
views, using 428–30

Project Communications Life Cycle 6–9

Project Documents .... *See* **Project Site**

Project Online vs Project Server 2019 Comparison 3–4

Project Online, explained 3–4

Project Options, setting 93–97

Project Properties, setting 87–88

Project Server 2019, explained 3–4

Project Site

accessing a Project Site 375–77
Documents library
    accessing 393–94
    attaching tasks to a document 412–19
    checking out a document for editing 402–3
    creating a new document 394–95
    creating a new folder 395–96
    editing a document 403–6
    e-mail alerts about document changes 411–12
    managing project Documents 393–412
    renaming a document 399–400
    sharing a document 401–2
    uploading a document 396–99
    uploading a template 396–99
    version history for a document 407–11
Issues
    attaching tasks to an Issue 412–19
    creating a new Issue 390–92

Index

  managing project Issues 389–92
   working with existing Issues 392
 Newsfeed, using 422–23
 OneNote notebook, using 419–20
 overview 375
 Risks
  applying a Risk view 383–84
  attaching tasks to a Risk 412–19
  autofiltering 384–85
  creating a new Risk 379–82
  deleting an existing Risk 386–87
  editing existing Risks 385–86
  e-mail alerts about changes to a Risk 387–89
  managing 378–89
  sorting 384–85
  viewing existing Risks 385–86
  working with existing Risks 383–89
 role-based permissions for Project Site access 375
 Tasks page, using 420–22
 viewing assigned Issues and Risks 392–93
 viewing Deliverables 242–44

Project Start Date, setting 85–87

Project Summary Task, displaying in a project 88–89

Project Update Life Cycle 301

Project Web App
 accessing 53–55
 Act as a Delegate page 495–96
 Approval Center page 302–20
 Assignment Details page 280–83
 Build Team page 451–52
 Capacity Planning page 159–61, 463–70
 creating an enterprise project from 70–77
 exporting a data grid 65–66
 features and functionality 55–66
 filters 59–60
 Force Check-in Enterprise Projects page 437
 groups 60–61
 Home page 53

Index

    Issues and Risks page 392–93
    manage Alerts and Reminders for your resources 487–89
    manage Alerts and Reminders for yourself 484–86
    Manage Alerts and Reminders page 484–86
    Manage Delegates page 491–94
    Manage My Resource's Alerts and Reminders page 487–89
    manipulating a data grid 61–64
    My Queued Jobs page 490–91
    overview 53
    personal settings overivew 483–84
    printing a data grid 64–65
    Project Center page 427–55
    Project Permissions page 438–41
    PWA Settings page 483–84
    Resource Assignments page 471–72
    Resource Center page 459–79
    Resource Requests page 473–76
    Rules page 314
    Tasks page 278–94
    Timesheet page 251–77
    views 57–61

Project Web App Login Account
    creating in Microsoft Project 29–33

Project, defined 3

Publishing a Project
    overview 6, 8, 228–29
    publishing a PWA project 453–54
    publishing after processing weekly task updates 328
    publishing an enterprise project 229–30
    setting custom permissions 230–34

PWA ..... *See* **Project Web App**

Q

Queue, defined 490
Quick Access Toolbar, understanding in Microsoft Project 13

R

Reporting
    adding tasks to the Timeline view in Microsoft Project 350
    arranging tasks in the Timeline view in Microsoft Project 350–51
    creating natural language queries in Power BI 370–71
    customizing a Dashboard Report chart in Microsoft Project 354–55
    customizing a Dashboard Report table in Microsoft Project 356–57
    filtering in Power BI reports 368–69
    formatting tasks in the Timeline view in Microsoft Project 351–52
    overview 349
    using Dashboard Reports in Microsoft Project 353–57
    using Power BI reports 359–71
    using the Project Online Content Pack for Power BI 359–67
    using the Timeline view in Microsoft Project 349–52

Rescheduling Incomplete Work in the Past **320–24**

Resource Assignments page **471–72**

Resource Availability, defined **464**

Resource Capacity, defined **464**

Resource Center
    accessing 459–60
    approving or rejecting Resource Engagement requests 473–76
    Capacity Planning page 463–70
    filtering enterprise resources 460
    Resource Assignments page 471–72
    Resource Requests page 473–76
    selecting and deselecting resources 461–62
    viewing resource assignments 471–72
    viewing resource availability 463–70
    views, using 460–61

Resource Center page
    subscribe to e-mails about Resource Engagements 489

Resource Engagements
    approval overview 473
    approving or rejecting pending Resource Engagements 476–79
    building a project team with Generic resources 100–103
    creating with Generic resources 103–8

creating with human resources 213–16
Resource Requests page 473–76
setting the Resource Utilization value 100
submitting 108–9
using with Generic resources to forecast resource utilization 99–109
viewing pending engagements 473–76
viewing resource availability prior to approval 476

Resource Leveling .... *See* **Leveling Overallocated Resources**

Resource Leveling dialog .... *See* **Leveling Overallocated Resources**

Resource Overallocations

clearing leveling results 210
leveling an overallocated resource 204–11
leveling overallocated resources 207–8
locating and analyzing 200–202
overallocation defined 199
setting leveling options 205–7
setting task Priority numbers 210–11
using a leveling methodology 204
viewing leveling results 208–10

Resource Requests page **473–76**

Revising a Project Plan **341**

Ribbons

Availability ribbon, Capacity Planning page 464
Design ribbon with the Chart Tools applied in Microsoft Project 355
Design ribbon with the Table Tools applied in Microsoft Project 356
Engagements ribbon, Resource Requests page 475
Format ribbon with the Chart Tools applied in Microsoft Project 355
Format ribbon with the Table Tools applied in Microsoft Project 357
Format ribbon with the Timeline Tools applied in Microsoft Project 351
Options ribbon, Resource Requests page 475
Options ribbon, Schedule PDP 448
Options ribbon, Timesheet page 259
Project ribbon, Schedule PDP 449
Projects ribbon, Project Center page 428
Report ribbon in Microsoft Project 353
Resources ribbon, Resource Center page 460
Task ribbon, Schedule PDP 443, 449
Team ribbon, Build Team page 452

Index

    understanding in Microsoft Project 13

Risks .... *See* **Project Site**

S

Saving a Project
    overview 6, 8
    saving an enterprise project 97–98

SharePoint .... *See* **Project Site**

SharePoint Deliverables .... *See* **Deliverables**

Single View in Microsoft Project, defined 17

Status Manager, setting for every task in a project 81–83

Subtasks, creating in Microsoft Project 118–21

Summary Tasks, creating in Microsoft Project 118–21

T

Tables, using in Microsoft Project 19–20

Task Calendars, setting in Microsoft Project 144–47

Task Dependencies, setting in Microsoft Project 128–34

Task Entry view, assigning resources to tasks 169–72

Task Mode Setting 113–14

Task Notes, setting in Microsoft Project 124–25

Task Planning Process
    adding task notes 124–25
    creating milestones 121–22
    creating summary tasks and subtasks 118–21
    creating the Work Breakdown Structure (WBS) 118–21
    deleting tasks 116–17
    entering new tasks 114
    estimating task Durations 147–48
    inserting tasks 115–16
    moving tasks 114–15
    setting Deadline dates 141–42
    setting Task calendars 144–47
    setting task constraints 137–40

    setting task dependencies 128–34
    Task Mode setting 113–14
    task planning process in a PWA project 449–51
    understanding 113
    understanding constraints and Deadline dates 136–44
    understanding missed constraints and Deadline dates 143–44
    understanding Planning Wizard messages about constraints 140–41
    using cell background formatting 125–27

Task Types **179–81**

Task Updates
    submitting 7, 270–71

Tasks page
    accessing the Assignment Details page 280–83
    adding a new proposed task 289–90
    adding a Team task 291–92
    adding tasks to the Tasks page 289–92
    adding yourself to an existing task 290–91
    methods for tracking task progress 278
    reassigning a task to a fellow team member 293–94
    removing a task 292–93
    reporting task progress 283–89
    tracking progress using Actual Work Done and Work Remaining 286–88
    tracking progress using Hours of Work Done Per Period 288–89
    tracking progress using Percent of Work Complete 284–86
    understanding 278–80

Team Task, defined **265, 291**

Timeline View
    adding tasks to the Timeline view in Microsoft Project 350
    arranging tasks in the Timeline view in Microsoft Project 350–51
    formatting tasks in the Timeline view in Microsoft Project 351–52
    reporting in Microsoft Project 349–52

Timesheet page
    accessing 251–52
    adding a new proposed task 263–64
    adding a new row 261–67
    adding a non-project line 266–67
    adding a personal task 267
    adding an Administrative line 266–67

Index

adding an existing task	262–63
adding Team tasks	265–66
adding yourself to a task	264–65
deleting a timesheet	274–75
entering time	260–61
reassigning a task to a fellow team member	268–69
recalling a submitted timesheet	273–74
removing a timesheet line	269
responding to a rejected timesheet	275
setting up for best use	255–58
submitting for approval	270–71
submitting planned PTO in the future	275–77
tracking time overview	251
understanding	253–55
using	259

Tracking Gantt View **335–36**

Tracking Task Progress

manually entering task progress in Microsoft Project	294–98
overview	251

V

Variance .... *See* **Variance Analysis**

Variance Analysis

actual vs. estimated variance	334–35
analyzing Cost variance	338–39
analyzing Date variance	335–37
analyzing variance	335–39
analyzing Work variance	337–38
formula for calculating variance	333–34
understanding	333
variance types	333

Views

using in Microsoft Project	15–18
using in Project Web App	57–61

W

Work Breakdown Structure
- creating in Microsoft Project 118–21
- defined 118

Workflow Stage Status page
- understanding 73–74
- using to create a new enterprise project 73–74

Working with a Project in Offline Mode **45–50**

Index